Communications in Medical and Care Compunetics

Volume 4

Series Editor

Lodewijk Bos
International Council on Medical and Care Compunetics
Utrecht, The Netherlands

For further volumes:
http://www.springer.com/series/8754

This series is a publication of the International Council on Medical and Care Compunetics.

International Council on Medical and Care Compunetics (ICMCC) is an international foundation operating as the knowledge centre for medical and care compunetics (COMPUting and Networking, its EThICs and Social/societal implications), making information on medicine and care available to patients using compunetics as well as distributing information on the use of compunetics in medicine and care to patients and professionals.

Rajeev K. Bali · Lodewijk Bos
Michael Christopher Gibbons · Simon R. Ibell
Editors

Rare Diseases in the Age of Health 2.0

Springer

Editors
Rajeev K. Bali
BIOCORE
Coventry University
Coventry
UK

Lodewijk Bos
International Council on Medical
 and Care Compunetics
Utrecht
The Netherlands

Michael Christopher Gibbons
Johns Hopkins Urban Health Institute
Baltimore
MD
USA

Simon R. Ibell
iBellieve Foundation
Toronto
Canada

ISSN 2191-3811 ISSN 2191-382X (electronic)
ISBN 978-3-642-38642-8 ISBN 978-3-642-38643-5 (eBook)
DOI 10.1007/978-3-642-38643-5
Springer Heidelberg New York Dordrecht London

Library of Congress Control Number: 2013944887

© Springer-Verlag Berlin Heidelberg 2014
This work is subject to copyright. All rights are reserved by the Publisher, whether the whole or part of the material is concerned, specifically the rights of translation, reprinting, reuse of illustrations, recitation, broadcasting, reproduction on microfilms or in any other physical way, and transmission or information storage and retrieval, electronic adaptation, computer software, or by similar or dissimilar methodology now known or hereafter developed. Exempted from this legal reservation are brief excerpts in connection with reviews or scholarly analysis or material supplied specifically for the purpose of being entered and executed on a computer system, for exclusive use by the purchaser of the work. Duplication of this publication or parts thereof is permitted only under the provisions of the Copyright Law of the Publisher's location, in its current version, and permission for use must always be obtained from Springer. Permissions for use may be obtained through RightsLink at the Copyright Clearance Center. Violations are liable to prosecution under the respective Copyright Law.
The use of general descriptive names, registered names, trademarks, service marks, etc. in this publication does not imply, even in the absence of a specific statement, that such names are exempt from the relevant protective laws and regulations and therefore free for general use.
While the advice and information in this book are believed to be true and accurate at the date of publication, neither the authors nor the editors nor the publisher can accept any legal responsibility for any errors or omissions that may be made. The publisher makes no warranty, express or implied, with respect to the material contained herein.

Printed on acid-free paper

Springer is part of Springer Science+Business Media (www.springer.com)

Never doubt that a small group of thoughtful, committed, citizens can change the world. Indeed, it is the only thing that ever has.

Margaret Mead

For our families and friends.

All proceeds from the sale of this book are to be donated to the iBellieve Foundation. (www.ibellieve.com)

Foreword

We continue to move very quickly through the period following the human genome mapping project, an international research effort to sequence and map all of the genes—together known as the genome. Newer methods of molecular diagnoses based on whole genome sequencing are utilized more frequently. Appropriate interpretation of the outcomes of sequencing efforts will become the norm for future diagnostic procedures. Expanding research, evaluation of potential therapies, and introduction of precision or personalized medicine interventions are now available for a few rare diseases. The rare diseases community anticipates the arrival of even more products into the market place for genomics identified subpopulation of both common and rare diseases.

Past research advances have made possible the molecular and enzyme replacement therapies for many diseases such as Gaucher disease and Fabry disease. Current research reveals a glimpse of the future with dramatic treatment effects from gene therapy and stem cell approaches to treatment. Regenerative medicine research offers hope to millions of patients worldwide. A greater emphasis will continue to be placed on the repurposing of investigational and approved products for uses other than the original intended patient population. High throughput screening facilities will provide access to products with potential application to other diseases through a more systematic approach to screening procedures for potential products for further development.

The patients, their families, and the patient advocacy communities have gained a more appropriate status as research partners to the traditional basic and clinical research and development efforts of the academic, federal, and biopharmaceutical research communities. Arrival and full utilization of the capabilities of the internet and world-wide-web approaches to information gathering, development, and distribution have provided ready access to reliable and useful information. Patient registries are considered useful tools to obtain a better understanding of the rare diseases across the lifespan. Developing an extended knowledgebase about rare diseases is now possible through a well-curated patient registry for an individual disease or a group of related disorders. The usefulness of a patient registry is dependent upon the quality of the data entered and the sustainability of a registry for a number of years to enable growth of the data sets for rare diseases.

More recent advances offered by social media and social networking provide the opportunity for gaining access to many more patients and families through crowd sourcing data gathering efforts unknown in the past. Rapid access to thousands of patients connected by common interests is now possible. Using these communication techniques and development of more useful patient registries will increase patient recruitment and entry into clinical trials. The goal remains to gather significant information from as many patients as possible on a systematic global basis to provide for pan-diseases analyses and to increase the knowledge-base for the individual rare disease. For the research investigator and the pharmaceutical industry, more ready access to patients should speed-up the opening of a clinical trial, accrual of patients, and completion of the investigation. Delays in recruiting sufficient participants for clinical trials have always been a major barrier to rare diseases research and development.

Increased research opportunities exist for most rare diseases based on the greater availability of patients worldwide to participate in clinical trials, the improved willingness of the research community to develop multinational research teams, the heightened interest of the biopharmaceutical industry in orphan products and rare diseases, and access to global markets. Multidisciplinary and multisite collaborative partnerships are required for research of rare diseases and the development of orphan products. The rare diseases community is responding to the needs and scientific opportunities offered by rare diseases and in particular, if there is potential usefulness in the more common diseases. The extent of the rapid response to these needs and opportunities will determine how quickly we eliminate the disparities in access to treatment of rare diseases by orphan products. The uniqueness of the book provides perhaps the most valuable research baseline of information—the patient perspective. The growing acceptance of this perspective resulted in the empowerment of the patient, family, and caregiver as a necessary and major partner in the clinical research environment. Individual patient vignettes provide a history of their experiences with a rare disease and how they were able to develop a treatment program to live with a rare disease, even when there are no apparent therapeutic interventions. Increased communication among all of the partners in the rare diseases community is required if optimal care is to be provided to the millions of patients around the world who have received a rare disease diagnosis. Even greater efforts and support are required for the large number of patients who have not received the appropriate diagnosis and remain as patients with undiagnosed diseases.

Stephen C. Groft
Director, Office of Rare Diseases Research
National Center for Advancing Translational Sciences
Department of Health and Human Services
National Institutes of Health
Bethesda, MD, 20892-4874, USA

Preface

> Rare diseases impact more people than AIDS and cancer combined.
> www.globalgenes.org

The Twin Aims of this book were simple: to be clear and to be useful. Clear and useful not only to an academic audience, the typical target group of this sort of text, but also to be of relevance to the most important people of all: patients. We have departed from convention somewhat in not just assembling another textbook, one written primarily for academics and healthcare professionals. Instead, we have elected to present chapters, stories, and vignettes in as accessible manner as we could manage. In this way, we hope that the book will act as a landmark and key resource in the field of rare and orphan diseases, useful not just to professionals but also to patients and their families. We are confident that this book meets a few essential requirements:

- It should meet a long overdue need—specifically, to act as a much needed resource (for academics, health professionals, patients, and their families) which combines essential elements of health, empowerment, new social media, and judicious management.
- Be relevant and accessible—we wanted this text to be different in that it combines the essential competencies and perspectives of clinicians, managers, and patients, all presented in an easy-to-read manner.
- Be globally inclusive—we have included chapters, cases, and vignettes from around the world, sadly proving the prevalence of orphan and rare diseases.

In trying to ensure relevance and usefulness for a wide readership, we instructed contributors to use as few clinical terms as possible (other than those deemed absolutely essential). To ensure accessibility for all stakeholders, we have included as many patient vignettes and case studies as we could. These examples may echo your own questions and concerns. Many chapters either include or end with a relevant insight into patient conditions and needs, often written by patients (or their advocates). These cases are at once intensely emotional, insightful, and inspirational. Most importantly: they are real.

The complexities associated with rare and orphan diseases require a complex response. Thankfully, this task is made somewhat easier thanks to the continued and rapid progress of technology; the phrase *Health 2.0* attests to this fact. The advent of new social media, such as *Facebook*, *Twitter*, *YouTube* (and similar outlets) has perhaps done more to *organically* enable "problem sharing" between affected disease communities than previous, targeted, initiatives. The ability to simply discover that there are "others like me" (via a web search) must give some hope and motivation. Connecting like-minded individuals together results in exchange of ideas, care practice, advice, and, most importantly, support. "People Power" is alive and well.

The multifaceted problems faced by the rare disease communities demand the respect of a similarly multi-layered response. Readers should be satisfied that the first such book of this type is edited by a team whose collective skills span medicine, integrated healthcare, knowledge sharing, and patient advocacy. Editors Bali, Bos, and Gibbons were particularly keen and subsequently pleased to convince a passionate patient advocate to join the team. Simon Ibell (founder and CEO of the *iBellieve Foundation*—a Canadian charity established to find a cure for Mucopolysacharridosis II (MPSII, or Hunter's Syndrome—and convener of the *Be Fair 2 Rare*™ public outreach campaign) provides the team with a powerful and credible voice when communicating with the rare and orphan disease communities. Simon's story features amongst the vignettes within the book. Any lingering doubts that this is merely "just another academic text" should be dispelled.

It should be noted that all proceeds from the sale of this book are to be donated to the iBellieve Foundation (www.ibellieve.com).

The book is split into four interlinked sections: *Rare and Orphan Diseases*, *Health 2.0*, *Patient Perspectives and Empowerment Issues* and *Closing Gaps: Promising research and future considerations*. Each section contains a set of chapters which, together, contain key definitions and concepts, applied research and development projects, opportunities and challenges in the field, and, as previously mentioned, patient-focused case studies for many chapters. Building on our collective, extensive and wide-ranging experience, we trust that we have produced a book which presents a consolidated perspective of the intricacies involved. We hope that readers enjoy the book and trust that it achieves what it set out to do: to convey an important message…simply.

August 2013

Rajeev K. Bali
Lodewijk Bos
Michael Christopher Gibbons
Simon R. Ibell

Acknowledgments

One of our key objectives for this book was *accessibility*. We hope we have achieved this by way of the numerous patient-focused case studies and vignettes and we are therefore indebted to the many kind folk who contributed to these, as listed below:

Patrice F. Band: *Autoimmune Polyendocrine Syndrome Type I (APS1)*
Jason Barron: *Recessive Dystrophic Epidermolysis Bullosa (RDEB): Sibling experiences*
Angela Covato: *Beta-Thalassemia Major*
Jocelyn Gardner: *Living with NOMID: Michael's Story*
Adrian (Ed) Koning: *Hope—Overcoming Fabry*
Marie Ibell: *A Giant of a Man: Simon Ibell (MPSII)*
Grainne Pierse: *The Blessings and Curse of Diagnosis—Myasthenia Gravis*
Deb Purcell: *The Journey of a Lifetime*
Jeneva Stone: *The Wilderness* and also *Route 125 (October 2009)*
Roy and Zezee Zeighami: *MPSIIIA(Sanfilippo)*

We of course say a big thank you to all chapter authors for their wise and expert input. Between the editors, we have had the great pleasure of either meeting in person or conversing with many of the contributors. Such encounters were invaluable in shaping the book and we would like to thank everybody for giving up so much of their time for these discussions. Their constructive contributions have made this book much more relevant than it would otherwise have been.

We thank our publishing team at *Springer*, in particular Christoph Baumann, for their helpful assistance during the publication process. We are grateful to our team of reviewers (and, in particular, to the speedy dedication of Dr. Jacqueline Binnersley) for their comments and suggestions.

Finally, we reserve particular thanks for Dr. Stephen C. Groft, Director of the *Office of Rare Diseases* at the *National Institutes of Health* (*NIH*) for his insightful foreword and accommodating discussions.

About the Editors

Rajeev K. Bali, B.Sc (Hons), M.Sc, Ph.D., PgC, SMIEEE is a Reader in Healthcare Knowledge Management at *Coventry University* (UK). His main research interests lie in healthcare and clinical knowledge management, management of change, and integrated healthcare. He founded and leads the *Knowledge Management for Healthcare* (KARMAH) research subgroup (working under the *Biomedical Computing and Engineering Technologies* (BIOCORE) Applied Research Group). He is well published in peer-reviewed journals and conferences and is regularly invited to deliver presentations and lectures internationally. He is an invited reviewer for several journals and publishers and has been involved with the authoring and editing of several well-received books. In addition to overseeing various projects in the UK, he is leading (or has led) several international development and research projects (or short courses) involving partners from the UAE, the USA, Canada, Malaysia, Singapore, Nigeria, Oman, and Finland.

Lodewijk Bos, MA is founder and president of the *International Council on Medical and Care Compunetics* (ICMCC), an international foundation operating as the knowledge center for medical and care compunetics (COMPUting & Networking, its EThICs and Social/societal implications). He is series editor of *Communications in Medical and Care Compunetics* (Springer-Verlag). He maintains one of the leading websites providing information on health IT-related news and science articles. He is co-Editor-in-Chief of the recently founded IUPESM/Springer journal *Health and Technology.*

M. Chris Gibbons, MD, MPH is Associate Director of the *Johns Hopkins Urban Health Institute* and an Assistant Professor at the *Johns Hopkins Schools of Medicine and Public Health.* He is an urban health expert and informatician who works primarily in the area of consumer health informatics where he focuses on using health information and communications technologies to improve urban healthcare disparities. His work is forming the foundation of the emerging field of Populomics which represents the fusion of the population sciences, medicine, and informatics. Prior to coming to the *Urban Health Institute,* Dr. Gibbons was a Senior Policy Fellow at the *Centers for Medicare and Medicaid Services (CMS).*

Simon Ibell is the founder of the *iBellieve Foundation* as well as the *Be Fair 2 Rare*™ campaign. A truly remarkable and inspirational individual, Simon is an advocate for Human Potential, a Role Model for Persons with Disabilities, and a crusader for the rare disease community. As one of 30 people in Canada (and approximately 2000 worldwide) born with a condition known as MPS II (mucopolysacharridosis) or Hunter Syndrome, living with MPS II has taught Simon that every moment of life is precious. He continues to learn how people perceive and handle disabilities and he uses this knowledge to lower the barriers and help bring about change. Receiving his degree in Sports Administration at the *University of Victoria* in June 2002, Simon is also a sought after speaker on human potential and overcoming obstacles. He has been awarded the *Spirit of Sport Story of the Year Award* at the 2003 *Canadian Sport Awards*, the *Queen's Golden Jubilee Commemorative Medal for Her Majesty Queen Elizabeth the Second* as well as the 2012 *St. Michaels University School Distinguished Alumni Award* and 2012 *University of Victoria Distinguished Alumni Award*. For more information on Simon, please visit: www.ibellieve.com, www.simonibell.com and http://nsb.com/speakers/view/simon-ibell.

About the Authors

Dr. Brendan Allison has been active in EEG research for about 20 years, most of which involved brain–computer interface (BCI) systems. He earned his Ph.D. in Cognitive Science in 2003 at *UC San Diego*, where he focused on BCIs based on visual attention (primarily P300) and imagined movement. He has since worked with several top researchers and institutes, including Prof. Wolpaw at the *New York State Dept of Health*, Prof. Polich at *The Scripps Research Institute*, and Profs. Pfurtscheller and Neuper at *Graz University of Technology*. He recently returned to his alma mater and is again with the Cognitive Science Dept. at *UCSD*. Dr. Allison's main interests involve making BCIs more practical, flexible, and robust. Despite considerable progress recently, most BCIs use simple interfaces that do not function well for disabled users outside of laboratory settings without ongoing expert support. These problems can be alleviated by combining BCIs with other communication devices and intelligent systems, such as hybrid BCIs, which is Dr. Allison's main research interest. He has also been active in addressing policy and infrastructural issues within the BCI community, such as surveying researchers and companies to assess their needs and analyzing and recommending specific directions.

Patrice Band lives in Toronto with his wife Jennifer, their daughter Julia, Julia's service dog Tory, Nibbles her rat, and Liberty her fish. When he is not spending time with his family and critters or playing hockey, he practises criminal law.

Jason R. Barron is Associate Director of Public Policy for the National Organization for Rare Disorders (NORD). His current focuses include a variety of operational activities that support NORD's efforts in advocacy, government affairs, and scientific affairs. Jason is a graduate of Dartmouth College, class of 2010. His previous work experience includes 2 years as a laboratory assistant in the Genetics and Molecular Biology Department of Dartmouth Medical School, as well as a 6-month internship as a junior legislative correspondent for health affairs in the legislative office of Patrick Kennedy, former Congressman of Rhode Island's 1st District. As the sibling of a young adult with Recessive Dystrophic Epidermolysis Bullosa, Jason brings his personal experiences as a rare disease patient advocate to NORD as he seeks to improve the lives of all affected by rare diseases.

Dr. Norman Barton's biomedical research career spans 30 years with leadership responsibility for clinical research programs in both the academic and industrial sectors. The unifying principle throughout has been the development of effective first-in-class therapeutics for genetic and metabolic disorders such as Gaucher, Fabry, and Niemann-Pick Disease, intractable gout, Duchenne Muscular Dystrophy and bronchopulmonary dysplasia in premature infants. Serving in the capacity of Chief, Clinical Investigations and Therapeutics Section at the National Institute of Neurological Disorders and Stroke, Dr. Barton conceived and executed the clinical development program that led to the approval of Ceredase as the first safe and effective enzyme replacement product for Gaucher Disease. Recognition for this contribution included the Meritorious Service Medal from the Public Health Service, the Outstanding Achievement Award from the National Gaucher Foundation and the Distinguished Alumnus Award from Pennsylvania State University. Dr. Barton earned a Ph.D. and MD at Pennsylvania State University. He was a resident in medicine at Albany Medical College Hospital, and a resident in neurology at Cornell University, the New York Hospital.

Dr. Cindy Bell joined *Genome Canada* in August 2000. From 2000 to 2008 she held the position of Vice-President, Genomics Programs in which she was responsible for providing policy and strategic advice on scientific and other aspects of *Genome Canada's* programs. This included overseeing and managing the peer review process used to establish the research program of *Genome Canada*. In her role as Executive Vice President, Corporate Development she provides leadership in the development and implementation of strategic initiatives and approaches to enhance *Genome Canada's* business model and secure funding to support genomics research in Canada. Prior to joining *Genome Canada*, Dr. Bell was a Deputy Director in Programs Branch at the *Canadian Institutes of Health Research* from 1994 to 2000. At CIHR she managed a number of research programs and was involved in policy development and implementation. From 1986 to 1994, Dr. Bell was a researcher at the University of California, Riverside. Her research focused on investigating the basic defect in the genetic disease, Cystic Fibrosis. She obtained her Ph.D. in Genetics from McGill University in 1986.

Dr. Raymond Bond completed a B.Sc (Hons) in Interactive Multimedia Design and a Ph.D. in Computing Science in 2007 and 2011 respectively. He teaches and carries out his research in the School of Computing and Mathematics. His research interests include health informatics, computerized electrocardiology, patient safety, and usability engineering.

Dr. Kym M. Boycott received her Ph.D. in Medical Genetics and subsequently studied Medicine and specialized in Medical Genetics at the University of Calgary, Canada. She is now a Clinical Geneticist at the *Children's Hospital of Eastern Ontario (CHEO)*, Investigator at the *CHEO Research Institute*, and an Associate Professor in the Department of Pediatrics at the University of Ottawa. Dr. Boycott's research, bridging clinical medicine to basic research, is focused on elucidating the molecular pathogenesis of rare inherited diseases using next-generation sequencing approaches. She is the Lead Investigator of the 'Finding of Rare Disease Genes in Canada' (FORGE Canada) project which is investigating the molecular etiology of almost 200 rare pediatric diseases.

Kyle Boyd completed a B.Sc (Hons) in Interactive Multimedia Design and an MA in Multidisciplinary Design in 2007 and 2009, respectively, from the University of Ulster. He is currently undertaking a Ph.D. in Leveraging Web 2.0 technology for Ambient Assisted Living. His research is to address social isolation in older carers of persons with dementia and chronic diseases. He is addressing both the barriers to use and usability issues of the current technology.

Angela Covato is the Managing Director of the *Canadian Organization for Rare Disorders (CORD)*. Since 2006, she has been actively involved in *CORD*'s advocacy work for a much-needed Canadian orphan drug policy, including an orphan drug regulatory framework. For the past 13 years, she has volunteered with the *Thalassemia Foundation of Canada*. She currently serves as the *Foundation*'s Research Grant Co-ordinator. The *Thalassemia Foundation of Canada*'s grant program provides medical research grants to new investigators in the field of Thalassemia. But most of all, she is the mother of an extraordinary teenager with a rare disorder.

Krissi Danielsson is a student at Lund University in Sweden, where she will complete her medical degree in 2015. During enrolment she has been active in research involving cancer and the PTEN gene and was the primary author of a review article on pancreatic cancer published in *Personalized Medicine* in early 2013. Krissi also holds a Bachelor's degree in Psychology from Excelsior College. Prior to studying medicine, Krissi worked as a freelance Internet content writer covering a range of topics, ranging from software-as-a-service to pregnancy-related issues and has been the author of four published consumer-oriented non-fiction books. After graduation, Krissi plans to pursue a career in oncology or medical genetics.

Dr. Remco de Vrueh works as senior adviser at *Schuttelaar & Partners*, a Dutch communications consultancy firm that strives to promote greater health and sustainability. As a personal initiative, he founded *Rare Disease Matters* at the beginning of 2012 to continue his research and teaching in the area of orphan drug development. Prior to this he worked for the *Netherlands Organisation for Health Research and Development* (ZonMw) as adviser to various (orphan) drug innovation programs. Between 2006 and 2011 he was active as Orphan Product Developer for the *Dutch Steering Committee on Orphan Drugs*. He was responsible for stimulating development of orphan drugs by the Dutch pharmaceutical industry. Apart from industry, he also actively interacted with academia and patient organizations. Remco has (co-)organized meetings, given presentations, provided training and teaching, and is the (co-)author of several scientific publications. In the same period he also fulfilled interim project management positions for the *European and Developing Countries Clinical Trials*

Partnership and the *Dutch Medicines Evaluation Board*. Before joining *ZonMw* he worked for seven years at *OctoPlus*, a Dutch pharmaceutical company. Remco de Vrueh holds a Ph. D. degree in Biopharmaceutical Sciences from the *Leiden-Amsterdam Center for Drug Research* in the Netherlands.

Dr. Mark P. Donnelly received both his Bachelor's degree (2004) and Ph.D. in Computer Science (2008) from the University of Ulster where he is currently a Lecturer with the School of Computing and Mathematics. He is a member of the institution's Smart Environments Research Group. His research interests include the application of connected health solutions to support independent living and remote monitoring of people with chronic conditions and, more recently, investigating the application of mobile technologies to support children with autism and their caregivers.

Jocelyn Gardner is a grandparent of a patient with Cryopyrin-Associated Periodic Syndromes. She is the President of the Canadian CAPS Network (CCN), whose mission is to improve the lives of all those affected by Cryopyrin-Associated Periodic Syndromes (CAPS) and related disorders. Amongst several objectives, CCN serves as a forum for bringing together patients and families, healthcare professionals, researchers, industry, funders, and policymakers to raise funds and promote health policies that improve the lives of those affected by CAPS and related disorders.

Dr. Rashmi Gopal-Srivastava is Director of Extramural Research Program in the Office of Rare Diseases Research (National Center for Advancing Translational Science) at the National Institutes of Health (NIH), Bethesda MD, USA. She oversees the national program on Rare Diseases Clinical Research Network (RDCRN). The RDCRN consists of 19 consortia and one Data Management Coordinating Center. She is responsible for developing and expanding the extramural research program to coordinate research activities on rare diseases across NIH in collaboration with other Federal Agencies, patient advocates, and other organizations. Prior to her current position she served as the program director for breast cancer SPOREs (Specialized program of Research Excellence) in the Office of Director, National Cancer Institute (NCI). Dr. Gopal-Srivastava received her Ph.D. in Microbiology and Immunology from the Medical College of Virginia, Virginia Commonwealth University, Richmond, Virginia in 1989 and received a Virginia Commonwealth fellowship. The same year she was selected for and awarded a Research Associateship from the US National Research Council of the National Academies of Science and joined the Laboratory of Molecular and Developmental Biology at the National Eye Institute, NIH and conducted research on regulation of gene expression for alphaB-cystallin (small heat shock protein). She has published several papers in peer-reviewed journals, written book chapters, and delivered invited oral presentations nationally and internationally. Dr. Gopal-Srivastava has received several honors and awards and is a member of a number of scientific committees.

About the Authors

Dr. Sylvie Grégoire is the former President of Shire Human Genetic Therapies (HGT), a division of Shire plc, the global specialty biopharmaceutical company. Headquartered in Lexington, Massachusetts, Shire HGT specializes in discovering, developing, manufacturing, and commercializing protein therapeutics primarily for the treatment of rare genetic diseases. Shire HGT employs 1,300 people in Massachusetts, and an additional 250 worldwide. Prior to joining Shire HGT, Dr. Grégoire served as Executive Chairwoman of the Board of IDM Pharma, a biotechnology company in California. Over the last 20 years, Dr. Grégoire has been the CEO of GlycoFi, held executive positions at Biogen Inc., and worked at Merck & Co. in the US and abroad. Dr. Grégoire received her Doctor of Pharmacy degree from the State University of New York at Buffalo, and her pharmacy degree from Université Laval, Québec City, Canada.

Stephen C. Groft, Pharm.D. is the Director of the Office of Rare Diseases Research (ORDR) in the National Center for Advancing Translational Sciences at the National Institutes of Health (NIH). His major focus is on stimulating research with rare diseases and developing information about rare diseases and conditions for healthcare providers and the public. To help identify research opportunities and establish research priorities, the office has co-sponsored over 1,200 rare diseases-related scientific conferences with the NIH research Institutes and Centers. Current activities include establishing patient registries for rare diseases, developing an inventory of available bio-specimens from existing bio-repositories, establishing a public information center on genetic and rare diseases, developing an international rare diseases research consortium, maintaining the Rare Diseases Clinical Research Network, and providing a special emphasis clinic with senior clinical staff for patients with undiagnosed diseases at NIH's Clinical Research Center Hospital. Steve received the B.S. degree in Pharmacy in 1968 and the Doctor of Pharmacy degree from Duquesne University in 1979.

Frank Grossmann, Dr. Med.vet. studied veterinarian medicine in Germany and conducted research at the ETH (Swiss Federal Institute of Technology) in Zurich. After supporting a start-up practice, he worked in management within the pharmaceutical industry. In addition to experience in dermatology, infections disease, nutrition, and orphan drugs, he is a recognized expert in sustainable businesses that deliver social impact. He is founder of a consulting company in pharmaceutical science, has been working as a guest lecturer in pharmaceutical science at the ETH/Zurich for many years and acts as a member of various organizations and boards. He is co-founder of Orphanbiotec, a social entity and is founder of Foundation Orphanbiotec, a Competence Center for Orphan Disease and Patient Empowerment.

Marie Ibell is the proud mother of Simon Ibell. Marie possesses over 30 years' experience in international business. She demonstrates strategic vision and possesses a portfolio of skills including governance, financial acumen, performance management, relationship-building, and the ability to champion change. She has 20 years' Board-level experience in organizations with significant budgets and complexity, a considerable reputation within the private and financial sectors and has experience of building alliances and working relationships with a range of stakeholders.

Adrian Francis (Ed) Koning, P. Eng is married to Marlene and together they have three young adult sons and one granddaughter. In early 2001, at the age of 43—and as a result of kidney failure—Adrian was diagnosed with Fabry disease, a life-threatening orphan disease. His life expectancy was reduced to between 45 and 50. A few months later, he was on dialysis, began enzyme replacement therapy on compassionate use, and was blessed with a live donor kidney transplant. From 2003 to 2010, he worked with others to secure access and funding of Enzyme Replacement Therapy (ERT) for all Fabry patients in Canada and was the first President of the *Canadian Fabry Association*. He also served as the Vice President of the *Canadian Organization for Rare Disorders* and the Secretary of the *Fabry International Network*. He continues to utilize his skills and abilities as a professional engineer to help improve the lives of those suffering from Fabry disease as well as support their families and caregivers. He continues to educate society and raise awareness about rare disorders and the need to ensure Canadians have access and funding to "orphan" drugs and therapies that are available in other parts of the world. He believes that Canada must not only adopt an orphan drug policy (ODP), but must also provide a comprehensive care program that includes the medical, psychological, and social needs of patients, families, and care givers who suffer the impact of living with a life-threatening rare disorder.

Caren Kunst is a healthcare consultant with a growing interest in digital healthcare. She is also a nurse and an experienced (homecare) mother of a 21-year-old son with the orphan disease Tracheoesophageal fistula (TOF)/Esophageal atresia (OA). She supports parents globally using such social media as *Facebook*, *Hyves* (a Dutch social network) and *NING* (an online platform for people to create their own social networks). Her aim is to organize effective and safe self-management in home and hospital situations.

Prof. Paul Lasko received his Ph.D. from the Massachusetts Institute of Technology in 1986 and joined McGill in 1990 after a postdoctoral period at the University of Cambridge. Using the Drosophila system, Dr. Lasko's research concerns regulatory processes that control gene expression at the levels of mRNA stability or translation, and that underlie germ cell or early embryonic development. He received the Award of Excellence from the *Canadian Society of Genetics* in 2004. At McGill he served as Chair of the Department of Biology from 2000 to 2011. He assumed his position at CIHR in May 2010 but maintains his research lab at McGill. Dr. Lasko has been highly active in research grant adjudication and served on CIHR or *Canadian Cancer Society* grant panels continuously since 1995. He has also worked extensively for the *Human Frontiers of Science Program Organization (HFSPO)* over the past 10 years, serving on its program grant panel from 2001 to 2005, and then as one of two Canadian representatives on the Council of Scientists. He chaired the HFSP Council of

Scientists from 2007 to 2010. Dr. Lasko also served as President of the *Genetics Society of Canada* from 2007 to 2010. As Scientific Director of the CIHR Institute of Genetics, Dr. Lasko oversees the Institute's strategic research funding initiatives, many of which involve fostering international partnerships. He is the incoming chair of the Executive Committee of the *International Rare Diseases Research Consortium*.

Dr. Gaye Lightbody received an M.Eng. (1995) and a Ph.D. (2000) in Electrical and Electronic Engineering from *Queen's University Belfast*. She worked in industry for *Amphion Semiconductor Limited* developing high performance FPGA and ASIC cores for image and audio processing before returning to academia in 2005. Since then she has been a Lecturer with the School of Computing and Mathematics in the *University of Ulster*. Her early research interests included high performance VLSI hardware design, FPGAs, and Adaptive Filtering. Progression into Biomedical Signal Processing followed, working in the area of automatic detection of Auditory Brainstem Responses for determining hearing threshold. Recent involvement in a project for development of Brain–Computer Interfaces has expanded her work in the medical computing domain. More recent activities have involved computer systems for supporting remote management and rehabilitation of children with autism. Dr. Lightbody has undertaken a range of teaching responsibilities from Advanced Computer Networks, Mathematics, Web Design, and Health Informatics. She received her PG Certificate in Higher Education in December 2012 and continues to perform a small level of pedagogic research activities alongside her teaching and scientific research activities.

Jimmy Lin, MD, Ph.D., MHS is the president of the *Rare Genomics Institute* (RGI), the world's first platform to enable any community to leverage cutting-edge biotechnology to advance understanding of any rare disease. Partnering with 18 of the top medical institutions, RGI helps custom design personalized research projects for diseases so rare that no organization exists to help. Dr. Lin is also a medical school faculty member at the Washington University in St. Louis and led the computational analysis of the first ever exome sequencing studies for any human disease at *Johns Hopkins*. He has numerous publications in *Science, Nature, Cell, Nature Genetics* and *Nature Biotechnology* and has been featured in *Forbes, Bloomberg,* the *Wall Street Journal, The Washington Post,* and *The Huffington Post.*

Amanda Lordemann is a senior at Washington University in St. Louis studying Philosophy-Neuroscience-Psychology. She joined the *Rare Genomics Institute* during the summer of 2012 as an intern with the patient advocacy team. She has been contributing to efforts in patient education, social media, marketing, and human resource management. After graduation, Amanda is interested in a career related to healthcare and medical research.

Dr. Alex E. MacKenzie received his MD and Ph.D. at the University of Toronto and specialized in Pediatrics at the University of Ottawa. He is currently a

Pediatrician at the *Children's Hospital of Eastern Ontario (CHEO)*, Senior Scientist at the *CHEO Research Institute*, and a Professor in the Department of Pediatrics at the University of Ottawa. Dr. MacKenzie's research is focused on the identification of treatment modalities for inherited pediatric disease using the neuromuscular disorder spinal muscular atrophy as a model for the repurposing of clinically-approved agents.

Dr. Paul McCullagh received a B.Sc (1979) and a Ph.D. (1983) in Electrical Engineering from *Queen's University of Belfast*. He is a Reader in the *School of Computing and Mathematics* at the the *University of Ulster* and is a member of the *Computer Science Research Institute*. He specializes in the teaching of data communications, computer networking, and health informatics. His research interests include Biomedical Signal and Image Processing, Data Mining, Brain Computer Interface, and Assisted Living applications. He is interested in advancing the education and professionalism of biomedical engineering and health informatics. He has worked on the EU FP7 *BRAIN* project for e-inclusion using the brain–computer interface, EPSRC SMART 2 for self-management of chronic disease, TSB NOCTURNAL for assisting people with dementia and ESRC *New Dynamics of Ageing*, *Design for Ageing Well* funded projects. He is a member of the *British Computer Society*, the *European Society for Engineering and Medicine* and the *UK Council for Health Informatics Professionals*.

Daniela M. Meier, Ph.D., worked in a leading position of a Swiss agency promoting innovation and entrepreneurship before she started her own business *Manda Idea Management*. She studied Modern History, Political Sciences, and Iranian studies at universities in Bern (Switzerland) and Oxford (UK).

Dr. Pierre Meulien was appointed President and CEO of *Genome Canada* in October 2010. Prior to this appointment, he served as Chief Scientific Officer for *Genome British Columbia* from 2007 to 2010. From 2002 to 2007, Dr. Meulien served as the founding CEO of the *Dublin Molecular Medicine Centre* (now *Molecular Medicine Ireland*) which linked the three medical schools and six teaching hospitals in Dublin to build a critical mass in molecular medicine and translational research. The Centre managed the Euro 45 Million "Program for Human Genomics" financed by the Irish government and was responsible for coordinating the successful application for the first *Wellcome Trust* funded Clinical Research Centre to be set up in Ireland. For over 20 years, Dr. Meulien has managed expert research teams with a number of organizations, including *Aventis Pasteur* in Toronto (Senior Vice President of R&D), and in Lyon, France (Director of Research). He also spent 7 years with the French biotechnology company *Transgene* in Strasbourg, France as a research scientist and part of the management team. Dr. Meulien's academic credentials include a Ph.D. from the University of Edinburgh and a post-doctoral appointment at the Institut Pasteur in Paris.

Laura Montini is a blogger and editor for *Health 2.0 News*. She received her B.A. in Journalism from the University of North Carolina at Chapel Hill. As a student, her reporting focused on several of North Carolina's large health systems as well as the state's internal debate on United States healthcare reform. She has spent much of the past 2 years in San Francisco, California editing *The Health Care Blog* and writing about technology for *Health 2.0 News*. She is passionate about good healthcare reporting and believes that healthcare is too important of an issue for the public not to understand it. Laura currently lives in San Francisco but is a New Jersey girl through and through.

Prof. Sara Newman holds a Ph.D. in Rhetoric and is a Professor in the Department of English at Kent State University, and a member of its Ph.D. program in Literacy, Rhetoric, and Social Practice. She is author of *Aristotle and Style and Disability and Life Writing: A Critical* History as well as articles in such journals as *History of Psychiatry*, *Rhetorica*, *Disability Studies Quarterly* and *Written Communication*. Her current research deals with the rhetoric of disability and medicine.

Prof. Chris Nugent is Professor of Biomedical Engineering at the University of Ulster. He received a Bachelor of Engineering in Electronic Systems and D.Phil in Biomedical Engineering both from the University of Ulster. His research within biomedical engineering addresses the themes of the development and evaluation of Technologies to support independent living. From a technological perspective his work has focused on the integration of mobile devices within smart environments coupled with the development of activity recognition systems. He has published extensively in these areas with work which spans theoretical, clinical, and biomedical engineering.

Grainne Pierse was born and raised in Edmonton, Alberta in Canada. She grew up swimming competitively and Irish dancing, both at a national level until the end of high school. At this point she had been recruited to swim with the University of British Columbia (UBC) swim team and moved out to Vancouver in September of 2008 to pursue this. She had two sisters compete with the UBC team before her and two more have followed since. In 2010, she was forced to take a year off from training and racing due to her myasthenia gravis but 3 months after her surgery she was back at it. Since then, she has won two national championships with her team and is a current Canadian record holder as a part of a 4 x 100 medley relay team. She is currently completing an undergraduate degree in Psychology with hopes of going on to do veterinary medicine.

Deb Purcell is a home-schooling mother to three awesome kids (Trey, Avery, and Sadie) and partner to an amazingly supportive and wonderful man, Ryan. She is an advocate for MPS and all rare diseases in Canada and around the world. When she is not scouring the Internet for the latest research, updates, and journal articles in MPS, or raising funds for research, you can find her reading, doing puzzles, and playing street hockey with her kids or watching them at soccer, hockey, gymnastics, and baseball. You can also find her on her yoga mat all over Vancouver and North America, at home, in hospital, at studios, or out in the sunshine. (www.treypurcell.com).

Dr. Francis (Fritz) Paul Rieger is Associate Professor in the Odette School of Business at the University of Windsor in Ontario, Canada. He has been a regular lecturer with the College of Business, Department of Management Studies at the University of Michigan-Dearborn, USA. Dr. Rieger began his academic teaching career as an Assistant Professor at Oklahoma State University in 1982. He came to the University of Windsor in 1984, where he has served as Director of the *Business Resource Centre* (1995–2000) and Director of the MBA Program (1997–2000). He has served as a Visiting Professor at Université du Québec à Trois-Rivières; Queen's University, Lyon Graduate School of Management, France, the University of Michigan, and the University of Electronic Science and Engineering China (People's Republic). Prior to his academic career, Dr. Rieger worked in the private sector in the Finance area as a Field Engineer for IBM, Stockbroker and Financial Analyst in Anchorage, Alaska, and for the Cree Indian School Board in Val D'Or, Quebec. Dr. Rieger carries out research and publishes on topics of business ethics, global business environment, strategic management and healthcare. Dr. Rieger has a B.Sc. (Physics Education) from Manhattan College, USA, an MBA from Columbia University USA, and Ph.D. (Management) McGill University QC, Canada. He has many published articles and a book.

Michael Seres was diagnosed aged 12 with the incurable bowel condition Crohn's Disease. After over 25 surgeries, and intestinal failure, he became only the 11th person to undergo a small bowel transplant in the UK at *The Churchill Hospital* in Oxford. Michael started blogging to chart his journey from Crohn's to a Bowel Transplant. http://beingapatient.blogspot.com is now syndicated on over 20 websites and has had received 55,000 views from transplant teams, medical students, and patients. He has developed an understanding as to how patient-to-patient interaction can be a powerful tool to assist in recovery. Michael mentors many patients and their families and is a published author and professional speaker. He uses social media to develop global online communities and devises social media strategies around patient engagement. He is the patient lead for #NHSSM, a facilitator for the *Centre for Patient Leadership* and is Digital Strategy Advisor to *The Patients Association* and the *Oxford Transplant Centre*.

Dr. Hannah Spring is a Senior Lecturer in Research and Evidence-Based Practice within the Faculty of Health and Life Sciences at *York St John University* in the UK. Prior to working at York St John, she held a variety of posts in Department of Health and NHS settings. By background, she is a clinical information specialist and is particularly experienced in working with health professionals in the primary and secondary healthcare, and academic sectors. As well as 10 years' teaching experience, she has worked in a variety of independent consultancy roles including working with *Nottingham University* as an independent reviewer and content provider for the *Intute Health and Life Sciences* database, and as a consultant information specialist for general practitioners. She has significant experience in research and information and knowledge management in the health and academic sectors, and have published widely in my specialist areas. She is currently editor for the *Learning and Teaching in Action* regular feature of the *Health Information and Libraries Journal*, and a member of the editorial board. Her specialist interest areas include systematic reviews and associated research methodologies, research development in health LIS professions, clinical librarianship in primary care and the allied health professions, evidence-based health practice, the impact of Internet and web 2.0 technologies on learning and information behavior, and information literacy and health.

Therese Stutz Steiger, Dr. med., is a physician specializing in prevention and public health. She has worked at the Swiss Federal Office of Public Health (BAG) dealing with new issues, health expertise, and non-contagious diseases (specifically cancer). She co-managed Online Services and Empowerment and implemented the eHealth Switzerland strategy initiated by the federal government. She works as an private consultant in public health, as a lecturer, teacher, and publisher on healthcare issues and is a member of several committees and advisory councils in the public and private sectors.

Hugh Stephens has been interested in the intersection between healthcare and social media for some years now both as a researcher and in its ability to change the way that we provide healthcare. Hugh is an MBBS/Ph.D. candidate studying at *Monash University* in Melbourne, Australia. Hugh is completing his Ph.D. at *The Alfred Hospital*, conducting research into resource allocation issues for deceased organ donation across Australia and is also conducting additional research into the use of social media within the health sector in Australia. For the last 3 years, Hugh has worked on the External Advisory Board of the *Mayo Clinic's Center for Social Media* based in the USA, encouraging health organizations and professionals worldwide to embrace social media technologies. He also works closely planning part of the program of the *International AIDS conference*, the largest health-related conference in the world, attracting 27,000 people in 2012 (Washington, DC).

Roy Sterritt is a faculty member in the School of Computing and Mathematics at the University of Ulster and a member of the University's Computer Science Research Institute. Internationally recognised as a leading innovator in Autonomic Computing and Communications (self-managing and self-healing computer-based systems) with currently 13 patents with NASA and 175 + publications in the field. He was the founding chair of the *IEEE Technical Committee on Autonomous and Autonomic Systems (TCAAS)*. Roy has also served on the *IEEE CS Publications board* and chaired the *Conference Publications Operations Committee (CPOC)*, the *Conference Publications Committee (CPC)*, the *Conference Advisory Committee (CAC)* and also served on the *IEEE CS Technical and Conferences Activities Board (T&C Excom)*.

Jeneva Ellen Stone is a writer and editor who lives with her family in the Washington DC area (USA). She is the recipient of fellowships from the MacDowell and Millay Colonies for her work in nonfiction. Her poetry and nonfiction have appeared in many literary journals, including the *Colorado Review*, *Poetry International*, *Pleiades* and *The Collagist*. Jeneva is currently working on a memoir about her son's rare disease.

Tracy VanHoutan was raised in Marshalltown Iowa, attended University of Iowa in Iowa City, and now lives in the Chicago-land area with his wife Jennifer and their three children, Noah, Laine, and Emily. Two of his three children (Noah and Laine) are affected by the Late Infantile form of a devastating childhood disease known commonly as Batten disease. Tracy has worked in the proprietary trading field trading various financial products for the past 16 years. Since his children's diagnosis, Tracy and his wife started a local organization called *Noah's Hope* to help raise awareness and funds for Batten disease. Tracy currently serves on the *Batten Disease Support and Research Association* board of Directors. Tracy regularly attends research conferences related to Batten disease and has been part of a small group of parents that actively fund peer reviewed projects on a regular basis. In addition to his Batten-related work, Tracy has become very active in the rare disease community within the United States. These efforts include testimony before the FDA on two occasions as well as speaking to a panel at the NIH on cell-based therapies for rare diseases. Tracy has also been very active in speaking to members of the United States Congress and organizing grassroots efforts to further common causes between all rare disease.

Dorothy Weinstein has a long tenure working in Washington DC on national, state, and local health policy. She has been employed at *Georgetown University*'s *Institute of Health Policy*, the *Association of American Medical Colleges*, the *American Diabetes Association*, the *Endocrine Society*, the *American College of Cardiology*, and the *National Health Council*. Her various positions in health policy have been broad based including research and writing, crafting legislation, and directing government relations departments at leading major non-profit health organizations. Her most recent activities are working on designing and now implementing healthcare reform legislation in the United States. The focus of her efforts is on patient engagement in the health care delivery process. Dorothy has a background in healthcare volunteer and philanthropic work at the *Children's National Medical Center* (CNMC) and at the *Prevention of Blindness Society* in Washington DC. Dorothy has a B.A. degree magna cum laude from the honors program in philosophy from the University of Maryland and an M.A. degree from Duke University in Public Policy. She is a member of Phi Beta Kappa and is a published author in the area of fetal tissue/stem cell research and environmental policy on sound and noise abatement.

Dr. David Whiteman joined Shire Pharmaceuticals 8 years ago as medical director of the Elaprase program, and principal medical director in the (then) newly developing department of Global Medical Affairs at Shire HGT. He provided medical leadership and expert input to the pivotal clinical trial and eventual approval of Elaprase for the treatment of Hunter Syndrome. Dr. Whiteman obtained his undergraduate degree (in Experimental Psychology and Physiology) and his medical degree, from Oxford University in the United Kingdom. Following general postgraduate medical training in Britain, he undertook a residency in Pediatrics in the USA (University of Connecticut Hospitals) and then a clinical and research fellowship in Medical Genetics and Metabolic Diseases at the University of Pennsylvania/Children's Hospital of Philadelphia in the early 1980s. Subsequently, he held a variety of positions in clinical and laboratory genetics and metabolism in medical schools and university hospitals throughout the United States. Dr. Whiteman is a Fellow of the American Academy of Pediatrics and a Fellow of the American College of Medical Genetics. He has served on numerous state and U.S. national committees related to metabolic disease management and newborn screening. Currently he is Senior Medical Director, Clinical Sciences in the Research and Development department at Shire Human Genetic Therapies in Lexington, Massachusetts, USA.

Dr. Durhane Wong-Rieger is President and CEO of the *Institute for Optimizing Health Outcomes*. She is also President of the *Canadian Organization for Rare Disorders (CORD)* and head of the *Consumer Advocare Network*, a national network to promote patient engagement in healthcare policy and advocacy. Internationally, Durhane serves as Chair of the Board of the *International Alliance of Patient Organizations*, Co-Chair of the *Health Technology Assessment International Patient/Citizen Involvement Interest Group* and on the Board of Directors of *DIA International*. She is a certified Health Coach and licensed T-Trainer with the Stanford-based *Living A Healthy Life with Chronic Conditions*. Dr. Wong-Rieger has conducted training, workshops, and evaluation for patient groups in Canada and internationally on all aspects of patient engagement and advocacy. She has served on numerous health policy advisory committees and panels, including the *Policy Dialogues for the Commission on the Future of Healthcare in Canada*, *Ontario Premier's Advisory Board on Organ Donation*, Health Canada's *Expert Advisory Committee on Vigilance of Health Products* and *Expert Advisory Panel on Special Access Programme*, and *Association of Family Health Teams of Ontario*. From 1984 to 1999, Durhane was professor of Psychology at the University of Windsor in Ontario, Canada. Durhane has a BA in Psychology from Barnard College in New York City and an MA and Ph.D. in Social Psychology from McGill University in Montreal. She is author of two books and many articles and is a frequent lecturer and workshop leader.

Roy and Zezee Zeighami are parent advocates for children suffering from rare disease. Their personal connection to the fight is their son Reed who suffers from MPS IIIA, *Sanfilippo Syndrome*. They have been active in the fight to speed treatment and improve upon the incentives provided by the Orphan Drug Act to treat rare diseases such as *Sanfilippo*. They live in Dallas, TX, USA with Reed and his older sister Aziza.

Contents

Part I Rare and Orphan Diseases

Why R&D into Rare Diseases Matter................................ 3
Remco L. A. de Vrueh

Vignette: Autoimmune Polyendocrine Syndrome Type I (APS 1) 21
Patrice F. Band

**Rare Diseases: How Genomics has Transformed Thinking, Diagnoses
and Hope for Affected Families**..................................... 27
Pierre Meulien, Paul Lasko, Alex MacKenzie, Cindy Bell and
Kym Boycott

Vignette: A Giant of a Man: Simon Ibell (MPS II) 39
Marie Ibell

Innovative Funding Models for Rare Diseases....................... 43
Amanda Lordemann, Krissi Danielsson and Jimmy Cheng-Ho Lin

**Rare Diseases: The Medical and the Disability Perspectives
in the Age of 2.0**... 51
Sara Newman

Vignette: Taking Control of Thalassemia 67
Angela Covato

Industry Perspectives on Orphan Drug Development................. 71
Sylvie Grégoire, Norman Barton and David Whiteman

Part II Health 2.0

Health 2.0: The Power of the Internet to Raise Awareness of Rare Diseases.. 83
Laura Montini

Vignette: Living with NOMID: Michael's Story...................... 97
Jocelyn Gardner

Health 2.0 and Information Literacy for Rare and Orphan Diseases.... 101
Hannah Spring

Social Media and Engaging with Health Providers................... 115
Hugh Stephens

Vignette: Hope–Overcoming Fabry................................ 123
Adrian (Ed) Koning

Empowering the Rare Disease Community: Thirty Years of Progress... 127
Jason R. Barron

Part III Patient Perspectives and Empowerment Issues

The Role of Social Media in Healthcare: Experiences of a Crohn's Disease Patient.. 139
Michael Seres

Vignette: The Blessings and Curse of Diagnosis: Myasthenia Gravis.... 145
Grainne Pierse

Noah's Hope: Family Experiences of Batten Disease.................. 149
Tracy VanHoutan

Using Technology to Share Information: Experiences of Oesophagus Atresia (OA) and Tracheaoesophageal Fistel (TOF).................. 163
Caren Kunst

Vignette: MPSIIIA (Sanfilippo)................................... 173
Roy and Zezee Zeighami

The Empowered Patient in the Health System of the Future........... 177
Frank Grossmann, Daniela M. Meier and Therese Stutz Steiger

Vignette: The Journey of a Lifetime 191
Deb Purcell

Personalized Medicine: A Cautionary Tale or Instructional Epic 195
Dorothy Weinstein

Part IV   Closing Gaps: Promising Research and Future Considerations

**Managing Communication for People with Amyotrophic Lateral
Sclerosis: The Role of the Brain-Computer Interface** 215
Gaye Lightbody, Brendan Allison and Paul McCullagh

Vignette: The Wilderness 237
Jeneva Stone

**Opportunities and Challenges for Supporting People with Vascular
Dementia Through the Use of Common Web 2.0 Services** 241
Kyle Boyd, Chris Nugent, Mark Donnelly, Raymond Bond and
Roy Sterritt

**Vignette: Recessive Dystrophic Epidermolysis Bullosa (RDEB):
Sibling Experiences** 263
Jason Barron

**Health Policies for Orphan Diseases: International Comparison
of Regulatory, Reimbursement and Health Services Policies** 267
Durhane Wong-Rieger and Francis Rieger

Vignette: Route 125 (October 2009) 279
Jeneva Stone

Rare Diseases Challenges and Opportunities 283
Rashmi Gopal-Srivastava and Stephen C. Groft

Epilogue 291

Part I
Rare and Orphan Diseases

Why R&D into Rare Diseases Matter

Remco L. A. de Vrueh

> *The real voyage of discovery consists not in seeking new landscapes, but in having new eyes*
>
> MARCEL PROUST

Abstract The total number of patients in Europe and the US suffering from a rare disease is estimated at 30 and 25 million, respectively. Moreover, rare diseases are not confined to Europe and the US, but affect people all over the world, and consequently represent a true global health issue. As such there can be no doubt that R&D into rare diseases matter. Specific orphan drug legislations across the globe have been introduced to stimulate the pharmaceutical industry to further develop and bring the necessary therapies to the market. A total of 400 of these products have made it to the market in the US, 70 have done so in the EU, and they are truly making a difference for specific patients suffering from a rare disease. However, for the majority of rare diseases no appropriate medical interventions or care exist. In this chapter will show that in the last decades considerable progress has nevertheless been made in rare disease understanding. I will show that translation of rare disease research into orphan drug development represents one of the most important steps towards alleviating the burden for patients suffering from a rare disease. Moreover, developing an orphan drug is certainly feasible, but also tough, not without risk and requires a great deal of persistence. In my view, the way forward to give a new stimulus to R&D into rare diseases is to ensure that countries across the globe join the fight against rare disorders. Apart from western world countries and Japan, other countries, like China, India and Turkey have to step into the arena thereby really making it a global fight against rare disorders. Recent data shows that this is exactly what is happening at the moment. Finally, if we want to move rare diseases to the next level we should not merely focus on stimulating

---

R. L. A. de Vrueh (✉)
Rare Disease Matters, Louis Elsevierstraat, 2332 PM Leiden, The Netherlands
e-mail: remco113@planet.nl

rare disease R&D in general but also focus on the specific needs at disease (class) level. Of course, this should be done in close interaction with all stakeholders, including patient organizations and learned societies.

Background

Rare diseases are a complex and heterogeneous mosaic of an estimated 6,000–8,000 conditions, many of which are of genetic origin and affect children at a very early age (Van Weely and Leufkens 2004; European Commission 2013b). A rare disease is, according to the European definition, a life-threatening or chronically debilitating condition from which not more than five affected persons per ten thousand citizens in the European Union (EU) suffer (Eurordis 2013). In the United States (US), a rare disease is defined as a disease that affects less than 200,000 inhabitants (Haffner et al. 2002). The majority of the estimated 6,000–8,000 rare diseases has a prevalence of less than 10 patients per 1 million inhabitants (less than 5,000 patients in the EU) (Aymé and Hivert 2011). The total number of patients in Europe and the US suffering from a rare disease is estimated at 30 and 25 million, respectively (Eurordis 2013). Moreover, rare diseases are not confined to Europe and the US, but affect people all over the world, and consequently represent a true global health issue. As such there can be no doubt that R&D into rare diseases matter.

Rare diseases exist in all disease classes and range from exceptionally rare diseases that occur in only a few individuals worldwide to more prevalent, but still considered rare diseases. Examples of more well-known rare diseases are cystic fibrosis (CF), haemophilia and phenylketonuria. However, the majority of rare diseases are less well-known. Who is familiar with diseases like Hunter syndrome, tyrosinemia type I or alkaptonuria?

Although rare diseases have been around for centuries, it was not until the early eighties that policy makers started to pay attention to the needs of patients suffering from a rare disease. Because of the rarity, the cost of developing and marketing a medicinal product to diagnose, treat or prevent a rare disease would not be recovered by the expected sales of the product under normal market conditions. In order to overcome this hurdle, in several jurisdictions specific legislation has been introduced to stimulate the research, development and bringing to the market of appropriate medication for rare diseases, so-called orphan drugs: US in 1983, Singapore in 1991, Japan in 1993, Australia in 1998, and the EU and Taiwan in 2000 (Franco 2012). The underlying principle of the legislation is that patients suffering from rare conditions should be entitled to the same quality of treatment as other patients. In general, the legislation aims to stimulate orphan drug research and development (R&D) through a number of regulatory and economic incentives, of which a market exclusivity period of seven and ten years in the US and the EU, respectively, is regarded as being the most important. Other incentives are a number of regulatory fee reductions and/or waivers and free scientific advice.

Since the introduction in 1983 of the US Orphan Drug Act, more than 2,500 products have been recognised as potential products for tackling a rare disease and have obtained an orphan designation (OD) in the US. More importantly, over 400 products have obtained a marketing approval and they target more than 200 rare diseases (Franco 2012; Braun et al. 2010). In contrast, in the decade prior to 1983 fewer than ten such products came to market (Haffner et al. 2002). Since the EU Regulation on Orphan Medicinal Products (OMP) came into force in 2000, more than 70 Orphan Medicinal Products (OMPs) have been approved for marketing and more than 1,000 medicinal products have received an OD in the EU (Franco 2012; COMP 2011).

The numbers are truly impressive and just like the US Orphan Drug Act (Haffner 2006), the EU Orphan Drug Regulation is highly appreciated for its role in creating a favourable orphan drug development environment and making therapies for patients with rare diseases available (COMP 2011). However, creating a favourable orphan drug development environment is only one part of the story. The other part is whether the introduction of specific orphan drug legislation has also stimulated research into rare diseases. Like regular drug development, the foundation of any orphan drug development programme is the availability of sufficient disease understanding. Only fundamental and clinical research into a rare disease (e.g. aetiology, diagnosis and genetics) can reveal the necessary drug targets (Griggs et al. 2009), which, in turn, can be translated into the discovery of potentially interesting drug leads and subsequent drug development (O'Connell and Roblin 2006). Development of an orphan drug is thus the actual translation of the findings from fundamental and clinical rare disease research, much of it publicly funded. More importantly, translation of rare disease research into orphan drug development represents one of the most important steps towards alleviating the burden for patients suffering from a rare disease.

Although rare disease research and translation of this knowledge into an orphan drug development programme is clearly crucial, it is impossible to assess whether the introduction of specific orphan drug legislation has had a positive impact on rare disease research. First, the introduction of specific orphan drug legislation has been accompanied by the introduction of various specific rare disease research programmes across the globe (see Aymé and Rodwell 2012 for an extensive overview). Moreover, through a number of research and information networks, like TREAT-NMD (2013) and Orphanet database (2013), infrastructure has been given more attention and has greatly facilitated the exchange of views, experience and collaboration in the area of rare diseases (Aymé and Rodwell 2012). Finally, like medical research in general, rare disease research continuously benefits from the growing knowledge of disease biology and genomics (Van Weely and Leufkens 2004).

Although it may be impossible to quantify the impact of the aforementioned initiatives on R&D into rare diseases, better understanding of the translational process is certainly warranted. An improved understanding of the dynamics surrounding rare disease research and the translational process into product development represents an important step towards enhanced orphan drug development and ultimately reducing the burden of rare diseases. In this chapter will share with you some of my findings in the area of rare disease research. Next, I will provide a more in-depth description of factors that have been identified as important in the translation of rare disease research into the orphan drug development programme. Finally, I will briefly touch upon the orphan drug development process itself.

Rare Disease Research

Jewish writings of the 2nd century AD provided circumstantial evidence for the existence of what is now known as haemophilia (National hemophilia foundation 2013). They describe a Rabbi ruling to exempt a woman's third son from being circumcised if his two elder brothers had died of bleeding after circumcision. Although other causes for his sudden death have been mentioned, a recent study suggests that the legendary Egyptian pharaoh Toetanchamon suffered from sickle cell anaemia, a rare blood disorder. The message here is that although we are only recently addressing the issue of rare diseases, these diseases have been around for centuries. However, it was not until the end of the 19th century that research into rare diseases really took off. Around that time a number of rare metabolic conditions, like alkaptonuria (Garrod 1902), Tay-Sachs disease (Tay 1881; Sachs 1887) and Gaucher disease (Gaucher 1882) were first described in scientific proceedings. Since then the number of rare diseases has been growing and has resulted in the 6,000–8,000 conditions we know today.

Bibliometric studies on rare diseases are limited (Esen et al. 2011; Carey 2010; Leshem et al. 2010; Escudero Gomez et al. 2005), and studies on research funding in the area of rare diseases are even more scarce. Reinecke et al. (2011) studied the funding of rare disease research in Germany. Although the authors revealed great difficulty in retrieving solid data, their pilot study showed enormous deficits and inequities in rare disease research. Although there were only a few, all the bibliometric studies revealed a growth in scientific output at an individual disease level. Al-Shahi et al. (2001) compared scientific output of rare neurological conditions with non-rare neurological conditions (total of 44 diseases). They showed that, based on the publication ratio (number of papers/disease frequency), rare neurological conditions like Creutzfeldt-Jakob disease (incidence 0.02 per 100,000) and Wilson's disease (prevalence 0.4 per 100,000) exhibited a disproportionately larger research interest than, for example, stroke and transient ischaemic attack (combined incidence 250 per 100,000 or migraine (prevalence 10,000 per 100,000). Of course, this is only relative and in absolute numbers more common neurological conditions easily outperform rare neurological ones.

To better understand rare disease research, Heemstra et al. (2009) performed a bibliometric study in which the scientific output of a large group of rare diseases was evaluated (following text adapted with permission (author's right)). The disease dataset consisted of 588 rare diseases, distributed over 3 prevalence classes, with 161 diseases in the 0.1–0.9 per 100,000 prevalence group, 248 in the 1–9 per 100,000 group and 179 in the 10–50 per 100,000 group. More than 60 % (N = 375) of the diseases included in the study belonged to the disease classes (ICD; N) of oncological (C00–D48; 59); endocrine, nutritional and metabolic (E00–E90; 87); nervous system (G00–G99; 85); or congenital (Q00–Q99; 144) diseases. For each disease, Heemstra et al. determined the number of publications in four time periods: 1976–1983, 1984–1991, 1992–1999 and 2000–2007. Figure 1 provides an overview of the average number of publications per disease for the disease classes for the consecutive time periods.

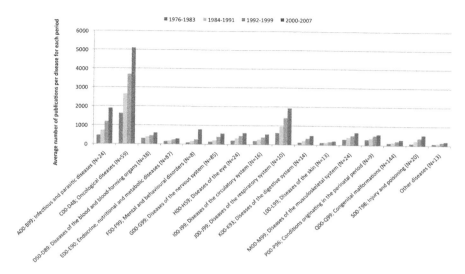

Fig. 1 Average number of publications per disease, by disease class and time period. The number of publications of each rare disease included in the study was determined with a search in PubMed (http://www.ncbi.nlm.nih.gov/pubmed/). For each disease, a PubMed search string was composed consisting of the disease name and synonyms of the disease mentioned in the Orphanet database (http://www.orpha.net). Possibilities for wrongful inclusion of a publication caused by a disease with the name of an author (e.g. Wilson disease) or a geographic region (e.g. Japanese encephalitis, West syndrome) were addressed by including Boolean NOT statements and PubMed search field tags for these terms, in a way comparable to that of Mendis and Mclean (2006). All searches were limited to English language articles and original research or case reports only. Reviews, comments and letters were excluded. Figure taken from Heemstra et al. (2009) with permission (author's right)

The average number of publications per disease increased over time for all 588 diseases in the study, from 330 publications in the period 1976–1983 to 1319 publications in the period 2000–2007, indicating a consistent increase in scientific output from 1976 to 2007. Comparison with the general increase of number of publications on overall biomedical research from 1976 to 2007 revealed that the observed increase is not statistically different from the general trend (data not shown). It appears that the introduction of the numerous research programmes and networks in the US, the EU and other jurisdictions has allowed our understanding of rare diseases to advance at a similar pace as more prevalent diseases. However, through its Framework programmes, the European Commission (EC) has funded numerous research proposals since the early 1990s. Between 2007–2010 a total of 50 proposals were funded, of which approximately 17 focussed on fundamental research and 8 on (pre-)clinical development of orphan drugs (Aymé and Rodwell 2012). A similar trend can be observed in the US where the National Institutes of Health (NIH) Office of Rare Diseases Research (ORDR) coordinates and authorises US investment in the development of diagnostics and treatments for patients with rare disorders. Programmes like the Undiagnosed diseases program (U.S department of Health and Human services 2013a) and the Rare Diseases Clinical Research Network (U.S department of Health and Human services 2013b) have been initiated to speed up R&D into rare diseases. Of course, the study by Heemstra et al. only focussed on research quantity and not quality, which is serious a limitation of the study.

What may be less encouraging is that a rather skewed distribution was observed with the highest average number of publications per disease observed for the disease class of oncological diseases. Apparently, rare cancer conditions seem to benefit more from available research funds (public and private) than all the other rare diseases. Some rare disease classes seem to hardly benefit from available research funds. One explanation could be that they all have to compete for the same research funds. Moreover, it is not uncommon that rare disease researchers have to compete with more prevalent disease researchers for the same funds (Aymé and Rodwell 2012; Van Weely and Leufkens 2004). Finally, research funds may have already been allocated to a specific disease class. This is certainly true for oncology. There is a considerable amount of public and private expenditure on oncology research, both in the USA and the EU, and a high-level transnational research infrastructure has evolved (European Cancer Research Managers Forum 2007; Eckhouse and Sullivan 2006). It is important to understand that rare cancer conditions have to compete for the same funds with more prevalent types of cancer, like colon and breast cancer.

There is sort of a "downside" to the growth in scientific output in the area of rare diseases. Through continuous research, in particular genetics, we are discovering that many rare disorders actually consist of various unique subtypes (e.g. Neuronal ceroid lipofuscinosis, mitochondrial DNA depletion syndrome, Congenital disorder of glycosylation). Each year about 250 new rare diseases or subtypes are described (Van Weely and Leufkens 2004). To illustrate the latter I have compiled a small dataset of around 350 low prevalence rare metabolic disorders (prevalence < 1/100,000 or < 5,000 patients), and shown when their clinical features were first described. As clearly depicted in Fig. 2, the data confirms the growth in the number of low prevalence rare (metabolic) disorders over time. What is interesting to notice is that the number of low prevalence rare metabolic disorders really started to go up around the time when the DNA structure was resolved (Watson and Crick 1953). Of course, the data is far too limited to determine a real cause and effect relationship. Over time there have been many pivotal discoveries in genetics that have made, and are making a difference. Just to name two: DNA sequenced for the first time (Sanger et al. 1977) and completion of the human genome project (Venter et al. 2001). Also, important discoveries in disciplines like biochemistry, cell biology and molecular biology have made considerable contributions (De Duve 1974). What may be worrying is that the number of rare disorders, in particular low prevalence rare ones, is growing at a much higher pace than therapy development. It is essential that the rare disease genetics research is accompanied by more translational research into disease aetiology and pathophysiology to provide the necessary targets and leads for an orphan drug development programme.

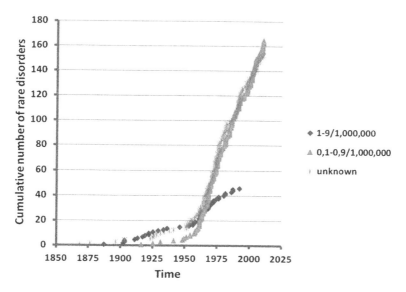

Fig. 2 Cumulative number of ultra-rare metabolic disorders over time (first clinical features described) for three prevalence categories

Translation of Research into Development

In principle, translational medicine describes the process of using the understanding of diseases to enable patients to access effective therapies. The translational medicine process consists of three main steps, although some also mention a fourth step (Homer-Vanniasinkam and Tsui 2012; Szczerba and Huesch 2012). When we focus on drug development, the first step (T1) is the actual translation of disease understanding into a phase I or first in man clinical study. A review of the summaries of opinion of the EU orphan designations reveals that the stage of development at the time of application varies from preclinical to clinical development (European Medicines Agency 2013). An orphan designation can thus be regarded as the first proof that studies are being conducted with the aim to develop an orphan medicinal product for a specific rare condition. The next step (T2) is the expansion of the clinical development programme towards the larger phase III trials and eventually approval. Clinical development of orphan drugs can be challenging and this will be discussed in the next paragraph. Finally, once a therapy has been approved, its use will have to be monitored and evaluated in the world-at-large (step T3).

For the majority of rare diseases no appropriate medical interventions or care exist. One explanation is that for many rare diseases there is little knowledge about the aetiology, pathophysiology and/or limited insight into their natural history (Van Weely and Leufkens 2004; Rare Disease Task Force 2009). Here, a clear need for more fundamental research exists; without it healthcare innovation and/or product development is not feasible. Even if a considerable disease understanding is available, the translation into product development or healthcare innovation may still be hindered. Reasons for a hampered translation may vary from lack of funding to the absence of a suitable preclinical model to establish a proof of principle. Finally, for some rare diseases, translation of research into product development or healthcare innovation has taken place, but product approval or implementation of healthcare innovation is hampered due to for instance the complexity of the disease. An example in this respect is cystic fibrosis. This chronic lung disease has been studied extensively for decades, however, the first medicine (ivacaftor; Kalydeco) to treat the underlying cause of cystic fibrosis was only approved recently (Perrone 2012).

Although knowledge is lacking, at an individual disease level several studies have appeared that describe the enormous progress that has been made from understanding the disease to the translation of this knowledge into a potential therapy. Take for example, lymphangioleiomyomatosis (LAM), a rare progressive lung disease which causes cystic destruction and eventually respiratory failure over one to two decades in young women (Henske and McCormack 2012). Over a period of fifteen years the genetic basis of LAM has been elucidated and has revealed targets to kill the LAM cells responsible for the clinical and pathological features of LAM. Moreover, clinical trials with Sirolimus, a product approved in the US to prevent rejection in organ transplantation, are ongoing (Bissler 2008). A similar story applies to cystinosis, an autosomal recessive disorder caused by mutations of the CTNS gene, which encodes a ubiquitous cystine-specific transporter (cystinosin) in the lysosomal membrane (Goodyer 2011). The disease itself has already been described in 1903. However, research in the last decade has revealed the causative gene and provided a better understanding of the natural history of the disease and its pathophysiology. It has allowed the development of cysteamine as a potential therapy for cystinosis (Gahl et al. 2007). More recently, the initial step has been undertaken to translate the disease knowledge into a proof of concept with stem cells (Syres et al. 2009). What is interesting about LAM and cystinosis is that both diseases are prime examples in which an important role is being fulfilled by patient foundations. Organizations like the LAM foundation, the LAM treatment alliance, the Cystinosis research foundation and the Cystinosis research network continue to raise considerable funds for LAM and cystinosis R&D. The focus of these organizations goes beyond supporting R&D. Like many patient foundations, they also aim to raise awareness and improve the research infrastructure (e.g. registries) (Mavris and LeCam 2012).

Studies that aim to enhance understanding of the role of translational research in the area of rare diseases at a higher aggregate level are limited. A study by Heemstra et al. (2008a) has shown the importance of pharmaceutical innovation for orphan drug discovery and development, and numerous analyses have emphasized that the big challenge in successful drug discovery and development lies in the translation of biomedical research into discovery and development of a successful product (O'Connell and Roblin 2006; Editorial-nature 2008). This translation incorporates the two-way process of using knowledge from basic research for the discovery and development of new methods for the treatment, prevention or diagnosis of diseases (O'Connell and Roblin 2006; Hall 2002), involving industry and regulators, as well as academia (Sanchez-Serrano 2006). Although the increase in biomedical research has failed to deliver the expected flood of new medicinal products (DiMasi et al. 2003; Lindsay 2003), the opposite may be true for orphan drugs. Nowadays orphan drugs make up about one third of all newly approved drugs and biologics in the US (Cote 2011). It is anticipated that this trend will continue, especially since there is a growing interest from larger pharmaceutical firms in orphan drug development, best exemplified by the introduction of specific rare disease units at Pfizer and GSK (Melnikova 2012).

Similar to medical research mentioned in the previous paragraph, in both the US and the EU, certain disease classes—in particular oncology—are associated with a high number of orphan designations and approvals, compared with disease classes with less orphan designations (Haffner et al. 2002; COMP 2011; European Commission 2013c). This skewed distribution of orphan drug development over the disease classes suggests that certain disease-specific factors favor the translation of rare disease research into orphan drug development. To better understand the translational rare disease research process, Heemstra et al. (2009) analyzed the influence of three major disease-specific factors on the chance for a rare disease to obtain at least one product with an orphan designation: disease class, research output and disease prevalence. The outcome of the study confirmed that successful translation of rare disease research into an orphan drug discovery and development programme is not only dependent on disease class, but also on rare disease research output and on disease prevalence.

The disease class of oncological diseases can serve as a valuable role model for other disease classes. Within oncology, an important boost was given to the translation of research by the 1971 National Cancer Act. This act provided the National Cancer Institute with not only the necessary funding and the mandate to support basic research, but also the application of the results of the research to reduce cancer incidence, morbidity and mortality (Haran and DeVita 2005). As already mentioned, since then, there has been considerable public and private expenditure on oncology research, both in the USA and the EU, and a high-level transnational research infrastructure has evolved (European Cancer Research Managers Forum 2007; Eckhouse and Sullivan 2006). Consequently, knowledge and understanding of oncology have advanced rapidly and have turned oncology into an attractive challenge for the pharmaceutical sector (Stratton et al. 2009). The observed differences between disease classes might also be explained by differences in the feasibility of identifying a suitable targets and drug leads (Overington et al. 2006). Within oncology, increased emphasis is given to targeted drug discovery (Benson et al. 2006). A notable success in this area has been the development of imatinib, the first tyrosine kinase antagonist, which was introduced as an orphan drug for the treatment of chronic myelogenous leukaemia (Capdeville et al. 2002).

What is equally important is that countries have national policies in place to stimulate the discovery and development of orphan drugs. To assess the impact of the EU Orphan Drug Regulation on the development of orphan drugs at a national level, Heemstra et al. (2008a) categorized the European orphan designations between April 2000 and December 2007 by country of origin. The authors defined country of origin as the country in which the company or institution was located that was leading the project; from preclinical work to initial clinical development of the particular product for the designated indication. The outcome of the study showed that the origin of designated orphan products is not homogenously distributed across the European countries. Moreover, the authors unveiled a strong relationship between orphan drug development and pharmaceutical innovation performance in Europe, and underlined the importance of innovation-based policies to enhance the development of orphan drugs in Europe.

In brief, those countries that harbor large amounts of pharmaceutical SMEs, and the countries in which companies spend larger amounts on pharmaceutical R&D and apply for more patents, do develop more orphan drugs. Interestingly, Heemstra et al. determined that scientific output in biomedical sciences plays a part in stimulating the development of orphan drugs as well, but not as large as innovation in pharmaceutical development. The quality of the science originating from a country appears to be of less importance than the quality of innovation in pharmaceutical development. It is therefore essential for countries not only to invest in the quality of their universities and schools, but also to encourage a climate that fosters innovation in pharmaceutical development. The same applies for the allocation of resources specifically aimed at orphan drugs; these should be allocated to drug discovery as well as drug development. There are examples that illustrate the impact of the allocation of resources on R&D into rare diseases: The first clinical proof of concept study of Alipogene Tiparvovec (Glybera®), the first gene therapy product approved in the EU in 2012 (Gallagher 2012), was partially funded through a translational research programme of the Netherlands Organisation for Health Research and Development (ZonMw 2005).

Orphan Drug Development

As already mentioned in the introduction, orphan drug regulations have proven to be an effective strategy to stimulate the pharmaceutical industry to develop therapies for rare diseases. Some argue that the legislation has allowed the start of a number of US-based biotechnology companies, like Genentech, Amgen and Genzyme, and consequently the translation of rare disease knowledge into numerous highly innovative rare disorder therapies (Haffner et al. 2002; Haffner 2006). Whether this is true is difficult to establish, however, it is clear that particular biopharmaceutical SMEs have embraced the concept of orphan drugs, and are generally considered to be the engine behind orphan drug development (Torrent-Farnell 2005). In the last five years the concept of orphan drugs has been the subject of numerous studies (Regnstrom et al. 2010; Heemstra et al. 2008b, 2011; Putzeist et al. 2012; Joppi et al. 2006, 2009, Kesselheim 2011).

Developing an orphan drug is fundamentally different from a regular drug. This is most apparent during the clinical development stage and concerns several factors: too small a number of patients, logistics, ethics (e.g. use of placebos), lack of validated biomarkers and surrogate end-points, poor diagnostics, limited clinical expertise and expert centres. Many companies focus on the success stories, and it is true that nowadays orphan drugs make up about one third of all newly approved drugs and biologics in the US (Cote 2011). However, the downside is that orphan drugs also excel in non-successful drug approval. In the US between 1998 and 2007 there have been 15 non-approved orphan NDAs (Heemstra et al. 2011). A similar trend was observed in the EU: a total of 43 orphan drug marketing authorization applications were withdrawn or refused (= non-approval) (Putzeist et al. 2012). What these numbers show is that developing an orphan drug is certainly feasible, but tough, not without risk and requires a great deal of persistence. Although indirectly, improved understanding of the orphan drug development process will not only stimulate innovation, but more importantly facilitate the availability of therapies for life-threatening and/or chronically debilitating rare disorders (Heemstra et al. 2008a; Eichler et al. 2008).

An overview of the outcome of the studies mentioned at the start of this paragraph has been the subject of a recent review (De Vrueh et al. 2013). In brief, three main factors were identified that (partially) explain the difference between approved and non-approved orphan drugs: the pivotal clinical trial stage, the size/experience of the sponsor and interaction with the regulatory agencies (Heemstra et al. 2011; Putzeist et al. 2012).

With regard to the pivotal clinical trial stage, although not always statistically significant, studies by Heemstra et al. and Putzeist et al. revealed the following success factors in the clinical development of orphan drugs. Most apparent was the importance of selecting and achieving a clinically relevant endpoint of the pivotal trial and identifying the most appropriate target population. In addition, the submission of sound dose finding data was also found to be important. Interestingly, other potentially important factors, such as clinical trial rigor (e.g. RCT versus open label) and the number of patients were not clearly identified as critical success factors. The latter was confirmed by recent reviews by Joppi et al. (2006, 2009) of approved orphan medicinal products in the EU in the first decade. They showed that quite a number of products have been approved based on uncontrolled pivotal trials with less than 100 patients included in the study (Joppi et al. 2006, 2009).The authors also expressed considerable criticism with regard to the quality of the registration dossiers and questioned the level of clinical development programmes. In their opinion *the number of patients studied, the use of placebo as control, the type of outcome measure and the follow-up have often been inadequate.*

Another aspect that was suggested as relevant in terms of influencing the likelihood of approval is the potential of the sponsor to carry out suitable clinical trials (Wastfelt 2006). Although not conclusive, the aforementioned studies indicate that successful orphan drug development is associated with previous experience of the sponsor in obtaining approval for another orphan drug (Heemstra et al. 2008b, 2011; Putzeist et al. 2012). Moreover, the outcome in each study revealed the size of a company as an equally important factor. Basically, the data are in line with the opinion expressed by experts: orphan drug development can be complicated for a variety of reasons and requires experience of developing and marketing an orphan drug to increase the likelihood of subsequent marketing approval. In particular, the pivotal clinical trial stage is complex and orphan drug development by inexperienced companies and/or SMEs can be hampered by a limited geographical outreach with poor access to patients and a lack of regulatory knowledge and experience in rare disease clinical trial design (Haffner et al. 2008). To overcome the gap of inexperience several strategies are being employed, either separately or combined. First, many former SMEs with one or more approved orphan drugs have brought on board management with the necessary experience at an early stage. Second, companies make use of the scientific advice service available at FDA, EMA, but also various regulatory agencies of the individual EU members. Third, small-and medium sized enterprises (SMEs) with limited experience are looking for a collaboration with an experienced partner to bring their product to the market.

Finally, the importance of regulatory dialogue during the (pivotal) clinical trial stage has been included in a number of studies (Heemstra et al. 2011; Putzeist et al. 2012; Regnstrom et al. 2010). Looking at all market authorization applications (non-orphan and orphan) in the EU in the period 2004–2007, Regnstrom et al. showed that requesting scientific advice at the EMA was not associated with successful drug approval although, complying with scientific advice was identified as predictive factor for successful approval (Regnstrom et al. 2010). The importance of compliance with scientific advice as an important predictive factor for successful drug approval was confirmed for orphan drugs by Heemstra et al. using US data (Heemstra et al. 2011). It is important to understand is that this does not mean that complying with scientific advice should be regarded as a reward for bluntly accepting the regulators' views.

The Way Forward

The occurrence of rare disorders does not stop at the US, Japan or EU border, but affects people all over the world. The way to give a new stimulus to R&D into rare diseases is to ensure that countries across the globe join the fight against rare disorders. Apart from western world countries and Japan, other countries, like China, India and Turkey have to step into the arena, thereby really making it a global fight against rare disorders. Recent data shows that this is exactly what is happening at the moment.

In particular, China is quickly becoming a medical research powerhouse. A recent literature search revealed that articles related to basic medical science and clinical research from China increased each year between 2000 and 2009 (Hu et al. 2011). In 2011, China was ranked fourth according to a number of medicine-related publications in the SCImago Country Ranking (in 2000 the rank was 18). This is a truly remarkable achievement, although some concern has been expressed with regard to the quality and trustworthiness of the published research. China's contribution to rare disease understanding is also growing. A comparison of

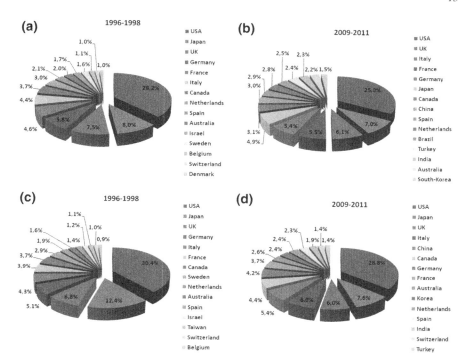

Fig. 3 Relative contribution of Top-15 countries to the total scientific output for 88 rare metabolic disorders (1996–1998: FIRST and 2009–2011: SECOND) and 84 rare nervous system disorders (1996–1998: THIRD and 2009–2011: FOURTH) PubMed was used to determine for every "SCImago" top-15 country the total scientific output. For every disease (prevalence 0, 1–50/100,000) a PubMed search string according to Heemstra et al. (2009) was prepared. Via concatenation search strings were combined into one PubMed search string

scientific output for 88 rare metabolic disorders and 84 rare nervous system diseases between 1996–1998 and 2009–2011 reveals that China now belongs to the top10 countries (see Fig. 3). The data reveals an increased scientific output for almost every country included in the study, however, the scientific output by Turkey, South-Korea, India, Brazil and especially China is increasing at a much steeper rate than the western world countries and Japan (data not shown).

China's growing role in (rare) medical research will certainly make a difference in our understanding of a number of rare diseases. However, to make a truly meaningful difference for the patient, disease understanding has to be translated into orphan drug development or some form of health care innovation. Therefore, it is important to know that China is not only making a difference in understanding disease, but is also focusing more and more on biopharmaceutical innovation (Rezzai et al. 2012). A recent publication by Rezzai et al. revealed that China has built close to 20 high-tech parks with a life sciences component in the last fifteen years. R&D expenditures as a portion of its gross domestic product has doubled between

2002 and 2007 (~ US$102 billion in purchasing power parity). Finally, between 1999 and 2007 approximately US$200 million of public funds was devoted to financing biotechnology companies.

With countries like China, Brazil, Turkey, South-Korea, and India also investing in R&D into rare diseases it is really turning into a global fight. This will certainly make a difference in the lives of rare disease patients around the world.

Conclusion

With more than 400 products approved in the US and 70 in the EU, the introduction of specific orphan drug legislation across the globe has truly stimulated the pharmaceutical industry to invest in the development and marketing of products for rare diseases. Although the legislation has been instrumental in this respect, what has been equally important is the availability of sufficient disease knowledge and the ability to transfer this knowledge into an orphan drug development programme. At a high aggregate level, data presented in this chapter shows that rare disease research has increased at a similar rate to overall medical research in the last decade.

However, the fact of the matter is that for the majority of rare diseases no appropriate medical interventions or care exist. One explanation is that for most rare diseases there is little knowledge about the aetiology, pathophysiology and/or limited insight into their natural history (Van Weely and Leufkens 2004; Rare Disease Task Force 2009). Even if considerable disease understanding is available the translation into product development may be hindered because of lack of funding or the absence of a suitable preclinical model to establish a proof of principle. Finally, product approval can be hampered due to the difficulties that are associated with the clinical development of an orphan drug.

What becomes clear is that if we want to move rare diseases to the next level we should not merely focus on stimulating rare disease R&D in general but also on the specific needs at disease (class) level. Of course, this should be done in close interaction with all stakeholders, including patient organizations and learned societies. In this respect, the disease class of oncological diseases can serve as a valuable role model for other disease classes. Public support has never been better. A recent special Eurobarometer that focussed on European awareness of rare diseases revealed that *Europeans have a relatively accurate understanding of what rare diseases are but detailed knowledge and awareness remain low* (European Commission 2011). Moreover, in the same survey, strong support for policy initiatives linked to rare diseases at both national and European level was expressed.

The first step in Europe, which is already underway, will be to link national efforts within a common European strategy for rare disease management, with the aim of levering national research resources on rare diseases through synergistic cooperation and preparation of joint strategic activities (Europlan 2013; European Commission 2013a). In the US, the National Institutes of Health are already providing support for specific, preclinical research and product development for rare and neglected diseases. To further maximise scarce resources and the coordination of research efforts, the EU and the US have recently joined forces and are the key players in the International Rare Diseases Research Consortium (2013). Through IRDiRC they want to achieve two main objectives: *to deliver 200 new therapies for rare diseases and means to diagnose most rare diseases by the year 2020*. This is certainly a goal I can live with.

References

Al-Shahi R, Will RG, Warlow CP. Amount of research interest in rare and common neurological conditions: bibliometric study. BMJ. 2001;323(7327):1461–2.

Aymé S, Hivert V, editors. Report on rare disease research, its determinants in Europe and the way forward, May 2011.

Aymé S, Rodwell C, editors. 2012 report on the state of the art of rare disease activities in Europe of the European Union Committee of experts on rare diseases, July 2012.

Benson JD, Chen YN, Cornell-Kennon SA, Dorsch M, Kim S, Leszczyniecka M, Sellers WR, Lengauer C. Validating cancer drug targets. Nature. 2006;441:451–6.

Bissler JJ, et al. Sirolimus for angiomyolipoma in tuberous sclerosis complex or lymphangioleiomyomatosis. N Engl J Med. 2008;358(2):140–51. doi:10.1056/NEJMoa063564.

Braun MM, Farag-El-Massah S, Xu K, Coté TR. Emergence of orphan drugs in the United States: a quantitative assessment of the first 25 years. Nat Rev Drug Discov. 2010;9(7):519–22.

Capdeville R, Buchdunger E, Zimmermann J, Matter A. Glivec (STI571, imatinib), a rationally developed, targeted anticancer drug. Nat Rev Drug Discov. 2002;1:493–502.

Carey JC. The importance of case reports in advancing scientific knowledge of rare diseases. Adv Exp Med Biol. 2010;686:77–86. doi:10.1007/978-90-481-9485-8_5.

Committee for Orphan Medicinal Products and the European Medicines (COMP), Westermark K, Holm BB, Söderholm M, Llinares-Garcia J, Rivière F, Aarum S, Butlen-Ducuing F, Tsigkos S, Wilk-Kachlicka A, N'Diamoi C, Borvendég J, Lyons D, Sepodes B, Bloechl-Daum B, Lhoir A, Todorova M, Kkolos I, Kubáčková K, Bosch-Traberg H, Tillmann V, Saano V, Héron E, Elbers R, Siouti M, Eggenhofer J, Salmon P, Clementi M, Krieviņš D, Matulevičiene A, Metz H, Vincenti AC, Voordouw A, Dembowska-Bagińska B, Nunes AC, Saleh FM, Foltánová T, Možina M, Torrent i Farnell J, Beerman B, Mariz S, Evers MP, Greene L, Thorsteinsson S, Gramstad L, Mavris M, Bignami F, Lorence A, Belorgey C. European regulation on orphan medicinal products: 10 years of experience and future perspectives. Nat Rev Drug Discov. 2011;10(5):341–9.

Cote T. Nat Rev Drug Discov. 2011;10:345.

De Duve C, de Barsy T, Poole B, Trouet A, Tulkens P, Van Hoof F. Lysosomotropic agents. Commentary. Biochem Pharmacol. 1974;23:2495–531.

De Vrueh R, Heemstra H, Putzeist M. Navigating orphan drugs through the regulatory maze: successes, failures and lessons learned 2013, in press.

DiMasi JA, Hansen RW, Grabowski HG. The price of innovation: new estimates of drug development costs. J Health Econ. 2003;22:151–85.

Eckhouse S, Sullivan R. A survey of public funding of cancer research in the European union. PLoS Med. 2006;3:e267.

Editorial-Nature. To thwart disease, apply now. Nature 2008;453(7197):823.

Eichler HG, Pignatti F, Flamion B, Leufkens H, Breckenridge A. Balancing early market access to new drugs with the need for benefit/risk data: a mounting dilemma. Nat Rev Drug Discov. 2008;7(10):818–26.

Escudero Gómez C, Millán Santos I, Posada de la Paz M. Analysis of Spanish scientific production in rare diseases: 1990-2000. Med Clin (Barc). 2005;125(9):329–32.

Esen F, Schimmel EK, Yazici H, Yazici Y. An audit of Behcet's syndrome research: a 10-year survey. J Rheumatol. 2011;38(1):99–103. doi:10.3899/jrheum.100335.

European Cancer Research Managers Forum: Investment and outputs of cancer research: from the Public Sector to Industry. http://www.ecrmforum.org (2007).

European Commision: National plans or strategies for rare diseases. http://ec.europa.eu/health/rare_diseases/national_plans/detailed/index_en.htm (2013a). Accessed 3 Mar 2013.

European Commission: Public Health. European Commission: Public health. Rare diseases–What are they? http://ec.europa.eu/health/rare_diseases/policy/index_en.htm (2013b). Accessed 20 Jan 2013.

European Commission: Register of designated orphan medicinal products. http://ec.europa.eu/health/documents/community-register/html/orphreg.htm (2013c). Accessed 2 Mar 2013.

European Commission: Special Eurobarometer 361: European awareness of rare diseases http://ec.europa.eu/health/rare_diseases/docs/ebs_361_en.pdf (2011). Accessed 3 Mar 2013.

European Medicines Agency. Rare disease (orphan) designation. http://www.emea.europa.eu/ema/index.jsp?curl=pages/medicines/landing/orphan_search.jsp&mid=WC0b01ac05800 1d12b (2013). Accessed 2 Mar 2013.

EUROPLAN: European project for rare diseases national plan development. http://www.europlanproject.eu/_newsite_986987/plans.html (2013). Accessed 3 Mar 2013.

EURORDIS: What is a rare disease? http://www.eurordis.org/content/what-rare-disease (2013). Accessed 20 Feb 2013.

Franco P. Orphan drugs: the regulatory environment. Drug Discov Today. 2012;. doi:10.1016/j.drudis.2012.08.009.

Gahl WA, Balog JZ, Kleta R. Nephropathic cystinosis in adults: natural history and effects of oral cysteamine therapy. Ann Intern Med. 2007;147(4):242–50.

Gallagher J. Gene therapy: Glybera approved by European Commission. BBC News Health website. http://www.bbc.co.uk/news/health-20179561 (2013). Accessed 2 Mar 2013.

Garrod AE. The incidence of alkaptonuria: a study in clinical individuality. Lancet. 1902;2(4137):1616–20. doi:10.1016/S0140-6736(01)41972-6.

Gaucher PCE. De l'epithelioma primitif de la rate, hypertrophie idiopathique de la rate sans leucemie (academic thesis). 1882. Paris, France.

Goodyer P. The history of cystinosis: lessons for clinical management. Int J Nephrol. 2011;2011:929456. doi:10.4061/2011/929456.

Griggs RC, Batshaw M, Dunkle M, Gopal-Srivastava R, Kaye E, Krischer J, Nguyen T, Paulus K, Merkel PA. Rare diseases clinical research network. Clinical research for rare disease: opportunities, challenges, and solutions. Mol Genet Metab. 2009;96:20–6.

Haffner ME. Adopting orphan drugs–two dozen years of treating rare diseases. N Engl J Med. 2006;354(5):445–7.

Haffner ME, Whitley J, Moses M. Two decades of orphan product development. Nat Rev Drug Discov. 2002;1(10):821–5.

Haffner ME, Torrent-Farnell J, Maher PD. Does orphan drug legislation really answer the needs of patients? Lancet. 2008;371:2041–4.

Hall JE. The promise of translational physiology. Am J Physiol Lung Cell Mol Physiol. 2002;283:L235–6.

Haran C and Vince DeVita. The view from the top. Cancer World. 2005;6:38–43.

Heemstra HE, de Vrueh RL, van Weely S, Büller HA, Leufkens HG. Orphan drug development across Europe: bottlenecks and opportunities. Drug Discov Today. 2008a;13(15–16):670–6. doi:10.1016/j.drudis.2008.05.001.

Heemstra HE, de Vrueh RL, van Weely S, Büller HA, Leufkens HG. Predictors of orphan drug approval in the European Union. Eur J Clin Pharmacol. 2008b;64(5):545–52. doi:10.1007/s00228-007-0454-6.

Heemstra HE, van Weely S, Büller HA, Leufkens HG, de Vrueh RL. Translation of rare disease research into orphan drug development: disease matters. Drug Discov Today. 2009;14(23–24):1166–73.

Heemstra HE, Leufkens HG, Rodgers RP, Xu K, Voordouw BC, Braun MM. Characteristics of orphan drug applications that fail to achieve marketing approval in the USA. Drug Discov Today. 2011;16(1–2):73–80.

Henske EP, McCormack FX. Lymphangioleiomyomatosis—a wolf in sheep's clothing. J Clin Invest. 2012;122(11):3807–16. doi:10.1172/JCI58709.

Homer-Vanniasinkam S, Tsui J. The continuing challenges of translational research: clinician-scientists' perspective. Cardiol Res Pract. 2012;2012:246710. doi:10.1155/2012/246710.

Hu Y, Huang Y, Ding J, Liu Y, Fan D, Li T, Shou C, Fan J, Wang W, Dong Z, Qin X, Fang W, Ke Y. Status of clinical research in China. Lancet. 2011;377(9760):124–5. doi:10.1016/S0140-6736(11)60017-2.

International Rare Diseases Research Consortium (IRDiRC). http://ec.europa.eu/research/health/medical-research/rare-diseases/irdirc_en.html (2013). Accessed 3 Mar 2013.

Joppi R, Bertele V, Garattini S. Orphan drug development is progressing too slowly. Br J Clin Pharmacol. 2006;61(3):355–60.

Joppi R, Bertele V, Garattini S. Orphan drug development is not taking off. Br J Clin Pharmacol. 2009;67(5):494–502.

Kesselheim AS, Myers JA, Avorn J. Characteristics of clinical trials to support approval of orphan vs nonorphan drugs for cancer. JAMA. 2011;305(22):2320–6.

Leshem YA, Pavlovsky L, Mimouni FB, David M, Mimouni D. Trends in pemphigus research over 15 years. J Eur Acad Dermatol Venereol. 2010;24(2):173–7. doi:10.1111/j.1468-3083.2009.03390.x.

Lindsay MA. Target discovery. Nat. Rev. Drug Discov. 2003;2:831–8.

Mavris M, Le Cam Y. Involvement of patient organisations in research and development of orphan drugs for rare diseases in europe. Mol Syndromol. 2012;3(5):237–43. doi:10.1159/000342758.

Melnikova I. Rare diseases and orphan drugs. Nat Rev Drug Discov. 2012;11(4):267–8.

Mendis K, McLean R. Increased expenditure on Australian health and medical research and changes in numbers of publications determined using PubMed. Med J Aust. 2006;185:155–8.

National Hemophilia Foundation. http://www.hemophilia.org/NHFWeb/MainPgs/MainNHF.aspx?menuid=178&contentid=6 (2013). Accessed 21 Feb 2013.

Orphanet Database. http://www.orpha.net (2013). Accessed 20 Jan 2013.

O'Connell D, Roblin D. Translational research in the pharmaceutical industry: from bench to bedside. Drug Discov Today. 2006;11:833–8.

Overington JP, Al-Lazikani B, Hopkins AL. How many drug targets are there? Nat Rev Drug Discov. 2006;5:993–6.

Perrone M. Drug approved to treat cystic fibrosis' root cause. Associated Press, January 31, 2012. http://www.huffingtonpost.com/2012/01/31/kalydeco-cystic-fibrosis-cause-drug_n_1244218.html (2012). Accessed on 20 Jan 2013.

Putzeist M, Heemstra HE, Garcia JL, Mantel-Teeuwisse AK, Gispen-De Wied CC, Hoes AW, Leufkens HG. Determinants for successful marketing authorisation of orphan medicinal products in the EU. Drug Discov Today. 2012;17(7–8):352–8. doi:10.1016/j.drudis.2011.10.027.

Rare Disease Task Force: Patient Registries in the field of rare diseases. http://www.eucerd.eu/?post_type=document&p=1218 (2009). Accessed 20 Jan 2013.

Regnstrom J, Koenig F, Aronsson B, Reimer T, Svendsen K, Tsigkos S, Flamion B, Eichler HG, Vamvakas S. Factors associated with success of market authorisation applications for pharmaceutical drugs submitted to the European Medicines Agency. Eur J Clin Pharmacol. 2010;66(1):39–48.

Reinecke M, Rommel K, Schmidtke J. Funding of rare disease research in Germany: a pilot study. J Community Genet. 2011;2(2):101–5. doi:10.1007/s12687-011-0045-1.

Rezaie R, McGahan AM, Frew SE, Daar AS, Singer PA. Emergence of biopharmaceutical innovators in China, India, Brazil, and South Africa as global competitors and collaborators. Health Res Policy Syst. 2012;10:18. doi:10.1186/1478-4505-10-18.

Sachs B. On arrested cerebral development with special reference to cortical pathology. J Nerv Ment Dis. 1887;14(9):541–54. doi:10.1097/00005053-188714090-00001.

Sanchez-Serrano I. Success in translational research: lessons from the development of bortezomib. Nat Rev Drug Discov. 2006;5:107–14.
Sanger F, Nicklen S, Coulson AR. DNA sequencing with chain-terminating inhibitors. Proc Natl Acad Sci USA. 1977;74(12):5463–7.
Stratton MR, Campbell PJ, Futreal PA. The cancer genome. Nature. 2009;458:719–24.
Syres K, Harrison F, Tadlock M, Jester JV, Simpson J, Roy S, Salomon DR, Cherqui S. Successful treatment of the murine model of cystinosis using bone marrow cell transplantation. Blood. 2009;114(12):2542–52. doi:10.1182/blood-2009-03-213934.
Szczerba RJ, Huesch MD. Why technology matters as much as science in improving healthcare. BMC Med Inform Decis Mak. 2012;12:103. doi:10.1186/1472-6947-12-103.
Tay W. Symmetrical changes in the region of the yellow spot in each eye of an infant. Trans Ophthalmol Soc. 1881;1:55–7.
Torrent-Farnell, ICORD 2005, see http://www.icord.cc/stockholm_2005.php?p=speaker_presentations (2005). Accessed 20 Jan 2013.
Treat-NMD. http://www.treat-nmd.eu/about/network/about-network/ (2013). Accessed 20 Jan 2013.
U.S department of Health and Human services: Office of rare diseases research of the national center for advancing translational research- Rare Diseases Clinical Research Network (RDCRN) consisting of Rare Diseases Clinical Research Consortia (RDCRC) and a Data Management and Coordinating Center (DMCC). http://rarediseases.info.nih.gov/Wrapper.aspx?src=asp/resources/extr_res.asp (2013b). Accessed 18 Jan 2013.
U.S department of Health and Human services: Office of Rare diseases Research of the National center for advancing translational research- Undiagnosed disease program http://rarediseases.info.nih.gov/Resources.aspx?PageID=31 (2013a). Accessed 18 Jan 2013.
Van Weely S, Leufkens H. Background paper: orphan diseases. In: Kaplan W, Laing R, editors. Priority medicines for Europe and the world—a public health approach to innovation. Geneva: World Health Organization; 2004.
Venter JC, et al. The sequence of the human genome. Science. 2001;291(5507):1304–51.
Wästfelt M, Fadeel B, Henter JI. A journey of hope: lessons learned from studies on rare diseases and orphan drugs. J Intern Med. 2006;260:1–10.
Watson JD, Crick FH. Molecular structure of nucleic acids; a structure for deoxyribose nucleic acid. Nature. 1953;171(4356):737–8.
ZonMw: Programma Translationeel Gentherapeutisch Onderzoek. Project Gene therapy for LPL deficiency. Start date 1 Aug 2005. http://www.zonmw.nl/nl/projecten/project-detail/gene-therapy-for-lpl-deficiency/samenvatting/ (2005). Accessed 2 Mar 2013.

Vignette: Autoimmune Polyendocrine Syndrome Type I (APS 1)

A.K.A. Autoimmune Polyendocrinopathy-Candidiasis-Ectodermal Dystrophy (APECED)

Patrice F. Band

Welcome to APS 1: An Exclusive Group

Our only child, Julia, suffers from an orphan disease known as APS 1. It is an autoimmune disease that causes dysfunction predominantly, but not exclusively, in the endocrine system. It is a genetic disorder associated with a defect

---

P. F. Band (✉)
15 Bedford Road, Toronto, ON M5R 2J7, Canada
e-mail: pfb@15bedford.com

in the *AIRE* gene. Currently, it is believed that APS 1 attacks one in 2,000,000 people: approximately 17 in Canada, and 160 or so in the United States. While it can attack many tissues in the body, it is best known as a disease that causes hypoparathyroidism (leading to hypocalcaemia), chronic candidiasis (inability to fight off yeast infections) and adrenal failure (by itself known as "Addison's Disease"). My wife Jennifer and I refer to those symptoms as the *trifecta*. But they represent only the minimum—the entrance criteria for membership in this small group. According to Julia's doctors at SickKids Hospital in Toronto, and our own ever-expanding understanding, she was hit by the disease very early and very hard.

The Long Road to Diagnosis

Julia enjoyed a very healthy first year of life and met all of her physical milestones early. Shortly after her first birthday, she began to suffer from fevers that continued for days and a strange rash that looked like a map of the world (later identified as giant urticaria). After a bout of pneumonia and gastro-intestinal blockages, tests showed that Julia had dysphagia (a problem swallowing that can lead to aspiration of food and fluids). She then got sicker. From that point, she lived the better part of her toddler-hood in hospital, undiagnosed until she was about two-and-a-half years old. She spent almost three months in the CCU (Critical Care Unit) and most of that on a ventilator due to serious lung infections related to chronic aspiration.

She was seen by scores of talented specialists at SickKids, a smaller group at the NIH in Bethesda, MD and many were consulted world-wide. None knew the answer. As one senior pediatrician told us in the early days, "sometimes, we need to wait and the disease will present itself". In Julia's case, APS 1 arrived wearing a disguise. Chronic lung infections, urticaria, prolonged fevers, gastro-intestinal blockages and oral thrush are not specific to APS 1. So it was not until hypoparathyroidism emerged and genetic tests were ordered that the diagnosis was made.

During that year-and-a-half, Julia was followed by over 10 departments at SickKids. Leukemia, lymphoma and degenerative muscle disease featured on the differential diagnosis. It was a terrifying period of limbo. As disease after disease was ruled out, APS 1 caused serious and irreparable damage to Julia's lungs, parathyroid, adrenal, salivary and tear glands. From this, further complications have cascaded.

To our surprise (and eternal gratitude), it was Julia's dermatologist who first mentioned the possible diagnosis. *Autoimmune Polyendocrinopathy Type I*. I asked her to repeat it. I wrote it down on a yellow sticky as though it was the winning number of the next lottery. And I kept it until the genetic results came back.

Such is the hunger for diagnosis. Any diagnosis.

Treatment at Last

After the diagnosis was made, we agreed to an aggressive trial of steroids. The results were dramatic and the long road to rehabilitation began. Now, almost seven years later, Julia's complicated medical regimen includes immune-suppression therapy and supplements, and she still has a g-tube because her dysphagia has improved but has not yet resolved. While she is frequently sick and often in hospital, she is otherwise a thriving, beautiful, bright and active girl. She is also wise beyond her years and her spirit is an intoxicating force. But most importantly, she is happy.

What I Have Learned...

Patient Care is like Warfare

Over the course of caring for Julia through the various stages of her illness, from undiagnosed and grave to identified and chronic, I have learned that patient care is like warfare. Some battles occur, of course, between the patient or her caregivers and the doctors and hospitals, but that is not what I mean. What I mean is that there is a common enemy-disease-against which many people fight using the various tools at their disposal. This war is waged by patients, their families, caregivers and countless healthcare professionals with differing backgrounds over many weeks, months and years. It is waged with high-tech weaponry: modern diagnostic instruments, pharmaceutical products and non-traditional supplements. In response, the disease engages in covert and guerrilla tactics.

As a caregiver, one has to engage in relentless efforts at advocacy, diplomacy and bargaining—usually in combination. And as with any lengthy war, battle fatigue can set in. Some become separated, never to be heard from again. Many stop fighting: some because they do not understand the cause, others simply from exhaustion. In this, caregivers have to focus their dwindling energy on ensuring that the goal is kept in sight.

Lost Health must be Mourned

Anyone who has been touched by serious illness understands intimately that lost health, like other bereavements, must be mourned. Rare diseases can take more time to diagnose than more common diseases because medicine tends to operate on the principle that "common things are common". Diagnosis is a process of elimination. So, in relative terms, the patient can lose more to a rare disease than to a readily discernible one. In the case of some rare diseases, the hardship of mourning the former "normal" self is compounded by the awareness that some health could have been salvaged with early diagnosis.

Rare Diseases can Cause Extreme Isolation

Serious illnesses isolate people. Julia has spent a significant part of her life in hospital, away from her peers, and so have we as parents at her bedside. Because she routinely suffers from pneumonia, she must frequently be kept in "isolation" within the hospital for infection control purposes. When that happens, she is kept separate from her peers in the community *and* in the hospital. At times like those, we are especially thankful to have Tory, Julia's loving COPE Service dog.

SickKids is, by any standard, a lovely building with fountains, several restaurants, and even a shopping area. But when you live there for months, it loses its veneer as you see the same exhausted parents shuffling up and down the halls, coffee in hand, around the clock. At those times, I have compared it to the island of misfit toys.

Disease also isolates people by making them (and their caregivers) different, or "other". Even when not physically isolated, you can often feel alone because family, friends and coworkers simply do not (and cannot) understand what you are going through. While it is almost an oxymoron, isolation can be felt together, as a couple or family. Fortunately, there are groups of people who band together under the banner of their particular disease. I always knew these groups existed in the world. They sold lottery tickets and marathons were devoted to them. But they did not exist in my world. They do now. For rare diseases, though, they are harder to find and their populations are both smaller and more diffuse. They are therefore less powerful at attracting attention and much needed research funds than their more "mainstream" counterparts.

We in the orphan disease camp can feel connected to one another on the internet, but we are literally countries apart in the real world. Nevertheless, these groups are an enormous wealth of knowledge. They empower us to bring information to the attention of health professionals. They also give us a sense of community, albeit on a virtual island of misfit toys.

Adapting to a New and Ever-Changing Normal

Sufferers of serious disease and their caregivers have to adapt to a "new normal" in ways that can be unimaginable to the healthy. Some are unable to. We have seen families permanently torn apart by life in the SickKids CCU. Others weathered the storm together, only to fall apart when relative calm was restored. In our case, APS 1 has tightened the bonds of our small and extended family.

APS 1 is poorly understood and its course is difficult to predict. It took hold of Julia's life when she was one year old—among the earliest recorded cases, we are told. No one can explain why Julia, who is cognitively normal (gifted, in fact), has dysphagia. Every time she has pneumonia, her team wonders whether it is infectious or auto-immune. Many of her high and countless fevers strike without warning and leave the same way. Others foreshadow hospitalization. APS 1 is also associated with other serious illnesses including autoimmune hepatitis, retinal disease and diabetes, to name just a few.

So for us, living in the "new normal" is a bit like walking through a minefield. But we do it hand-in-hand-in-hand. While Jennifer and I are in a constant state of "high-alert", Julia is miles ahead of us. She loves to hear idiosyncratic things about herself as a baby. She was born on her due date. She did not sleep that night. She refused to be swaddled and had to have at least one free arm. Recently, she said of those things that "maybe I was trying to tell you and Mama that I had a rare disease". I think Darwin himself would wonder at Julia's adaptability!

Rare Diseases: How Genomics has Transformed Thinking, Diagnoses and Hope for Affected Families

Pierre Meulien, Paul Lasko, Alex MacKenzie, Cindy Bell and Kym Boycott

Abstract The latest methods for sequencing DNA have already revolutionized our approach to the management of very rare diseases. It is now possible—using whole genome sequencing and whole exome sequencing of genomes—to diagnose phenotypically complex monogenic diseases in a significant number of cases. A pan-Canadian initiative launched in 2011 (FORGE) has developed a highly sophisticated and productive pipeline that has identified the causative genes for 67 % of cases studied. This has obvious impacts that are immediately actionable for families and their caring physicians as well as implications for implementing models of personalized medicine. It also promises to impact profoundly on our understanding of pathway biology and could accelerate the speed at which we develop medicines for both rare and common diseases.

Introduction

Due to the unprecedented technological advances in our ability to "read" our DNA (our personal code of life) the time is rapidly approaching when we will have our personal genome sequenced and available for a variety of health-related interrogations. The first human genome was sequenced at a cost of $3 billion and took

---

P. Meulien (✉) · C. Bell
Genome Canada, 2100-150, Metcalfe Street, Ottawa, ON K2P 1P, Canada
e-mail: pmeulien@genomecanada.ca

P. Lasko
CIHR Institute of Genetics, McGill University, 3649 Promenade Sir William Osler, Montreal, QC H3G 0B1, Canada

A. MacKenzie · K. Boycott
Children's Hospital of Eastern Ontario Research Institute, University of Ottawa, Ottawa, ON K1H 8L1, Canada

thousands of scientists over 10 years to complete (International Human Genome Sequencing Consortium 2001, 2004; Venter et al. 2001). Less than 10 years after the first genome, any one of the many established genome sequencing centers in the world (of which there are 3 in Canada) can do the job in a few days for around $5,000. What other area of science and technology has undergone such a rapid evolution where the cost of a significant operation has dropped by around 1 million fold within a ten year period?

No wonder then that the biomedical world is abuzz with a profusion of potential applications for this now accessible technology. How will the enormous amounts of information generated through high throughput DNA sequencing be analyzed, by whom, who will own the data and how on earth will we integrate this "new world" of medicine into an already stressed healthcare system? In order to answer these questions we need to understand what our personal genome can tell us about our individual health status at any one time and to what extent this information can inform us with regards to our susceptibility to certain diseases later in life. As importantly, we have to understand what our genome sequence cannot now tell us about our destiny and what it shall never be able to. The degrees to which our genes impact our health differ greatly depending on the condition or disease in question. As usual, with any biological system there is a wide spectrum of situations; the end of each spectrum is illustrated in Fig. 1.

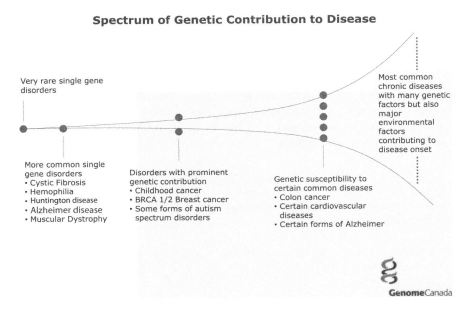

Fig. 1 The spectrum of genetic contribution to disease spans a range from very rare diseases with strong genetically determined outcomes to more common chronic diseases with more complex genetic and environmental etiology

At one end are the single gene disorders that are relatively rare in any given population (and the subject of this chapter). These include the majority of extremely rare diseases that affect perhaps one in several hundreds of thousands of individuals, and fewer more prevalent ones like cystic fibrosis, certain forms of bleeding disorders (hemophilia) and Huntington disease that affect one in a few thousand. For these diseases, the genetic component is the main (if not unique) driver of the disease. No matter what environmental factors are at play, if someone is unlucky enough (and, yes, it is a matter of which cards you have been dealt with to go through life's journey) to have a certain mutated gene for one of these disorders, then he or she will most likely have the disease.

At the other end of the spectrum, there are the more common chronic diseases where many genes may cooperate together to confer susceptibility to a disease. The disease will, however, manifest itself only if environmental components (including lifestyle choices) are added to the mix. An example is type 2 diabetes which is driving healthcare costs to unsustainable levels in most developed countries. There is a genetic component to type 2 diabetes but the disease will express itself preferentially in those who, for example, do not take regular exercise, whose nutrition is suboptimal, and who consume alcohol at above average levels (Moore and Florez 2008).

We will see later on how discoveries at one end of the spectrum can inform how we cope with complex diseases at the other end but, for now, let us concentrate on the very rare genetic diseases that affect from one in 50,000 to one in one million people. The causative mutations are referred to as *highly penetrant*, meaning that they almost always cause disease. Given that a single gene is involved they are referred to as monogenic or Mendelian diseases, the latter term reflects that the vast majority of these diseases respect the first laws of genetics laid down by the Austrian Augustinian monk Mendel (1865). Mendel worked for 8 years, using the garden pea as model system, performing the fundamental experiments on which much of modern genetics is based.

In this chapter we will describe the outcome of a two-year program of research from a pan-Canadian network that brings pediatric clinicians, clinical geneticists, scientists and high throughput genome centers together. This project (called FORGE, see below) has already had a major impact in rare disease research. As we will show, FORGE has given Canada a front line role internationally in the quest to solve the mysteries of rare diseases, many of which have been under investigation for decades without much progress toward their understanding. The new knowledge emanating from this initiative will not only help the affected families directly but will, we believe, have profound effects on how we establish an evidence based approach to personalized health in Canada.

Technology

The rapid development in DNA sequencing and bioinformatic analyses of genomes has facilitated truly revolutionary progress in the rare disease field; the work summarized below was not conceivable even 5 years ago from both a financial and technological viewpoint. So what have been the main technological breakthroughs that have enabled this paradigm shift? The so called next-generation DNA sequencing technologies were developed shortly after, and as a direct result of, the Human Genome Project (HGP) (International Human Genome Sequencing Consortium 2001, 2004; Venter et al. 2001), a ten-year $3 billion effort that profoundly altered our fundamental understanding of the genome at a number of levels. Given the complexity of the human organism, it was predicted by many that there would be over 100,000 (and perhaps several hundred thousand) genes in our genome, far greater than the current estimate of 22–25,000 functional genes (just a few more than a nematode worm). It would thus appear that the far greater complexity of humans as compared to a nematode worm derives from the 95 % of the genome outside the protein coding region. Second, when human genomes were compared with one another, it became clear that the number of repeated stretches of DNA sequence, (i.e. copy number variants or CNVs) was much greater than anticipated (Feuk et al. 2006). From a practical and technical point of view it also became clear that, although very robust, the traditional Sanger DNA sequencing methodology which had been the gold standard for several decades (and remains useful today for many specific uses) had limitations in terms of scalability and cost. If the analyses of hundreds or thousands of human genomes, and the development of reference sequences for several key species in the living world with complex genomes such as wheat, conifer trees or salmon were to be undertaken, new approaches were clearly needed.

Several groups thus started to work on developing scalable technologies capable of determining short stretches (or reads) of DNA sequence on a massively parallel scale, as well as the powerful computational based bioinformatic tools to assemble them, making alignments with the reference human genome sequence (Bentley et al. 2008; McKernan et al. 2009). This field exploded from 2007 onwards; in the following 4 years the capacity to sequence DNA increased by 1,000 fold per sequencing run! Thus in 2007 it cost about $500,000 to sequence one human genome and in 2011 it was possible for $10–20,000—a 25–50 fold reduction. Exomes (the portion of the genome which encodes protein and is reflected in the number of genes) could be sequenced for $3,000 meaning this technology was now available for the analysis of small cohorts such as those used to elucidate the causal mutations underlying rare monogenic diseases. Both Genome Canada and the Canadian Institutes of Health Research (CIHR) immediately saw value in initiating a partnership. The vision was to bring existing high throughput sequencing platforms and large scale genomics research supported by Genome Canada together with the extensive expertise and patient resources found in the large network of CIHR funded scientists and Canadian clinicians; the result: a very productive partnership.

Since 2000, Genome Canada has invested heavily in both the technologies and large scale genomics projects changing the landscape of Canadian science in this realm. Because of the decade-long sustained investment, Canada could hit the ground running when the opportunity came to tackle rare diseases. The genome sequence focused Science and Technology Innovation Centres (STICs) in Canada (one at McGill University in Montreal, one at the Hospital for Sick Children in Toronto and one at the BC Cancer Agency in Vancouver) had been at the cutting edge in the ten years since the HGP and were thus primed for this opportunity. In addition, because of their inherent skill and the fact that the Canadian health care system is organized around public single payer systems funded by provincial governments, the existing pediatric networks, medical geneticists and CIHR funded clinician scientists were rapidly able to provide a large number of extremely well phenotyped patients and families affected with rare, presumably monogenic diseases.

Recent Progress in Rare Disease Research in Canada

Collectively, rare diseases affect approximately 500,000 children in Canada with an estimated annual cost to the health care system measured in the billions of dollars. Although the cause of a monogenic disease is simple (i.e. a single gene), the clinical manifestations are sometimes so complex that a clear diagnosis is very challenging. Affected families may thus spend years visiting many disciplines of medical practice, undergo a myriad of clinical testing involving blood draws, tissue biopsies, sophisticated (and expensive) imaging technologies often with an inconclusive result. This long and frequently non-productive journey is referred to as the diagnostic odyssey. However, the new genomics-based approach promises that very soon, a patient presenting to the genetics clinic with features of a rare genetic disease will have a rapid, comparatively inexpensive and accurate molecular diagnosis, a true revolution in the care of patients and families affected by these disorders. In Canada we have been gaining insight into this future reality through a rare disease initiative called FORGE Canada (Finding of Rare Disease Genes in Canada) initiated in April 2011.

FORGE is led by Drs Kym Boycott (Children's Hospital of Eastern Ontario Research Institute, University of Ottawa), Jan Friedman (Children's and Women's Hospital, University of British Columbia) and Jacques Michaud (CHUM Sainte Justine, University of Montreal). It is supported by Genome Canada, the Canadian Institutes of Health Research, Genome British Columbia, Genome Quebec, the Ontario Genomics Institute and the McLaughlin Centre, Toronto. The early success of this initiative is due to four main strengths (1) the scientific strength and inclusive nature of the team leadership, (2) the network of clinicians who have access to superbly defined clinical phenotypes and family history data reflective of a publically funded health system, (3) intimate links with the Genome Canada funded Science and Technology Innovation Centres which provide the latest cutting edge high throughput DNA sequencing and bioinformatic analysis of human genomes, and (4) a flexible funding model that allowed, in the first instance, CIHR and Genome Canada to launch an innovative call for proposals.

From the outset, it was decided (on a Canada-wide basis) which of the 350 diseases proposed by the more than 150 FORGE members would have the most chance of benefitting from the genome sequencing technology—defined as arriving at a molecular etiology for a particular disease. The inclusion and exclusion criteria were configured by a steering group of Canadian investigators, clinical geneticists and genomics experts. The result was a pipeline of approximately 200 disorders that were primarily subjected to whole-exome sequencing analysis with a small subset of disorders undergoing a whole-genome sequencing approach. Currently over 100 genes have been identified as causal for the different diseases studied. About half of these genes are novel and never before associated with disease; others frequently broaden our understanding of the clinical presentation of a given disorder. The proportion of hits (number of disorders solved relative to the total analyzed) is a remarkable 67 %—the highest we are aware of. The utility of genomics-based diagnostic approaches can be demonstrated in patients who fall into three distinct categories:

1. Patients with diseases for which a phenotype has been described and genes are known but are nonetheless undiagnosed usually because their clinical features are atypical. In such cases the molecular diagnosis frequently broadens the established phenotype for the disease. These patients often undergo the diagnostic odyssey described above.
2. Patients with diseases for which a phenotype is known but the causal gene is not.
3. Patients with previously undescribed disorders for which there is neither a name nor a gene.

Category 1—Expansion of a phenotype associated with a disease gene:

The diagnostic odyssey is best captured in the story of two brothers from rural Canada who for over a 5 year period went from specialist to specialist, from hospital to hospital—undergoing brain scans, muscle biopsies and metabolic tests, only to be informed every time that a diagnosis was not forthcoming. With the FORGE based genome sequencing solution of the genetic riddle in 2011, the family's life has changed significantly. The two affected siblings exhibited hearing difficulties early on and then motor neurological symptoms leading to one brother being more or less confined to a wheel chair by the age of 15 years. Both brothers have now been definitively diagnosed with D-bifunctional protein deficiency (McMillan et al. 2012). They have a previously undescribed mild form of the disease (in most cases, survival beyond 2 years is unusual), a result of the type and distribution of the two recessive mutations making a diagnosis based only on clinical and biochemical assessment virtually impossible. Now the family can concentrate on managing their lives knowing exactly the cause and prognosis of their disease.

Category 2—Gene discovery for a known phenotype:

A FORGE team led by Dr. Jacques Michaud analyzed genomes from a French Canadian family affected with a subtype of Joubert syndrome (JBTS)—a rare autosomal recessive neurological disorder with a distinctive diagnostic brain malformation. Over 20 clinical variants have been described and the genetic etiology is known for just over a half. For the JBTS present in the French Canadian population, no clear causal association had been made with specific known genes even though the syndrome was first described in Quebec families over 40 years ago by Dr. Marie Joubert (a Quebec pediatric neurologist). Using individuals from eleven unrelated but clinically well documented families, including members of the family originally described by Dr. Joubert, Michaud and colleagues sequenced the exomes of fifteen individuals and discovered that mutations in the gene *C5ORF42* was the cause of this syndrome in this geographic region. Since the publication of this work (Srour et al. 2012a, b) the same gene has been associated with cases of JBTS in Saudi Arabia and is now proposed as being a relatively common cause of JBTS world-wide.

A second example of this category is Floating Harbour Syndrome (FHS) (Pelletier and Feingold 1973), a rare condition characterized by short stature, delayed bone maturation and distinctive facial appearance. The unusual name reflects its first description by investigators from both Boston Floating Hospital and Harbor General Hospital (Torrance, CA). Many cases are sporadic although a few parent to child transmissions have been documented suggesting that FHS is an autosomal dominant disorder. Despite the general recognition that FHS is a distinct syndrome—in over 25 years little progress has been made relative to its underlying genetic cause. In a FORGE study (Hood et al. 2012) led by Dr. Kym Boycott, 13 unrelated patients were identified and exome sequencing in 5 immediately revealed that the gene *SRCAP* was responsible. Targeted Sanger sequencing revealed that the same gene was mutated in the 8 other patients. Interestingly, the gene product of *SRCAP* is involved in chromatin remodeling and another gene involved in this biological process (encoding CREB-binding protein) has been shown to be mutated in a similar rare disease—that of Rubenstein-Taybi syndrome. As an anecdote, Dr. M Feingold (the clinician who originally described FHS in 1973) on hearing about the discovery prior to publication, called Dr. Boycott in amazement at the finding and could not believe that this had been elucidated using DNA sequencing technology. A comprehensive review of the genotype-phenotype correlation in FHS, in collaboration with Dr. Feingold, involving over 50 patients, will be reported shortly.

Category 3—Gene discovery for a novel phenotype:

Microcephaly-capillary malformation (MIC-CAP) syndrome was described for the very first time in two patients by FORGE clinicians in 2011 (Carter et al. 2011). Once recognized as a distinct clinical entity, several additional patients were quickly reported internationally. MIC-CAP syndrome is a severe disorder characterized by microcephaly, intractable epilepsy, profound developmental delay and multiple small capillary malformations of the skin. The FORGE team led by Dr. Kym Boycott analyzed exome data from five patients with MIC-CAP syndrome and identified novel recessive mutations in *STAMBP*, a gene encoding the deubiquitinating (DUB) isopeptidase STAMBP (*STAM-b*inding *p*rotein) that plays a key role in the recycling of cell surface receptors (McDonell et al. 2013). Within a very short period of time, the team had moved from a single Canadian patient, to an internationally recognized syndrome to a gene which implicated a new area of biology to progressive neuronal loss.

Many more of the FORGE successes published to date can be found in the reference section (Samuels et al. 2013a, b; Moffatt et al. 2013; Fernandez et al. 2012; Schuurs-Hoeijmakers et al. 2012; Koenekoop et al. 2012; Lynch et al. 2012; Rivière et al. 2012; Bernier et al. 2012; Doherty et al. 2012; Lines et al. 2012; Gibson et al. 2012; Majewski et al. 2011). These examples illustrate the power of next generation DNA sequencing in elucidating the causes of rare genetic diseases. It has been suggested that over the next eight years the underlying cause of almost all Mendelian disorders will be solved. The only way to achieve this ambitious goal is to encourage international collaboration at a massive scale. In a first step towards this the International Rare Disease Research Consortium, IRDiRC, (www.irdirc.org) was created. Initially driven by the European Commission and the NIH, this initiative now involves over 30 public and private funders from France, Italy, Germany, the Netherlands, Spain, the EU, UK, USA, Australia, China and Canada. Indeed one of us (PL) will be assuming the chairmanship of the executive committee of IRDiRC in April 2013, and another of us (KB) co-chairs one of three scientific committees that advises the executive. This reflects Canada's leading role in this field.

Families with rare diseases are immediately impacted by the results of this new approach; molecular insight influences how they live their lives going forward knowing exactly what has caused their disease, what the prognosis might be, how to best manage complications and informs reproductive decision making. While knowing what biological pathway is perturbed may not lead to a cure for the families in the short-term it will hopefully pave the way for best practice guidelines and novel interventions for future patients. Indeed Beaulieu et al. (2012) have proposed the development of a strategic tool box and preclinical research pathway for inherited rare diseases. This may well lead to targeted therapeutic interventions using, for example, repurposing of already approved drugs so that patients can benefit as soon as possible after diagnosis. This is indeed the perfect model for personalized medicine.

Lastly, insight into rare genetic diseases can contribute to much more common diseases. Thus, the insight into the molecular etiology of rare diseases will not only help the affected families but also will contribute a wealth of knowledge to human biology shedding light on how we are structured and function in both health and disease, rare and more common. For example, the discovery that mutations in the *NOTCH2* gene is responsible for the rare Hadju-Cheney syndrome (a disease exhibiting dramatic deficiencies in bone formation and degeneration) could provide great mechanistic insights into much more common forms of osteoporosis and may eventually give rise to better treatments for this complex disease (Majewski et al. 2011).

Implications for Personalized Medicine

The integration of new technologies such as genomics into complex health systems represents a challenge for healthcare providers to fully embrace. The main reasons for this are:

1. The relative lack of good economic models for health technology assessment necessary for payers to see the value of proactive integration of genomics into the healthcare system.
2. A limited receptor capacity within a healthcare system that is not optimally adapted to the efficient translation of new technologies as they mature.
3. The lack of education and training of health care professionals in the field of genomics.
4. The lack of robust harmonized health information systems needed to integrate high density data sets (omics) with detailed clinical phenotypic data while making them readily accessible to the end user.

Nonetheless and notwithstanding these barriers, we believe the work on rare diseases in Canada paves the way for more general integration of genomics technologies into the health system over time. Rare diseases represent the first true test case for personalized medicine and will, in our opinion, create the model of intervention for more common diseases in the future. We are learning through the rare disease program what genome sequencing can enable and, as importantly, what some of the limitations will be. Being able to stratify patient groups according to genomic profile will allow more targeted clinical assessments to be carried out and should give rise to more efficient drug development processes and ultimately more effective therapies. We are also learning how to deal with those discoveries which were unanticipated yet have clinical significance, the so called incidental finding; when and under what circumstances these should be communicated to patients and their families. Indeed one of the key integrated parts of the FORGE program is a study on what legal, ethical and social implications should be considered when whole genome sequencing is used to determine the root cause of a specific condition. How individuals are consented for these studies is critical as the ramifications of discovering incidental findings of a clinically actionable nature can be far reaching for both individuals and their families. But it is very early days for personalized medicine: we are at the very beginning of the application phase and so, for some, the potential benefits to patients and to the system are hard to imagine. As Arthur Kornberg (Nobel laureate who discovered the enzyme that replicates DNA) used to say: "the future is invented, not predicted".

Acknowledgments We would like to thank Alain Beaudet, President of CIHR and Thomas Caskey, who served as Chairman of the Board of Genome Canada for several years, both of whom were key players in the conceptualization of the rare disease initiative in Canada. We would like to thank the Government of Canada for their continued support of the Canadian Institutes of Health Research and Genome Canada.

References

Beaulieu CL, Samuels ME, Ekins S, McMaster C, Edwards A, Krainer A, Hicks GG, Frey BJ, Boycott KM, MacKenzie AE. A generalizable pre-clinical research approach for orphan disease therapy. Orphanet J Rare Dis. 2012;7:39.

Bentley DR, Balasubramanian S, et al. Accurate whole genome sequencing using reversible terminator chemistry. Nature. 2008;456:53–9.

Bernier FP, Caluseriu O, Ng S, Schwartzentruber J, Buckingham KJ, Innes AM, Jabs EW, Innis JW, Schuette JL, Gorski JL, Byers PH, Andelfinger G, Siu V, Lauzon J, Fernandez BA, McMillin M, Scott RH, Racher H, FORGE Canada Consortium, Majewski J, Nickerson DA, Shendure J, Bamshad MJ, Parboosingh JS. Haploinsufficiency of *SF3B4*, a component of the pre-mRNA spliceosomal complex, causes Nager syndrome. Am J Hum Genet. 2012;90:925–33.

Carter MT, Boycott KM. A new syndrome with multiple capillary malformations, intractable seizures, and brain and limb anomalies. Am J Med Genet A. 2011;155:301–6.

Doherty D, Chudley AE, Coghlan G, Innes AM, Lemire EG, Rogers RC, Mhanni A, Ishak GE, Jones SJM, Zhan SH, Fejes AP, FORGE Canada Consortium, Triggs-Raine B, Zelinski T. Mutations in the G protein signaling modulator 2 gene, *GPSM2*, cause the brain malformations and hearing loss in Chudley-McCullough syndrome. Am J Hum Genet. 2012;90:1088–93.

Fernandez BA, Green JS, Barrett B, Macmillan A, McColl S, Fernandez S, Rahman P, Mahoney K, Pereira SL, Scherer SW, Boycott KM, Woods MO, FORGE Canada Consortium. Adult siblings with homozygous *G6PC3* mutations expand our understanding of the severe congenital neutropenia type 4 (SCN4) phenotype. BMC Med Genet. 2012;13:111.

Feuk L, Carson AR, Scherer SW. Structural variation in the human genome. Nature Rev Genet. 2006;7:85–97.

Gibson WT, Hood RL, Zhan SH, Bulman DE, Fejes AP, Moore R, Mungall AJ, Eydoux P, Babul-Hirji R, An J, Marra MA, FORGE Canada Consortium, Chitayat D, Boycott KM, Weaver DD, Jones SJM. Mutations in *EZH2* cause weaver syndrome. Am J Hum Genet. 2012;90:110–8.

Hood R, Lines MA, Nikkel S, Schwartzentruber J, Beaulieu C, Nowaczyk MJM, Allanson J, Kim CA, Wieczorek D, Moilanen JS, Lacombe D, Gillessen-Kaesbach G, Whiteford ML, Robledo C, Quaio DC, Gomy I, Bertola DR, Albrecht B, Platzer K, McGillivray G, Zou R, McLeod DR, Chudley AE, Chodirker BN, Marcadier J, FORGE Canada Consortium, Majewski J, Bulman DE, White SM, Boycott KM. Mutations in *SRCAP*, SNF2-related CREBBP activator protein, cause floating-Harbor syndrome. Am J Hum Genet. 2012;90:308–13.

International Human Genome Sequencing Consortium. Initial sequencing and analysis of the human genome. Nature. 2001;409:860–921.

International Human Genome Sequencing Consortium. Finishing the euchromatic sequence of the human genome. Nature. 2004;431:931–45.

Koenekoop RK, Wang H, Majewski J, Wang X, Lopez I, Chen Y, Li Y, Fishman G, Ren H, Schwartzentruber J, Solanki N, Traboulsi E, Cheng J, Nageeb M, FORGE Canada Consortium, Keser V, Mardon G, Fu Q, Chen R. Mutations in *NMNAT1* cause Leber congenital amaurosis and identify a new disease pathway for retinal degeneration. Nature Genet. 2012;44(9):1035–40.

Lines MA, Huang L, Schwartzentruber J, Douglas S; Lynch DC, Beaulieu C, Almeida MLG, Zechi-Ceide RM, Gener B, Gillessen-Kaesbach G, Nava C, Baujat G, Horn D, Kini U, Caliebe A, Alanay Y, Utine GE, Lev D, Kohlhase J, Grix AW, Lohmann DR, Hehr U, Böhm D, FORGE Canada Consortium, Majewski J, Bulman DE, Wieczorek D, Boycott KM. Haploinsufficiency of a spliceosomal GTPase encoded by *EFTUD2* causes mandibulofacial dysostosis with microcephaly. Am J Hum Genet. 2012;90:369–377.

Lynch DC, Dyment DA, Huang L, Nikkel SM, Lacombe D, Campeau PM, Lee B, Bacino CA, Michaud JL, Bernier FP, FORGE Canada Consortium, Parboosingh JS, Innes AM. Identification of novel mutations confirms *PDE4D* as a major gene causing acrodysostosis. Hum Mutat (Epub). 2012;31(1):97–102.

Majewski J, Schwartzentruber JA, Caqueret A, Patry L, Marcadier J, Fryns JP, Boycott KM, Ste-Marie LG, McKiernan FE, Marik I, Esch HV, FORGE Canada Consortium, Michaud JL, Samuels ME. Mutations in *NOTCH2* in families with Hadju-Cheney syndrome. Hum Mutat. 2011;32:1114–7.

McDonell LM, Mirzaa GM, et al. Mutations in STAMBP, encoding a deubiquitinating enzyme, cause microcephaly–capillary malformation syndrome. Nature Genet. 2013; 45(5):556–562.

McKernan KJ, Peckam HE, et al. Sequence and structural variation in a human genome uncovered by short read, massively parallel ligation sequencing using two-base encoding. Genome Res. 2009;19:1527–41.

McMillan HJ, Worthylake T, Schwartzentruber J, Gottlieb CC, Lawrence SE, Mackenzie A, Beaulieu CL, Mooyer PA; FORGE Canada Consortium, Wanders RJ, Majewski J, Bulman DE, Geraghty MT, Ferdinandusse S, Boycott KM. Specific combination of compound heterozygous mutations in 17β-hydroxysteroid dehydrogenase type 4 (HSD17B4) defines a new subtype of D-bifunctional protein deficiency. Orphanet J Rare Dis. 2012;7(1):1–9.

Mendel G. Experiments in plant hybridisation. 1865. http://www.mendelweb.org.

Moffatt P, Ben Amor M, Roschger P, Klaushofer K, Schwartzentruber JA, Paterson AD, Hu P, Marshall C, FORGE Canada Consortium, Fahiminiya S, Majewski J, Beaulieu CL, Boycott KM, Rauch F. Metaphyseal dysplasia with maxillary hypoplasia and brachydactyly is caused by a duplication in RUNX2. Am J Hum Genet 2013;92(2):252–258.

Moore AF, Florez JC. Genetic susceptibility to type 2 diabetes and implications for anti-diabetic therapy. Annu Rev Med. 2008;59:95–111.

Pelletier G, Feingold M. Case report 1. In: Bergsma D, editor. Syndrome identification. White Plains: National Foundation-March of Dimes; 1973. p. 8–9.

Rivière JB, Mirzaa GM, O'Roak BJ, Beddaoui M, Alcantara D, Conway RL, St-Onge J, Schwartzentruber JA, Gripp KW, Nikkel SM, Worthylake T, Sullivan CT, Ward TR, Butler HE, Kramer NA, Albrecht B, Armour CM, Armstrong L, Caluseriu O, Cytrynbaum C, Drolet BA, Innes AM, Lauzon JL, Lin AE, Mancini GM, Meschino WS, Reggin JD, Saggar AK, Lerman-Sagie T, Uyanik G, Weksberg R, Zirn B, Beaulieu CL, FORGE Canada Consortium, Majewski J, Bulman DE, O'Driscoll M, Shendure J, Graham JM Jr, Boycott KM, Dobyns WB. De novo germline and postzygotic mutations in *AKT3*, *PIK3R2* and *PIK3CA* cause a spectrum of related megalencephaly syndromes. Nat Genet. 2012;44:934–40.

Samuels ME, Gallo-Payet N, Hasselmann C, Magne F, Patry L, Chouinard L, Schwartzentruber J, Rene P, Sawyer N, Bouvier M, Djemli A, Delvin E, Huot C, Eugene D, Deal CL, van Vliet G, Majewski J, Deladoey J, FORGE Canada Consortium. Bioinactive ACTH causing glucocorticoid deficiency. J Clin Endocrinol Metab 2013a;98(2):736–742.

Samuels ME, Majewski J, Alirezaie N, Fernandez I, Casals F, Patey N, Decaluwe H, Gosselin I, Haddad E, Hodgkinson A, Idaghdour A, Marchand V, Michaud JL, Rodrigue MA, Desjardins S, Dubois S, Deist FL, Awadalla P, Raymond V, Maranda B. Exome sequencing identifies mutations in the gene TTC7A in French-Canadian cases with hereditary multiple intestinal atresia. J Med Genet 2013b;50(5):324–329.

Schuurs-Hoeijmakers JH, Geraghty MT, Ben-Salem S, de Bot ST, Nijhof B, van de Vondervoort II, van der Graaf M, Nobau AC, Otte-Heller I, Vermeer S, Smith AC, Humphreys P, Schwartzentruber J, FORGE Canada Consortium, Ali BR, Al-Yahyaee SA, Tariq S, Pramathan T, Bayoumi R, Kremer HP, van de Warrenburg BP, van den Akker WM, Gilissen C, Veltman JA, Janssen IM, Vulto-van Silfhout AT, van der Velde-Visser S, Lefeber DJ, Diekstra A, Erasmus CE, Willemsen MA, Vissers LE, Lammens M, van Bokhoven H, Brunner HG, Wevers RA, Schenck A, Al-Gazali L, de Vries BB, de Brouwer AP. Mutations in *DDHD2*, encoding an intracellular phospholipase A1, cause a recessive form of complex hereditary spastic paraplegia. Am J Hum Genet. 2012;91:1073–1081.

Srour M, Hamdan FF, Schwartzentruber JA, Patry L, Ospina LH, Shevell MI, Désilets V, Dobrzeniecka S, Mathonnet G, Lemyre E, Massicotte C, Labuda D, Amrom V, Andermann E, Sébire G, Maranda B, Rouleau GA, FORGE Canada Consortium, Majewski J, Michaud JL. Mutations in *TMEM231* cause Joubert syndrome in French Canadians. J Med Genet. 2012a;49:1–6.

Srour M, Schwartzentruber J, Hamdan FF, Ospina LH, Patry L, Labuda D, Massicotte C, Dobrzeniecka S, Capo-Chichi JM, Papillon-Cavanagh S, Samuels ME, Boycott KM, Shevell MI, Laframboise R, Désilets V, FORGE Canada Consortium, Maranda B, Rouleau GA, Majewski J, Michaud JL. Mutations in *C5ORF42* cause Joubert syndrome in the French Canadian Population. Am J Hum Genet. 2012b;90:693–700.

Venter JC, et al. The sequence of the human genome. Science. 2001;291:1304–51.

Vignette: A Giant of a Man: Simon Ibell (MPS II)

Marie Ibell

It is said that we cannot know joy without pain. One of my greatest joys was December 6, 1977, the day my son, Simon, was born. It was almost two years later before I felt the pain of hearing that Simon had a rare and incurable disease: MPS II or Hunter Syndrome. A few minutes of doctor's explanation and my world came crashing down; I was powerless to express the simplest words, I wanted to scream. As I left that meeting with the doctors, there was a darkness that was lonely and intense. In those moments, I realized I was allowing the fear of my son suffering or

---

M. Ibell (✉)
Suite 612, 225 Davenport Rd, Toronto, ON M5R 3R2, Canada
e-mail: simon@ibellieve.com

dying to potentially stifle his wonderful spirit and that spirit would be the determining factor in his life: his defiance.

The fear of Simon's suffering was constant and all-consuming. Each day I had to remind myself that Simon had a life to live, and he needed to live without sympathy or question. He needed a daily dose of confidence and the encouragement to go in search of his dreams. Thus, the next time the doctors discussed his physical problems—short life span and the advice that he avoid all contact sport—my response and approach was very different. By this time Simon—the charismatic, funny, seven year old with a positive attitude—loved being around his friends and playing street hockey and many other games. From the doctor's office, we went directly to the sports store and bought soccer boots, hockey pads (goalie), skates and stick, badminton racquet and several other items for all round involvement, camaraderie, and joy. Simon became the *Mooredale* goalie. His father, an avid and former soccer player, coached the soccer team and Simon made many friends at the *Granite Club*'s badminton and bowling. Friends vied to spend time with him, to be on the same team. All this was important for Simon's confidence, his physique, and to keep all the muscles moving and in shape.

Simon was twelve when we were informed that a bone marrow transplant would help. Seattle was the foremost site and thus we moved to Victoria on Canada's west coast. Within a very short period after our arrival in Victoria, it was established that a bone marrow transplant would not help MPS II, only MPS I—a very different condition. Victoria seemed like a good place to raise children and Simon and his sister had settled in their new schools. Simon made many friends at the school but he was also subjected to some extreme bullying and difficult times in grades seven and eight. In the early days of computer animation, four boys placed the school picture of Simon in a cage, complete with animals feasting on him with the caption *"SMU School not for the deformed"*. Simon's attitude and response to this ugly bullying had a significant impact on this private school. The headmaster wanted to expel these boys but Simon insisted that they stay, that the school help them as any new school would not know they had a problem, and could not help.

The respect for Simon grew in every respect and aspect. He was involved in many sports, music, and drama and was excelling academically. On Sports day that same year, Simon was completing his second lap of the four lap run when all the other boys had past the finish line. The entire school surrounded the track clapping and chanting and even the youngest said they learned a valuable lesson to "never give up". Simon suffered a great deal, which for me was unbearable. The bullying felt like a betrayal, and I felt the pain for both of us. But, seeing how it united these children and how Simon endured the pain and did not let it destroy his trust in people helped me grow emotionally: my son became my teacher and the bullies admired him.

Each experience, each pain served as an education for Simon. He learned how to handle the worst of the abuse; it broadened his understanding of people, helped build his own character and purified his heart and soul to the point that he was sought as a major source of help for friends and those with problems. Also, he was well recognized as the manager of the senior basketball and rugby teams and travelled with the team.

During his university years, Simon continued his devotion to sport and became manager of his university basketball team. He took a year out before enrolling in his Master's program to organize a Bike ride to create awareness for MPS II. In 2002, Simon and many of his friends and supporters cycled 500 miles down Vancouver Island. Many top athletes, friends, corporations and broadcasting entities participated, gave their support and followed his progress. The ride was successful and Simon and his team reached their goals. Shortly after the bike ride, Simon had a few surgeries before moving back to Toronto to participate in the MPS II trial, which required travel to the University of North Carolina every week for almost 18 months. A few weeks into the trail Simon had a severe reaction, which was devastating for the entire family and friends. Thankfully, the results of the much awaited drug allergy test proved negative; he could continue the weekly travel. After the drug was approved by the FDA and the Canadian Government, Canada's Provincial government, Ontario, failed to give financial support for this extremely expensive treatment. Simon went forward confidently and took on the provincial government to secure support for children suffering from MPS II. This long and intense battle revealed Simon's vivid love and consideration for the MPS II children and their parents.

Standing at four foot eight inches, Simon has and continues to attract people's curiosity, blatant stares and questions from adults and children alike. Externally, Simon always appears to handle people's prying, but I think his internal dialogue becomes unsettled. It certainly affects the circulation of my energy and field of my very existence. Yet, in his world of basketball giants, friends in the professional, NBA, and college ranks give no thought to his short stature, other than poking friendly fun at him and many comments that he makes up for it in sheer brain power. Such is his professional status that he has held consulting positions in very reputable corporate entities and, after the medical trial, took a position at a global sports charity, which again kept him in the field he loved: sport. He has had the confidence to create unlimited wealth in his relationships with effortless ease and to experience success in every endeavour. Understanding that the power of his attention lay with the children that suffer from this rare and terrible condition, Simon made a conscious choice to leave *Right To Play* and to form his own foundation to garner support for MPS II.

Today, Simon is at the helm of the *iBellieve Foundation* which he founded in 2010. He is a reservoir of creativity and pure potentiality. His heart is intuitive and takes everything into account and is precise within the limits of rational thought and what he can do to retain the research and recognition for MPS II. There is a deep bond between Simon and the people he values. He has suffered a lot, but he believes trials and tribulations help build character: what we sow is what we reap. He believes happiness is life supporting and life sustaining and his next big goal to find a partner and have children of his own.

Innovative Funding Models for Rare Diseases

Amanda Lordemann, Krissi Danielsson and Jimmy Cheng-Ho Lin

> *I found in running businesses that the best results come from letting high-grade people work unencumbered*
>
> WARREN BUFFET

Abstract In the field of human genetic diseases, approximately 7,000 different diseases account for approximately 10 % of the total disease prevalence. For rare disease research, it is a very difficult task for a central, top–down entity to create and fund research for so many different diseases. With the traditional research model, therapies have been developed for less than 500 of these diseases, meaning that more than 9 in 10 diseases do not have any type of therapy. Current solutions to this problem have leveraged a bottom-up approach to allow for many stakeholders to contribute from their perspectives, a method often referred to as "crowdsourcing". Here, we propose the crowdsourcing model for both research as well as funding as a suitable way forward for rare disease research. Spurred on by the dual advances in crowdfunding and genomics, we at the Rare Genomics Institute are trying to combine these aspects to advance understanding and progress towards a cure for all 7,000 rare diseases. Because these projects are unique and often do not fit into existing grants, a web page for each patient is created and the funds collected go directly to the research for that patient. This chapter discusses how crowdfunding provides a powerful alternative funding mechanism to complement the existing infrastructure.

A. Lordemann · K. Danielsson · J. Cheng-Ho Lin (✉)
Rare Genomics Institute, 4100 Forest Park Avenue, Suite 204, St. Louis, MO 63108, USA
e-mail: jimmy.lin@raregenomics.org

Introduction

Human genetic diseases approximate the Pareto distribution, also known as the power law probability distribution. Often referred to in shorthand as the 80–20 principal, this distribution describes the phenomenon where a small number of common diseases (approximately 20 %) accounts for a large proportion of total diseases (approximately 80 %). As a result, there are a large number of diseases that are rare in occurrence that lie in the long tail, which can be approximated crudely as 80 % of diseases accounting for 20 % of the population. While the 80–20 number is just a gross approximation, the actual figures for rare diseases are as such: approximately 7,000 different diseases account for approximately 10 % of the total disease prevalence.

As shown in areas such as online retail, microfinance, marketing and social networks, these long tail problems require new models and solutions. For rare disease research, it is a very difficult task for a central, top–down entity to create and fund research for so many different diseases. With the traditional research model, therapies have been developed for less than 500 of these diseases, meaning that more than 9 in 10 diseases do not have any type of therapy.

To aid in finding cures, there have been scores of different disease advocacy groups and foundations that have been created to support existing efforts to understand these diseases and cures. Traditional disease non-profit organisations have grown to become powerful allies in the search for cures for these rare diseases, not only by fundraising, but also creating research networks, recruiting for clinical trials, and other activities. Venture philanthropy is another area that has recently added to the needed funding and innovation needed in this area.

If we examine the different solutions for the different long tail problems that exist, a common thread appears. Instead of creating larger and larger top–down structures, these solutions have leveraged a bottom up approach, to allow for the many to contribute from the bottom. This method is sometimes called crowdsourcing. For example, in online retail, Amazon allows many small niche vendors to create their own stores to sell their goods and also allows user and consumer contributions for a review system of the products and vendors. In the example of Wikipedia, with the large task of trying to curate and create content for all human knowledge, the organization has permitted many contributors to participate and add to the knowledge database as they please.

Here, we propose the crowdsourcing model for both research and funding as another solution for rare disease research. What if each of the hundreds of millions of people affected were able to contribute to the research effort? What if each of their projects was a combination of efforts from their collective social networks as well? When combined with the developments in genomics towards personalized medicine and the dropping costs of large throughput research, we are at the first time in history where we can truly tackle each rare disease one at a time with small crowdsourced and crowdfunded projects. For this chapter, we focus on using crowdsourced methods for raising funds for rare disease research.

What is Crowdfunding?

Fundraising via the collection of individual donations is nothing new, but moving fundraising efforts online can mean reduced time commitments and administrative costs. When compared with offline fundraising, going online enables easier targeting of interested audiences and potential contributors, and Internet technology simplifies the administrative side. A particular online trend that holds great promise for science is that of crowdfunding.

Inspired by "crowdsourcing," the development of technologies via code contributions from an online programming community, crowdfunding is the collection of monetary contributions from an interested online "crowd" toward business-related, artistic, or scientific projects.

The term "crowdfunding" is credited to Michael Sullivan, who in 2006 created a project called FundaVlog that collected donations to support video blogs.[1] Today, hundreds of crowdfunding platforms exist with varying features for project owners and donors. Basically, project creators submit ideas to a team at the chosen crowdfunding platform company. Approved pitches are added to the crowdfunding venture's website, and thereafter, donations are accepted through the website. Some platforms operate on an "all-or-nothing model", releasing funds only if/when a project meets its funding goal and refunding donors if projects fail to reach their goals. Some platforms offer tangible rewards to donors based on the amounts donated. In most cases, the project owner is legally required to use the funds for the stated project, and donors are kept informed about the project's progress.

Examples of Crowdfunding Ventures

Two examples of popular crowdfunding platforms are DonorsChoose (www.donorschoose.org) and Kickstarter (www.kickstarter.com).

Predating the term crowdfunding, DonorsChoose is a 501(c)3 charity launched in 2000 by a social studies teacher for the purpose of collecting funds for educational projects. Site visitors make tax-deductible donations towards specific classroom projects submitted by teachers, such as field trips or purchases of science supplies. The site once even supported the rebuilding of a Missouri school ravaged by a tornado in 2011. To date, DonorsChoose has funded nearly 340,000 projects and collected over $170 million in donations.[2]

Kickstarter, in contrast, is a platform for funding creative projects, such as independent films, art projects, or computer games. As of this writing, the company has taken in more than $450 million in pledges from over 3 million people for

---

[1] http://socialmediaweek.org/blog/2011/12/a-social-history-of-crowdfunding/#.URaOcDn-SQI.

[2] http://www.donorschoose.org/about/impact.html.

over 35,000 projects since its launch in 2009.[3] According to the company's blog, Kickstarter was behind 10 % of the independent films on the slate for the Sundance film festival in 2011 and 2012.[4]

Crowdfunding for Science

Because the availability of funding has always been a determining factor in whether projects can proceed, the growing crowdfunding trend holds considerable potential for the scientific community. Three examples of platforms targeting this niche are Petridish (www.petridish.org), Microryza (www.microryza.com), and SciFund Challenge (www.scifundchallenge.org).

Petridish is a scientific funding platform from which, depending on the project, donors can earn rewards such as T-shirts or videos, dinners with prominent scientists, acknowledgments in journals or participation in field projects. As of February 2013, 32 projects have been completed via Petridish funding. Some examples include a project to track Pacific killer whiles by GIS,[5] a study of communication in Bonobo monkeys[6] and an effort to excavate fossils of whales found in Virginia's Carmel Church Quarry.[7]

Named after *Mycorrhizae* fungi that live in soil and support plant ecosystems,[8] Microryza is a company that launched in April 2012 with the hope of supporting research that might otherwise remain unfunded. As of this writing, the site has successfully funded ten small projects with individual costs ranging from $1,150 to $7,000, and is currently collecting funds for another seven works.[9] One project to study how email spammers harvest email addresses collected over five times more money than had been needed for the project to proceed.[10]

Using the platform Rockethub (www.rockethub.com), SciFund Challenge is a project that was launched in July 2011 and aims to inspire scientists to interact with the public to fund research rather than relying on grant applications. During time-limited rounds, researchers blog about their projects via the SciFund Challenge site and post on social networking sites.Inspired donors then contribute to projects via Rockethub. As of December 2012, SciFund Challenge has completed three separate campaign rounds that have funded over 100 research projects with total donations of over $250,000.[11]

---

[3] http://www.kickstarter.com/help/faq/kickstarter%20basics?ref=home_learn_more.

[4] http://www.kickstarter.com/blog/100-million-pledged-to-independent-film.

[5] http://www.petridish.org/projects/tracking-killers-gis-mapping-of-pacific-killer-whales.

[6] http://www.petridish.org/projects/the-language-of-wild-bonobos.

[7] http://www.petridish.org/projects/saving-fossil-whales-in-virginia.

[8] https://www.microryza.com/faq.

[9] https://www.microryza.com/discover.

[10] https://www.microryza.com/projects/how-do-spammers-harvest-your-e-mail-address.

[11] http://scifundchallenge.org/blog/2012/12/15/scifund-in-3-rounds/.

Crowdfunding for Healthcare Ventures

In addition to supporting research, crowdfunding has also made inroads as a method of financing healthcare ventures. The public has traditionally shown great interest in contributing to health-related causes such as cancer or heart disease, and platforms now exist to channel this philanthropic mindset.

One example is MedStartr (www.medstartr.com), which aims to crowdfund solutions to healthcare problems, which may be anything from iPhone apps for cancer survivors to startup funds for new medical practices. MedStartr was launched on July 4, 2012, and as of February 2013, has successfully funded 18 separate projects with goals ranging from $20 to $45,000,[12] and has dozens more projects actively in the works. For example, one successfully funded project aims to provide gynecological care to poor women in the Dominican Republic, while another aims to develop reconstructive bras for breast cancer patients.

Advances in Personalized Genomics for Rare Diseases

While crowdfunding developed, there was a parallel advancement in personalized genomics. After the sequencing of the human genome in 2001 and the development of next generation sequencing, we now enter the so-called post-genomic age. While the first human genome cost billions of dollars, there has been a million-fold decrease in price to only thousands. It is now possible to think about sequencing genomes for patients who have rare diseases—possibly funded by crowdfunding.

One of the first promising examples of rare disease research by an individual is the story of Beatrice Reinhoff. Hugh Reinhoff, a physician with a strong background in clinical genetics, took this idea in 2003 to try to improve his daughter's condition. His daughter Beatrice was born with some of the symptoms of Marfan's Syndrome and some of Beals' Syndrome, yet she did not have either syndrome.[13] Concerned about her failure to thrive, Reinhoff started looking into genetic sequencing for answers. Through research and discussions with other doctors in the field, Reinhoff found himself focused on TGF-β, a growth factor important in development, where a defect could lead to abnormal bone or muscle growth.[14] He decided he would sequence the relevant genes of his daughter's genome to see what had happened, and discovered that there was indeed a change—but a puzzling one that was not where it was expected. In Beatrice, a mutation in the TGF-β

---

[12] http://www.medstartr.com/explore/successful.

[13] http://www.nature.com/news/2007/071017/full/449773a.html.

[14] http://www.hopkinsmedicine.org/news/publications/hopkins_medicine_magazine/hopkins_medicine_magazine_fall_2012/the_bea_project.

pathway has been found. This is an example of how research is now possible even for only one patient.

A second story shows how, with genomic studies, the research may even result in immediate cures. The Medical College of Wisconsin (MCW) was recently successful in sequencing the exomes of Nic Volker, resulting in the eventual cure of his genetic disorder.[15] Young Nic had a mysterious disease that caused to painful holes leading from his intestine to his skin, causing fecal matter to escape from the intestine through the holes. In a desperate attempt to try to find an answer, or even a cure, the MCW decided they would sequence some portion of his genome. In 2009, when this happened, it would have been prohibitively expensive to sequence his entire genome, so they decided to sequence portions of some of his exons, the part of the genome that possess the "instructions" for making proteins. Through much work and comparison, the doctors involved narrowed the gene in question down to XIAP, a gene associated with inflammatory bowel disease, which had similar characteristics to Nic's symptoms. To confirm, they made sure the mutation had never before been documented, since Nic's disease was unknown before he came along; it did not appear in medical literature, and out of 2,000 other human genomes, not a single one possessed this mutation. It must have been the reason. After a bone marrow transplant, Nicholas is currently cured of his genetic disease, and lives a happy normal life.

Crowdfunding as Applied to Rare Disease

Spurred on by the dual advances in crowdfunding and genomics, we at the Rare Genomics Institute are trying to combine these aspects to advance understanding and push forward towards a cure for all 7,000 rare diseases. When families approach us, we work with them to help create personalized experiments. Our current work focuses on using genomics to identify the causative variant(s) for the disease in question. We partner with clinicians and academic researchers at 18 of the top research universities in the US, including Harvard, Yale, Johns Hopkinsand Stanford. The research is then performed at these different sites. Because these projects are unique and often do not fit into existing grants, the families raise funds from their friends, family, and strangers through crowdfunding. A webpage for each patient is created and the funds collected go directly to the research for that patient. Donations range from $5 to hundreds of dollars that come from the patient's hometown or someone across the globe that had heard about their story. When the sequencing is accomplished, the researchers and the patients meet together with RGI and the researcher and one of the members of the RGI team help the patient and their family understand and interpret the results.

---

[15] http://www.pulitzer.org/archives/9180.

Crowdfunded Personalized Genomics Projects Success

In the past year, we have connected with hundreds of different families. We have setup projects for several dozen of their children and have started to see the results of these projects. Robert and Maya are two of our success stories.

Robert started out as a normal baby, doing ordinary baby activities. After his first birthday, however, in 1998, he had a dramatic medical breakdown over a few days and was left with some major disabilities, such as mobility, speech, and feeding impairments. As of now, he has to be in a wheelchair and is fed through a tube. He cannot really communicate his thoughts, other than by raising one hand for yes and the other for no, but it is obvious to his parents and others in his life that he is intelligent, alert, and social. Through investigation into his disorder, it was suggested that perhaps he had a genetic mutation. Robert's mother, Jeneva Stone, then stumbled across the Rare Genomics Institute. Within a couple of months, Robert was one of the first patients of RGI to start his crowdfunding campaign. Hundreds of people donated for his research, with donations ranging from $10 to $500, to reach the $7,500 mark for the sequencing to proceed. Once the project got the go-ahead, Robert got his blood drawn for the sequencing and his family sat down to wait. Nine months later, Robert's results came back; he had two extremely rare mutations, one on each of his PRKRA genes. Mutations in the PRKRA gene lead to dystonia 16, a syndrome that leads to gait issues and general twisting or repetitive movements. Robert is the ninth case that has been reported. His family is happy to finally have a definitive answer about their son. They now know the types of drugs that will work best for Robert, that he most likely has normal intelligence—and that he probably will continue to deteriorate. However, his mother said to *Johns Hopkins Magazine*, "We can prepare ourselves and our family, and also provide Robert with the appropriate types of emotional and life support. Information is a source of power."[16]

Maya is an adorable little girl who loves her dog, her parents, and the school bus. However, Maya has always been different. Suffering from global developmental delays, Maya has undergone multiple operations, has a hard time speaking, and has difficulty hearing. She is currently able to speak through a "talker," a device that pronounces the words when she presses the appropriate buttons. Despite visiting countless physicians, her condition had remained unexplained for years. Doctors agreed that "something genetic" was responsible for her condition, yet six genetic tests—each screening for a myriad of known genetic defects—yielded no definitive explanation. Maya's family connected with the Rare Genomics Institute to try genome sequencing. Her fundraising goal of $7,500 was met and exceeded within six hours, and the research project could start right away. The funds for Maya and her parents facilitated full exome sequencing to hunt for the disease gene. Less than one year into the project, researchers found what they believe to be the culprit behind Maya's illness: a gene active in fetal development and early

---

[16] http://hub.jhu.edu/magazine/2012/winter/jimmys-kids.

childhood. This is a scary and uncertain prospect for her family, as it is a mutation that has never been documented in medical literature. However, with more research and investigation, perhaps a treatment could be found for Maya, or her parents could get more information about the type of assistance that would help.

Conclusion

The stories of Robert, Maya, Beatrice, and Nic are but the first of a flood of discoveries and successes in rare disease research. Crowdfunding provides a powerful alternative funding mechanism to complement the existing infrastructure. Recent advances in crowdsourcing and crowdfunding are creating revolutions in microbiome research, with two large consortia projects occurring. More and more crowdfunding platforms are now available and some have partnered with university research projects, such as Consano and Indiegogo. While this field is still young, there is great possibility and promise for this alternative funding model to address these long tail diseases. It is an exciting time to be in rare disease research and we look forward to the day that we can find cures for all 7,000 rare diseases.

Rare Diseases: The Medical and the Disability Perspectives in the Age of 2.0

Sara Newman

> *If a better system is thine, impart it; if not, make use of mine*
> HORACE

Abstract Rare diseases are marginalized within the realm of medical and scientific research. This chapter addresses recent twenty-first century initiatives to improve the visibility and hence research on rare diseases by examining how they are conceptualized and represented in contemporary health care communities. Such conceptualizations are framed in terms of two models, the medical model, which prevails in the medical and scientific literature, and the social/disability studies perspective, typically found in advocacy materials. These models profoundly shape how research on rare diseases is conducted, how individuals with rare diseases are treated, and how the public perceives these issues. Because of the social model's focus on the patient, this chapter argues for incorporating significant aspects of that model into current medical thinking and education.

Introduction

Rare diseases are by definition a marginalized phenomenon within the spectrum of issues with which medicine and science deal. As such, individuals who experience rare diseases are, in a sense, twice disadvantaged; they not only cope with conditions which complicate their lives, often severely, but these conditions also receive relatively little attention from the medical community, a circumstance which limits diagnosis and treatment on various levels. Recently, however, the medical/scientific community has recognized this deficit and initiated research and advocacy efforts centered on rare diseases. This chapter addresses these twenty-first century

---

S. Newman (✉)
Department of English, Kent State University, P.O. Box 5190, Kent, OH 44242-0001, USA
e-mail: snewman@kent.edu

initiatives by examining how the very notion of rare disease is conceptualized and represented in contemporary health care communities.

As the discussion demonstrates, such conceptualizations are framed in terms of two models, the medical model, which prevails in the medical and scientific literature, and the social model, typically found in advocacy materials. These models not only shape how research on rare diseases is conducted but also how individuals with rare diseases are treated and how the public perceives those individuals. Because of the social model's focus on the patient, especially as manifested in online advocacy sites, this chapter argues for incorporating significant aspects of that model and those manifestations in scientific thinking; that merger can be facilitated if the curricula of graduate programs in medicine and science include the social model of illness and language use.

Health and Disability in the Medical and Social Models

Although no two human bodies are precisely the same, their parts and functions can be understood in terms of certain expectations, norms which reflect the state of medical knowledge of the given time and place in which they are held. Once applied, such normative knowledge influences how a given culture understands disease and health. Until the end of the eighteenth century, for example, Western cultures evaluated health predominantly in humoral terms. Without direct access to the internal living body at work, this perspective understood illness by examining perceived correspondences between outer features and inner health and character; thus, the external melancholic, or depressed demeanor and facial features reflected an inner excess of the melancholic bile (Lavater 1789). Historical and technological differences aside, some physical variations inevitably complicate the lives of those affected. Indeed, the percentage of problematic physical conditions across recorded history ranges somewhere between 10 and 25 % (Michelle and Asch 1988). At present, the term used to describe those problematic physical conditions or illnesses is disability.

In the twentieth and early twenty-first centuries, disabilities are typically evaluated by means of the medical model (Mishler 1981). Based on observable data, this model enables scientists to develop categories relating to the operation of the human body and/or the natural world. Because the model relies on the inductive scientific method, its data and categories are presumably objective; when used to diagnose, treat and/or cure physical problems, the categories function as pathological measures of normality and abnormality (Clarke et al. 2010). Such measures are evident, for example, in the mental health rubrics of the *Diagnostic Statistic Manual of Mental Disorders* (*DSM*) (American Psychiatric Association 1994) and the statistical compilations on which evidence-based medicine operates. By framing individual instances of illnesses within broader statistically derived units, this approach allows scientists to quantify illnesses into broad-based yet specifically-focused categories. Although the categories enhance diagnosis and treatment, they mask aspects of individual illnesses. Within the resulting (bio) medical model, then, each rare illness is but one "tree" within the "forest" of rare types; the experience of the patient and the specifics of each condition are hidden within the many names and types. Accordingly, the (bio) medical model accentuates a fundamental and problematic characteristic of rare diseases, their lack of cultural presence and the deficit of scientific research on them.

Recently, advocacy and academic communities have countered this (bio) medical model of health and disability by offering the social model (Linton 1998). As its proponents point out, even facts derived through the scientific method are not entirely objective when examined without context. That the criteria for many disorders have changed within the *DSM*'s successive editions discloses how scientific knowledge changes as well as how it interacts with social norms. Homosexuality, Asperger's Syndrome, and Narcissistic Personality Disorder have all appeared in an earlier addition of the *DSM*; none is included in the latest version, the *DSM-V*, although their symptoms may appear under different diagnoses.

To recognize the variations in human bodies, the social model examines each physical disability in terms of difference and rehabilitation rather than the (bio) medical model's aims of normalization and cure (Siebers 2006). To those ends, the social model distinguishes between "impairment" and "disability". An impairment is a physical fact, but a disability is a social construction; immobility impairs while a building without a ramp disables (Shakespeare 2006). Moreover, the social model insists that the voice of the patient belongs to any discussion of a condition (Couser 1994). Such an inclusive approach recognizes the individual's personal experience while chipping away at the objective veneer of the medical model. As such, it is especially appropriate for rare diseases whose individual manifestations are often lost in the (bio) medical model's statistical perspective (Siebers 2008).

To recapitulate, the (bio) medical model supports the current normative understanding of illness which underpins how members of the medical and scientific communities are taught to understand their disciple and, accordingly, how they conduct evidence-based research on rare diseases. Aimed at diagnosing, treating, and/or curing these conditions, this model influences how rare diseases are medically framed and perceived in medical contexts and beyond; understood within a quantified aggregate, specific rare diseases are beyond the ken and care of most world citizens. Unaware that this model supports the work, research proceeds in ways, as discussed below, which contradict the goals of those efforts. In contrast, advocacy groups use the social model to consider rare diseases within their broader contexts; by including the voices of those with rare diseases, and offering these materials online, these sites educate a wider public about the ways in which actual people experience these often diverse and complicated conditions.

The Medical Model and Rare Diseases

A Mid-Twentieth Century Example

William Bean's standard 1967 study, *Rare Diseases and Lesions: Their Contributions to Clinical Medicine*, offers a typical mid-twentieth century medical perspective. To begin, the work is "Dedicated to the Victims of Rare Diseases" whose "quiet courage is a lesson in patience... with the hope that someday we may control or eradicate the biological troubles they so unhappily exemplify" (Bean 1967). From the outset, Bean characterizes rare diseases as beyond the patients' control; dependent on her/his physician, that patient is a victim who suffers patiently (pun intended) with a broadly conceived condition. Such paternalistic language is not only used to identify the patient with the illness s/he might "exemplify", but that identity involves eliminating or regulating the "unhappy" "troubles" suffered without considering how the patient's experiences them or what their specific conditions are.

This sentiment continues in the "Forward" by Victor A. McKusick, MD who offers "four reasons that rare conditions are, or should be, of interest to physicians" (Bean 1967). In addition to their values for advancing science and helping the "victim", "finally, they are fun. They introduce variety into the humdrum of the physician's daily routine. They keep his powers of observation from undergoing atrophy" (Bean 1967). Although this reason for pursuing research might appeal to the physician reader, the patient may not share that perspective. By characterizing the study of rare diseases as clinical fun, even indicating concern for the physician's (rather than the patient's), physical health, McKusick further reduces the sense of the patient's experience; thereby, that voiceless patient is dehumanized and marginalized as indeed her/his rare disease is by definition.

The text moves on with Bean indicating that rare diseases are too numerous to name in their entirety. "Any thought of making this book all-inclusive is defeated by the rapid recovery of new diseases and the necessity of keeping the survey down to a manageable size. I had no such intention, anyhow" (Bean 1967). Bean's comment renders the notion of examining individual rare illnesses unworthy and/or unnecessary. Moreover, his statement about "rapid recovery" is not clear. Does he mean that the illnesses are incidental to the patient or that some types of illness have been eradicated; regardless, specific characteristics of illnesses are unspecified and insignificant (although rare diseases are, Bean acknowledges, extremely diverse). Continuing, he dismisses the approach of nineteenth century physicians. "Gould and Pyle have usurped the field in *illustrating* the outrages of deformity and freaks. I have resisted the temptation to make this a gallery of the unfortunate or ugly" (Bean 1967). On the one hand, Bean is a cutting-edge scientist and will not merely illustrate outward ugliness but instead will categorize inner problems. On the other hand, he justifies his explicit aim of looking at the whole, inner and outer, by his incomplete survey thereby reifying the marginal, deviant status of rare diseases, patients and their bodies, as did Gould and Pyle in their outdated terminology (Adams 2001).

Next, Bean (1967) provides an overview of past and present concepts of medicine and disease causation: the ontological, platonic, anthropologic-biographic, ostrich, and punitive. In contrast to these subjective approaches, he notes, his is objective. "Firm belief in any a priori system of medicine tends to disqualify the mind for correct observation, since everything is seen through the astigmatic eye of bias" (Bean 1967). Bean speaks to the objectivity of his medical model as if that objectivity is unassailable; in fact, he has again undermined this non-biased approach in his preceding discussion, which shapes his attitude about rare diseases as much as Gould and Pyle's work. In Bean's defense, hindsight is almost always 20/20.

The body of the text addresses various classes of rare diseases. Each chapter consistently presents the facts in language which distances diseases from the patient and the patient from the physician. For example, the first chapter covers "Bleeding from the Gut in Rare Disorders with Diagnostic Lesions of the Skin and Membranes":

> Every careful report of a large number of patients with bleeding from the alimentary canal contains an unhappy category, "Cause unknown." This may range from a small percentage to about a quarter of the total, depending on the nature of the clinical experience, the skill and honesty of the physician and other variables. It is toward the reducing of this group, even by a little, that a consideration of rare diseases which cause such bleeding is emphasized here. (Bean 1967)

Thus described and following the medical model, this rare disease type is all about the physician's expertise and its application to reducing numbers. Following that model, too, the patient has no role in these efforts; s/he is a victim who depends on the physician for cure and even a description of the illness s/he experiences. When Bean refers to a qualitative, human matter, a feeling, he does so to reveal how difficult determining causation is from the physician's perspective; specifically, the "unhappy category" (the same feeling he mentioned in his dedication) prevents the scientist from knowing the rare disease's cause. Such knowledge is important. But, Bean never mentions how bleeding from the gut presents in the patient, let alone how it feels. Despite the genuine concern Bean offers for rare disease patients, his linguistic choices obscure the experience of the patient in diseases which are defined by their individuality.

In one sense, Bean's language is factual, as it must be; yet, the facts are represented incompletely by his own admission, not only avoiding the complex categories beyond the scope of the book but also avoiding any representation of the patient's perspective. Ironically, having rejected other more subjective approaches than his own, Bean's work is not entirely objective; in his medical modulated voice, the patient is a victim who depends entirely on the doctor's "skill and honesty" to improve the situation surrounding his/her deviant body, even if only minimally in the case of rare diseases. In his voice, too, rare diseases are one rather than many.

The Early Twenty-First Century Biomedical Model and Rare Disease

The language of Bean's medical model resonates in a 2010 collection on rare disease epidemiology (Paz and Groft 2010). Here too, rare diseases are represented by presumably objective facts which locate individual illnesses within broad categories. Now reflecting the twenty-first century (bio) medical model within the state-of-the-art evidence-based approach to medicine, the language defines rare diseases as an aggregate by means of highly statistical, technical terminology and passive voice (rather than the author's voice evident in Bean's work). Such language, which typifies contemporary scientific discourse, maintains apparent objectivity by distancing doer from what is done (Gross 1990).[1] No doctor or scientist is present as framed in the passive voice, and no one but the expert can understand the technical terminology. Framing the language numerically also increases the sense that the rare diseases represent deviance by locating that deviance not in a physical body but in an abstract statistical norm. The articles apply these characteristics, manifesting the biomedical model, in their focus on broad populations yet specific types, creating arguments about what rare diseases are, what causes them, and how they present in patients that speak to prevalence rather than individuality.

The articles consistently define rare diseases by means of numbers, contributing to the sense that the aggregate is a monolithic entity. In the first chapter, for example, the authors note that "estimates approaching or even exceeding 7,000 conditions have been expressed" (Groft and Paz 2010). As stated, rare diseases constitute a whole without apparent parts. Although that whole is provided as a discrete number, 7,000, it is nonetheless an estimate, an inexact data point. In addition, the use of the single number elides the distance between the estimate and its environment. Of course, quantitative data is essential to scientific research on rare diseases. No one wants to try a treatment based on untested generalities or anecdotal evidence (Gross 1990). But, such information does not disclose alternative material, either numerical or contextual, and is not, therefore, universally clear or complete, as its statement here suggests.

As several articles point out, rare diseases are framed as an aggregate in part to support research initiatives: when aggregated, rare diseases constitute a bigger and, therefore, more fundable, category than any one of its many manifestations alone. In addition to its clear benefits for rare diseases, this approach has negative side. As Chap. 1 indicates, for instance, because rare diseases are understood as an aggregate, they are difficult to categorize. As a result, "the burden of most of these diseases remains invisible to the system" (Groft and Paz 2010). Ironically, and despite this acknowledgement, the quantitative, objective language reinforces the invisibility of rare diseases—and thus an abstract association with deviance—by characterizing them as a statistical, disembodied aggregate rather than by acknowledging their many types and diverse symptoms. Studying the aggregate depends in turn on prevalence, a concept involving frequency, incidence, commonness, and pervasiveness. As the authors of Chap. 3 put it:

> Prevalence is one of the most popular epidemiological measures and is defined as the probability that an individual in a population will be a case at time t. Generally speaking, the target population for this measure should be the population at risk but, owing to the

---

[1] In active voice, the subject, or doer of the action, is represented as doing the action—as being the subject of the statement; in the passive voice, the subject of the sentence appears at the end of the statement, seeming to be the object of the action. For instance, "she read the book" is in active voice with the subject at the beginning. Stating it is the passive, "the book was read by the woman" places the subject at the end.

> difficulty of ascertaining the latter, the general population is regularly used as a denominator. Indeed, another more practical definition of prevalence is the proportion of a population that has any given diseases at some specific point in time—usually called point prevalence. A second prevalence measure is the period prevalence, which is the probability that an individual in a population will be a case anytime during a given period of duration. (Paz et al. 2010)

Measured in graphic, mathematical nomenclature—points and periods and units—the passage frames prevalence vis-à-vis time and numbers rather than particular symptoms. Like the use of the aggregate, the technical language distances data from the world, the physical body specifically, and requires expertise to understand. This distancing effect is also evident in the use of equations to define prevalence, for example, "Prevalence proportion = Incidence rate x Duration of the disease" (Paz et al. 2010). Again, the language's sense is that rare diseases represent differences from statistical norms. Because epidemiological rare disease research depends on quantitative data, numbers and their prevalence, it excludes the experience of patients. As described in Sect. 5.6, on "What Else Matters? The Place of Personal Experience", the patient is an add-on concern.

> We turn finally, albeit briefly, to two elements of evidence-based medicine (encompassed in its definition) that often get forgotten. These are the expert opinions of the treating physician relating to the individual patient and—perhaps most importantly—the opinions and wishes of the patient. As illustrated above, most new treatments in the early phases of clinical development are probably worse than placebos. This is a sad fact but a realistic one. Of course, every patient will have a different perspective on treatment options and what matters to them. Some of us will clutch at any straw of hope and others will feel the emotional and physical burdens of an experimental toxic treatment (possibly after several earlier options have failed) are too much to bear. A patient suffering with a life-threatening disease, might argue that nothing can be worse than the inevitable disease prognosis. Put in slightly more scientific terms of benefit/risk assessment, if survival is the efficacy endpoint, then almost any and all adverse effects tend to be of secondary importance to mortality. (Day 2010)

Although this section considers the patient's personal experience, it does so "briefly" so as to focus on the physician's knowledge. As described, patients face treatments so punishing that their perspectives are rendered subjective and, therefore, not useful to rare disease research. In the end, the physician is the only expert here; the patient's perspective is dismissed as subjective, useless, and deviant vis-à-vis scientific information. It is also significant that the passage reframes discussion of patient experience "in slightly more scientific terms", as if to reinforce the evidential superiority of the scientific voice and data. Elsewhere, patients' thinking is characterized as emotional and therefore neither objective nor useful in slightly different terms.

> The severity of the disease often results in a limited life span so that the prevalence (total number of cases) remains low. This also implies a disproportionate distribution of young patients with rare diseases. The combination of rarity, severity and children makes this a particularly emotive topic. Again, little sense appears of a person within the objective data. Patients' wishes, therefore, may often over-ride the data. To what degree should this be respected? The easy answer is "always" but in some cases those wishes cannot be respected.... patients may need to be protected against their own over-enthusiasm.

The understanding of risk is generally poor and similar risks are interpreted differently depending on the context—both by patients and professionals. Hope in desperate situations is important but the distinction between hope and expectation is blurred. (Day 2010)

After mentioning that the person is lost in the "objective data", the passage first notes the likelihood that the rare disease patient will die young and then comments that early death contributes to low disease prevalence. Given the nature of this "emotive topic", the scientist reader is cautioned against using such irrational patient information in scientific decision-making. Ironically, the patient is thereby lost in the data. Moreover, patients' experiences are characterized as "wishes", that is, as desires that suggest fantasy. Described from the scientific community's perspective, patients are desperate and cannot think straight. Although wishes should always be respected they are not always; final authority is placed firmly in the control of the rational scientists. The authors of Chap. 1 offer yet a different take on patient subjectivity. Having already identified the dearth of research on rare diseases, they note:

For some rare diseases, it is not lack of information, but information overload that can be overwhelming to patients and their families. It is important with multiple sources presenting information to the patients or their families to remain aware that not all patients are capable of accepting or absorbing the same amount of information and at the same pace as others. Recognizing variability in perceptions of the disease and desire to learn more about a rare disease occurs at different rates for everyone. Family members and friends have to be prepared to meet the patients where they are or where they want to be intellectually or psychologically with respect to their disease and not where others believe they should be. (Groft and Paz 2010)

Previously, these authors indicated that scientists need more information to reduce the prevalence of rare diseases. Here, however, they state that the same impoverished amount is too much for many overburdened patients to handle. That line of reasoning suggests that individuals with rare diseases suffer from as unaddressed cognitive deficits, another manifestation of their bodily deviance from a norm.

Interestingly, patient experience is not only expressed in terms of aggregated data but also in terms of surveillance. In epidemiological terms, surveillance involves passive or active manifestations, rooted in direct or indirect contact with the patient. In other words, surveillance is about getting good, i.e., non-biased, information to crunch rather than looking at individual patient experiences (Richesson and Vehik 2010). Although this term, surveillance retains the objectivity of inference, its cultural implications are quite clear and are its perspective on the patient as a data point (Foucault 1974). Similarly, the patients' perspective is subsumed in the quantitative language of assessment when Chap. 23 addresses rare diseases patients' quality of life or HRQOL:

HRQOL assessment in patients with rare diseases can help to identify health needs.... Reliability as well as content, criterion, and construct validity, and also responsiveness should be taken into account in selecting the instrument to be used assessing patients with rare diseases. The use of a proxy-report may be essential in some cases where the patient is cognitively impaired or unable to communicate.... Given the impact of rare diseases on the quality of life of both patients and carers, it is likely that interest in its measurement will continue to increase among professionals, patients and the general public. Improving the quality of life of people with rare diseases should be one of the most important goals of any health care intervention or multidisciplinary approach. (Rajmil et al. 2010)

From this scientific perspective, quality of life issues involve numbers, their reliability, and the patterns they form in aggregate. Such numbers presumably rise above the patient's subjective experience and speak for everyone. So too are other elements of the patients' worlds excluded from the scientific data. When family enters the discussion, the articles separate them from the research and researcher. The family's role is often invoked in terms of family-management and within "the community" as stakeholders rather than particular experiences. Similarly, the patient advocate is distanced from the medical community:

> Frequently, those involved with larger numbers of patients in their practice or research protocols recognize the expression of a rare disease may vary from patient to patient. In many instances, it is the active patient advocacy group leader who describes the differences in patients. Appropriate epidemiological studies are required to confirm the opinions offered by clinicians, patients, and families. (Paz et al. 2010)

The passage dismisses advocacy work from the realm of science. In fact, the authors state that discussion of difference is utterly invalid without support by epidemiological research; in so doing, the article explicitly separates the social model from scientific work. True, as the authors state just below, the stated scientific goal is "to enhance social awareness and visibility" (Paz et al. 2010). But, the language used to that end, as well as the facts brought to bear on it, have the opposite effect. The work of considering the patient is deferred to advocacy organizations rather than scientists and characterized as lacking legitimacy on its own.

To recapitulate, epidemiological inquiry into rare diseases serves its disciplinary goal of treating rare diseases across many populations. To that end and enhancing funding initiatives, its language, as represented in the articles considered here, defines rare diseases monolithically, characterizing them in terms that call attention to their types and patterns; as such, rare diseases constitute an abstract category of statistical deviance from a monolithic norm. When human behavior, physical actions, or opinions are evoked, as in Bean's work, they are deemed too subjective to consider and separated from scientific research and decision-making. The existence of this common discourse reveals that scientific and medical language is learned and applied in ways which instill and perpetuate particular conceptions of rare diseases; these biomedical conceptions have significantly improved treatment of rare diseases but without necessarily increasing their visibility as much as envisioned.

The Social Model and Rare Disease

While the epidemiological research attempts to reduce the prevalence of rare diseases, advocacy discourse aims at helping individuals with rare diseases live with their physical bodies. Speaking from the social model of illness, these sites include the experiences of individuals and locate them within the broader healthcare context. In so doing, the site makes rare diseases visible from various perspectives and to more people globally. These elements are evident in the website of EURODIS, a European advocacy organization for rare diseases (EURODIS). Following the social model, the site defines terms fully, providing multiple perspectives on rare diseases and on individual manifestations. In particular, EURODIS relies not on numerical data alone, nor does it consider rare diseases in the aggregate. Instead, EURODIS presents the "paradox of rarity" and clarifies misperceptions associated with it. First, the site presents the European definition of rare diseases as "**less than 1 in 2,000**" (EURODIS).[2] As it explains:

> Despite the rarity of each rare disease, it is always surprising for the public to discover that according to a well-accepted estimation, "about **30 million people have a rare disease** in the 25 EU countries", which means that 6 % to 8 % of the total EU population are rare disease patients. This figure is equivalent to the combined populations of the Netherlands, Belgium, and Luxembourg. (EURODIS)

Rather than offering one number, the discussion compares several statistical sets on the presence of rare diseases in European countries. As the multiple facts suggest, rare diseases are a whole constituted by parts, each one of which merits attention. EURODIS makes this point repeatedly, for example, when it states that "diseases are rare, rare diseases are many" and that it is "**not unusual to have a rare disease**" (EURODIS). By connecting part to whole, the site demonstrates that rare diseases are individually few yet affect many as a whole. To clarify this paradox of rarity, the site also addresses any "misperception and confusion" (EURODIS) that the audience may hold true about rare disease terminology and practices. In addition to defining such concepts as orphan drugs and diseases, EURODIS focuses on diagnosing specific manifestations of bodily difference rather than on identifying statistical deviance. In the discussion, as indicated, it uses numbers but combines them with patient experiences in a broader health-care context.

> A survey by Eurordis (EurordisCare2)[4] focusing on diagnostic delays for rare diseases, has revealed that, for Ehlers Danlos syndrome, 1 out of 4 patients waited for more than thirty years before being given the right diagnosis. 40 % of patients participating in the survey received a wrong diagnosis before being given the right one. Among them:
> - 1 out of 6 underwent surgical treatment based on this wrong diagnosis;
> - 1 out of 10 underwent psychological treatment based on this wrong diagnosis.
>
> (EURODIS)

From a diagnostic perspective, the site's goal is avoiding and "correcting mistakes" rather than reducing the prevalence of illnesses. By focusing on diagnosis,

---

[2] As is not evident, the site's visual elements contribute significantly to its efforts to be direct and locate parts within wholes in their appropriate contexts. A picture of an individual in a wheelchair appears at the outset; and evident above, the site uses red numbering to emphasize certain information. Moreover, the site uses red throughout to highlight significant elements. These red areas are bolded in this text.

the site can describe the effects of rare diseases on real people as well as on the health-care system. Thus, described, those effects reveal how these bodies are individually different rather than statistically deviant. In clarifying misperceptions about diagnosis, the site also considers the physician's expertise. As EURODIS notes, the physician is not the only expert on rare diseases. To support this point, the site reports that:

> Up to 50 % of patients have suffered from poor or **unacceptable conditions of disclosure**. In order to avoid face-to-face disclosure, doctors often give the terrible diagnosis by phone, in writing - with or even without explanation – or standing in the corridor of a hospital. Training professionals on appropriate ways of disclosure would avoid this additional and unnecessary pain to already anguished patients (8). (Gross 1990)

Here, the physician is not all-knowing. The use of the striking statistic, 50 % of patients have problems with disclosure, gives clinical value to individual patient experiences. From this perspective, personal data are not subjective and dismissible matters nor do they detract from the authenticity of the discussion. Instead, these experiences contribute to the goal of improving care. In addition to offering multiples perspectives, both through statistics and patient experience, EURODIS allows patients to speak for themselves, demonstrating the significance of their particular perspectives to clarifying misperceptions. For instance:

> From the Agrenska Center in Sweden, we can quote the following reactions from patients and families who have participated in the Family Program:
> - We finally get a true perspective on our children's disability;
> - We now feel "normal";
> - Exchange of experience is as important as expertise knowledge. (Gross 1990)

All knowledge is valuable, and all people can benefit from acquiring it. As the patients themselves disclose, effective treatment acknowledges the patient as a full, if differently bodied, member of the world community. As part of this bigger picture, money matters affect patients not in terms of funding research that considers numerical deviance but on coping with bodily difference:

> these expenses are born exclusively by the families, thereby generating an **additional inequality** between rich rare disease patients and poor rare disease patients. Travel costs to specialised centres are high in terms of time off work and financial cost. Furthermore, the anxiety is amplified because usually only one parent can travel whilst the other looks after other children or has to work......As a consequence, **while expenses increase dramatically, incomes are considerably reduced**. In the case of an adult rare disease patient who is well enough to be able to work, the work hours must be adapted to allow for medical visits and appropriate care. In terms of logistics, much remains to be done to ensure real equality between a disabled and a healthy citizen. It is well accepted that impairment leads to a disability if the environment and regulations do not take into account the special needs of people with impairments to participate in society. The impairment is a part of our being. The **disability comes from outside by disabling factors.** (EURODIS)

The costs of treating rare diseases are many and extraordinary; ignoring this non-medical factor marginalizes patients in yet another way. As the site again makes explicit, non-medical elements inform medical matters and belong to patient treatment. Ignoring these elements, cost in this case, reduces the patient once more to an existence as a statistical deviance rather than to living as different but accommodated with the world. If there is any aggregate here, it is the aggregate of forces working toward helping individuals with rare diseases. Often, rare diseases affect "**the whole family of a patient** in one way or another" (EURODIS). From EURODIS's perspective, the family experience is part of the bigger health-care picture to which the research contributes. This means that patient, scientist, physician, and advocacy organization must work together "co-producing a knowledge base" (EURODIS). Thus, the site links its advocacy with the health-care system, public policy, and the international community:

> **A global approach to rare diseases enables the individual rare disease patient to escape isolation. Appropriate public health policies** can be developed in the areas of scientific and biomedical research, industry policy, drug research and development, information and training of all involved parties, social care and benefits, hospitalisation and outpatient treatment. In order to foster clinical research, the public funding of rare disease clinical trials should be promoted through national or European measures. Healthcare professionals, public health experts and policy makers cannot apply traditional responses and prioritisation to greater need. This approach is not valid for rare diseases and is not ethically sustainable. (EURODIS)

Rather than separating science and society, EURODIS advocates for bringing together their current work, work that is now "scattered and fragmented" (EURODIS). The global approach addresses current systemic problems, personal (isolation) and in treatment (lack of knowledge). Human and social matters are central to the goal of helping individuals with rare diseases. Again, scientific data does not exist to validate individual experiences but the two kinds of data work together. EURODIS discusses rare diseases from the patient perspective, using language which centers on bodily difference and rehabilitation rather than on bodily deviation from a norm. By using multiple statistics to highlight rather than hide aspects of rare diseases, the site embraces the paradox of rarity, its many parts, within a whole. And, by including patient experiences, the site demonstrates that patient experiences are essential in the effort to deal with these conditions.

Implications

As the preceding discussion demonstrates, both the epidemiological and advocacy accounts are working to increase the visibility of rare disease research. To that end, rare disease scientists, epidemiologists specifically, typically propose research based on the prevalence of an aggregate. Aided by new medical technologies, this evidence-based research looks deeply into the sources, causes, and spread of rare diseases; by accumulating data, this research is helping to reduce the numbers of these conditions worldwide. The significant of these efforts cannot be underestimated. Still, that approach limits how well it can serve its goal. Ironically, aggregated data hides particular rare diseases and patient experiences of them. Such data—focused on identifying deviance—disenfranchises those whose numbers are already low when they are considered as individual illnesses and, at the same time, casts their bodies as deviant. But, it is not the numbers per se which are problematic but the way those statistics are presented as universal and context-less.

In contrast, EURODIS focuses on the patient and integrates his/her experience into the account with statistics. By presenting numbers and experiences, this approach contextualizes the part in the whole. And, by clarifying misperceptions, the site demonstrates that human experiences and reactions to difference are not merely subjective and expendable but significant to the dataset. Because they approach the problem from different models and to different ends, the two kinds of efforts cannot increase visibility as well as if they were combined. Although data about prevalence is essential to improving the lives of individuals with rare diseases, they will remain invisible if they are understood as an aggregate of very diverse conditions, that is, if the paradox of their rarity is not made clear. And no matter how inclusive it is, advocacy alone cannot treat the physical symptoms of the diseases nor explain their development and course to support future efforts. Both models must be part of efforts to improve the health-care situations of individuals with rare diseases.

To combine (bio) medical and social models requires scientists to acknowledge the strategic argument about the aggregate inherent in their efforts; they must also accept that individual experience is not antithetical to good science and that disease is a matter of bodily difference rather than deviance. Accordingly, the scientific and medical literatures should recognize that quality of life issues can enhance their data and goal of reducing numbers. By understanding that language matters, that feelings and experiences constitute valid information—that facts can be misleading, and that scientific practices perpetuate values—physicians and scientists can render rare diseases more visible without compromising their research goals.

In practice, this merger requires a major overhaul in scientific and medical education, one incorporating the social model of illness and discussions of language in scientific and medical curricula. Current research in writing studies is considering how scientists are professionalized as they progress through graduate school. These efforts have demonstrates that writing actually mediates scientific practice at the same time that most scientists take this mediation for granted (Takayoshi et al.). Such writing research calls for more explicit instruction in and attention to writing in graduate programs in the sciences. This instruction should consist not of lessons in grammar and spelling but in how language choices affect cultural conceptions of science and the human body.

Similarly, although medical schools have begun to realize the importance of communication in doctor/patient relationships, they are less aware of the role writing and communication play in professionalization (Awad-Scrocco 2012). To address the bedside manner, so to speak, many medical schools include courses in medical humanities and bioethics as well as in multicultural communication (Fadiman 1997). Some schools are also developing M.A. programs in these areas to provide health-care professionals with in depth understanding of and experience with applying these models. These efforts must be continued and combined with explicit discussion of the medical and social models of illnesses and of role such language plays in shaping cultural perceptions about disease and health.

Finally, this merger of models requires a major overhaul in cultural thinking about science, medicine, and the body. For this purpose, the EURODIS site demonstrates the benefits of online delivery in shaping cultural concepts of health and illness. Such a presence not only disseminates ideas to a broader public but blends words and images, in black, white, and color. In so doing, these messages embody the inclusiveness of the social model of illness and, specifically, the paradox of rare diseases—the site speaks of many different kinds of body rather one deviant type. This kind of presence, then, is one of the significant contributions the age of 2.0 makes to rare diseases.

References

Adams R. Sideshow USA: freaks and the American cultural imagination. Chicago: University of Chicago Press; 2001.

American Psychiatric Association. Diagnostic statistic manual of mental disorders (DSM-IV), 4th ed. Rev. ed. Washington, D.C.: American Psychiatric Association; 1994.

Awad-Scrocco D. An examination of the literate practices of resident physicians and attending physician preceptors in a resident-run internal medicine clinic. Kent: Kent State University; 2012.

Bean WB. Rare diseases and lesions: their contributions to clinical medicine. Springfield: Charles C. Thomas Publisher; 1967.

Clarke A, Shim J, et al., editors. Biomedicalization: technoscience, health, and illness in the US. Durham: Duke University Press; 2010.

Couser G. Recovering bodies: illness, disability, and life writing. Madison: University of Wisconsin Press; 1994.

Day S. Evidence-based medicine and rare diseases. In: de la Paz MP, Groft SC, editors. Rare diseases epidemiology. Advances in experimental medicine and biology, vol 686. Berlin: Springer; 2010. Chapter 3.

EURODIS. Rare Diseases Europe. http://www.eurordis.org/.

Fadiman A. The spirit catches you and you fall down. Among child, her American doctors, and the collusion of two cultures. New York: Farrar, Straus and Giroux; 1997.

Foucault M. The birth of the clinic: an archaeology of medical perception. Sheridan Smith AM, trans. Vintage Books; 1974/1994.

Groft SC, de la Paz PM. Rare diseases—avoiding misperceptions and establishing realities: the need for reliable epidemiological data. In: de la Paz PM, Groft SC, editors. Rare diseases epidemiology. Advances in experimental medicine and biology, vol. 686. Berlin: Springer; 2010. Chapter 1.

Gross Alan. The rhetoric of science. Cambridge: Harvard University Press; 1990.

Lavater JC. Essays in physiognomy: designed to promote knowledge and love of mankind. London: John Murray; 1789.

Linton S. Claiming disability: knowledge and identity. New York: New York University Press; 1998.

Michelle F, Asch A. Disability beyond stigma: social interaction, discrimination, and activism. J Soc Issues. 1988;44(1):3–21.

Mishler E. Social contexts of health, illness, and patient care. Cambridge: Cambridge University Press; 1981.

Paz MP de la, Groft SC, editors. Rare diseases epidemiology. Advances in experimental medicine and biology, vol 686. Berlin: Springer; 2010.

de la Paz PM, et al. Rare diseases epidemiology research. In: de la Paz PM, Groft SC, editors. Rare diseases epidemiology. Advances in experimental medicine and biology, vol. 686. Berlin: Springer; 2010. Chapter 2.

Rajmil L, Perestelo-Pérez L, Herman M. Quality of life and rare diseases. In: de la Paz PM, Groft SC, editors. Rare diseases epidemiology. Advances in experimental medicine and biology, vol. 686. Berlin: Springer; 2010. Chapter 15.

Richesson R, Vehik K. Patient registries: utility, validity and inference. In: de la Paz PM, Groft SC, editors. Rare diseases epidemiology. Advances in experimental medicine and biology, vol. 686. Berlin: Springer; 2010. Chapter 6.

Shakespeare T. The social model of disability. In: Davis L, editor. The disability studies reader. London: Routledge/Taylor and Francis Group; 2006.

Siebers T. Disability theory. From social constructionism to the new realism of the body. In: Davis L, editor. The disability studies reader. London: Routledge/Taylor and Francis Group; 2006.

Siebers T. Disability theory. Ann Arbor: University of Michigan Press; 2008.

Takayoshi P, Newman S, van Ittersum D. Teaching scientific writing at the graduate level; paper in process.

Vignette: Taking Control of Thalassemia

Angela Covato

My daughter, Cassandra, was diagnosed with Beta-Thalassemia Major, a genetic blood disorder, at 6 months old. For the first 5 months of her life, she was a happy-go-lucky baby, always smiling and moving her arms around. Then, one day I noticed a difference. I went to a party and saw another baby her age there, laughing and so happy. I thought about how Cassandra was usually as happy, but that lately, she cried more often and just seemed lethargic.

A. Covato (✉)
Canadian Organization for Rare Disorders (CORD), 151 Bloor Street West, Suite 600, Toronto, Ontario M5S 1S4, Canada
e-mail: angela@optimizinghealth.org

Earlier that morning, I suspected that something was wrong so I took her to the local hospital because it was very early and no clinics were open. They told me that nothing was wrong. One doctor said it was normal for babies to have a fever sometimes when there was no particular reason for it. He then had the nerve to tell me to, "go home, because there are other patients that have more serious issues than her". I took his advice—he was a doctor after all—and went home. At the party, looking at the other baby, I knew in my heart—the doctor was wrong.

After the party, I went directly to the hospital where Cassandra was born. The intake nurse asked me a series of questions, including asking about her father's and my own ethnic backgrounds. We are both Italian. Immediately, the nurse determined that Cassandra had a urinary tract infection (UTI) but she also suspected a genetic blood disorder called Thalassemia. I was confused but also happy that we were closer to a diagnosis so that the hospital could treat her.

Once the doctors confirmed that she had a UTI, they put a catheter in her. Still, to this day, the thought of my 6-month old baby having a catheter put in brings tears to my eyes. What they told me next was even more of a shock: Her hemoglobin was 49. A normal 6-month old's hemoglobin is between 110 and 120! She needed to be transfused and she needed it immediately. The guilt I felt at that moment for letting it get so low is indescribable.

I told the doctor that I was afraid. The tainted blood scandal in Canada (thousands of people were being infected with HIV and Hepatitis C because of inadequately screened blood) was still in the back of everyone's minds and certainly mine. The doctor assured us that there was no option. She needed blood or she would die. However, because her hemoglobin was so low she could only receive the minimal amount or she could go into cardiac arrest.

When they transfused her, what I saw was amazing—within 20 min, I saw my quiet, tired, baby transform back into the happy child, with rosy cheeks, that loved to move her arms around and smile so much, once again. The guilt weighed on me like a brick, but I decided--from that moment on—I would do everything in my power to make sure no one ever doesn't give her exact the care she needs and deserves.

We were then referred to *The Hospital for Sick Children* in Toronto. Because Cassandra received blood products, they needed a DNA sample from both me and her father, to confirm the diagnosis. Once the hemotologist confirmed that she had Thalassemia, I felt alone. I've always had family or friends that could relate to something I was going through—this time was different. The clinic gave me the name of a parent that I could speak to so I called her and she explained what it was like to be a parent of a child with a chronic, life-long condition and that although Desferal, the treatment to remove the excess iron she received during transfusions, was hard, as long as we stuck to the treatment, Cassandra had a good chance of having a good life. To this day, I feel indebted to this mother. The time she took with me meant more than she will ever know.

At the next transfusion, the hematologist explained the genetic reasons for Cassandra having Thalassemia. I felt like my whole world was shattered and I felt such incredible guilt. Her parents, the people that would do anything to protect her, passed along this genetic trait to her. Over time, I learned to accept that she needed monthly transfusions and saw how they gave her more energy and life. What was truly hard for me was that she needed Desferal too. Having to poke my own child with a needle, and cause pain, was so hard to comprehend. What was even harder was knowing that I had to sit my beautiful 1-year-old down and explain that she had to be poked with a needle for 5 nights a week.

I looked up information on Thalassemia and saw the severe bone deformities and secondary conditions that can occur and was so upset; I felt like I could barely breathe. The only thing I knew was despite my guilt, I needed to keep pushing ahead—I needed to fight for Cassandra. Soon, I started to talk to the other parents in the clinic and began to develop a network of support. I also met a few adult patients that not only became a source of information for me, but my friends. I started volunteering my spare time to the *Thalassemia Foundation of Canada*. I knew this was the only thing in my power to make sure Cassandra received the best care possible. I also found what I needed—someone to talk to and the support system I craved, as a parent.

As part of the *Foundation*, we advocated with the government to make Exjade, an oral medication that could hopefully work as well as Desferal, available to all patients in Canada. When Exjade was finally approved, most patients switched from Desferal to Exjade immediately. I was cautiously optimistic, so, I waited. I waited almost 2 years, all the while doing my own research and attending conferences.

Finally, when Cassandra was 9 years old, her physician mentioned that Cassandra was one of the few patients still on Desferal. She didn't say it out of judgment; just simply a fact. I knew this was the time. I sat Cassandra down and explained that there was this new drug. What I could not promise was that once she started it that it would work, or that she wouldn't need to go back to the needles and Desferal. My brave little girl said she understood but was ready to try.

Cassandra is now 14 years old and takes Exjade every night. It truly has made a difference in our lives. No longer does it take time to mix the medication—and even better, she no longer needs to have needles or be hooked up to a pump for 10 hours every night. Transfusions have become part of the routine of our lives. Cassandra is now a beautiful, active teenager who is the flyer on her cheerleading team, plays soccer and volleyball and is one of the first people you will see join a speech competition—she loves the limelight, and it loves her back.

I'm proud that the baby who seemed so lifeless, at 6 months old, is not only thriving, but overcoming and shining. Last year, she was called down to the school office, as all of the kids that have medical conditions were. She was confused and thought "why are they calling me down there? There is nothing wrong with me!" She is right—there is nothing wrong with her. She has a blood disorder but it doesn't define her. What I hope that every parent realizes from our story is this: whether your child has Thalassemia or any other disease, they want what every kid wants: to run and play, and be part of the crowd. Don't treat them any different than you would a child without that disease.

The harsh reality is that we won't always be here for them. As hard as it is as a parent to slowly let go, we must do it, so they can become stronger and eventually slowly spread their wings. Cassandra, at nine years old, learned to do her own needles. At first I was hesitant when she asked, but soon saw the confidence it gave her to take her own healthcare upon herself. You must also, as a parent, build yourself up. Try, (although it is hard, and I still struggle to this day), to let go of the guilt and reach out to others around you. When you feel weak, networking with others that understand can make you stronger. Your child can benefit too by spending time with other children with the same disease.

Most of all, advocate. Advocate for your child. No one will look out for your kid like you. You have to do it. What led up to this point is now history, and out of your control, but you can control the future for your child and ensure that they receive the best care and treatment possible. That is what is in your power, as a parent.

Industry Perspectives on Orphan Drug Development

Sylvie Grégoire, Norman Barton and David Whiteman

> *The secret of business is to know something that nobody else knows*
> ARISTOTLE ONASSIS

Abstract This chapter presents perspectives from the pharmaceutical industry concerning the development of orphan drugs. This includes outlining orphan drug development in biotechnology, the various factors necessary for commercializing orphan drug candidates, phases and trials of clinical development and factors which may affect commercialization of a product. The chapter concludes with implications for patients and families.

Background

Interest and activity in orphan and rare diseases by the pharmaceutical (and, in particular, by the biotechnology) industry has grown enormously over the past 30 years. Many observers trace the development of this emphasis to the passing of the Orphan Drug Act in the United States in 1983, and, subsequently, to similar legislation developed in many other countries (Talele et al. 2010). In reality, interest in orphan and rare diseases has been driven not only by specific legislation, but also by collaborative partnerships involving interested physicians and medical centers; disease, or disease-group, specific patient organizations; biotechnology and pharmaceutical companies; politicians and health care agencies and providers.

The Orphan Drug Act established a formal pathway to designate treatments for specific diseases as orphan products, along with fiscal incentives (including

S. Grégoire · N. Barton · D. Whiteman (✉)
Shire Human Genetic Therapies, Lexington, MA, USA
e-mail: dwhiteman@shire.com

periods of market exclusivity; specific tax credits; and administrative fee exemptions) for their development. These are over and above protection provided by patents specific to the drug. Appropriate safety and efficacy data, through the Investigational New Drug (IND) process, however, remained as the basis for their phased development and ultimate marketing approval.

The standards of safety and efficacy required of orphan drugs are, and should be, no different than those for all therapies. To be approved, all drugs (Orphan and non-Orphan) must "demonstrate substantial evidence of effectiveness/clinical benefit" (Code of Federal Regulations 2012a) specifically defined as the impact of treatment on how a patient feels, functions or survives and improvement or delay in progression. Evidence of effectiveness is defined as "Evidence consisting of adequate and well–controlled investigations on the basis of which it could fairly and responsibly be concluded that the drug will have the effect it purports" (Code of Federal Regulations 2012b).

Orphan Drugs for Rare Diseases

Orphan drug development programs have also been able to be considered (alongside more conventional drugs) for accelerated approval: "FDA may grant marketing approval for a new drug product on the basis of adequate and well-controlled clinical trials establishing that the drug product has an effect on a surrogate endpoint that is reasonably likely to predict clinical benefit" (Code of Federal Regulations 2012c). Such an accelerated approval requires post marketing verification study (for instance, outcome studies, registries or specific additional trials) to verify and describe its clinical benefit. Post marketing studies must also be adequate and well-controlled, and conducted "with due diligence".

With marketing approval, a previously designated orphan drug receives 7 years of exclusive US marketing rights for the indication for that specific rare disease. Talele et al. (2010) summarize the impact of those innovations on the development of therapies for rare inborn errors of metabolism in the United States over the preceding 26 years. Heemstra et al. (2008, 2009) provide a perspective on the effects of similar legislation in the European Union.

Orphan disease definitions are often prescribed in legislation or regulation and vary with geography. In the United States, an orphan disease is defined as one that affects less than 200,000 of the national population. In the European Union (EU), the definition is a disease that affects fewer than 5 per 10,000 population. In Japan, less than 50,000 of the national population. In the EU and in Japan, however, designation as an orphan disease carries a 10 (rather than a 7) year period of market exclusivity upon authorization. Using the United States orphan drug definition, 6,000–8,000 rare diseases affecting 7 % of the population exist, 4/5ths of which have a clear genetic basis, and 70 % of which have a prevalence of less than 100,000 people. More than 2200 molecules have so far been designated as orphan drugs and more than 362 orphan drugs have been approved since 1983.

History of Orphan Drug Development in Biotechnology

Early examples of biotechnology products designed to treat orphan diseases began with products extracted from, and in many cases modified from (for instance by the alteration of sugar molecule side chains that affected transport across cell membranes) biological tissues. The next wave of products was driven by the need to eliminate tissue as the primary manufacturing source and was enabled by advances in molecular genetics and bio-engineering. An example is recombinant human growth hormone (GH), developed by *Genentech* for the treatment of growth hormone deficiency. This was driven by the need to eliminate the use of cadaver-derived pituitary growth hormone production. The cadaver-derived pituitary hormone carried the attendant risk of transmitting prion related diseases, such as Creutzfeldt-Jakob disease. Genetic engineering technology was well-enough developed for a quantum leap in therapeutics which eliminated this cadaver-related risk.

Likewise, *Genzyme Corporation* with followed Ceredase (alglucerase), derived from human placental tissue, with Cerezyme (imiglucerase), a product of recombinant bioengineering. The demonstration that this could be done, and could be translated into a successful, commercialized, therapeutic program (which included not only the provision of the drug, but a group of related services for patients and educational programs for health care providers) opened up the possibility of other recombinant human enzyme replacement therapies (Barton and Brady 1997). There followed laronidase (Aldurazyme, *Genzyme*) to treat mucopolysaccharidosis (MPS) I: idursulfase (Elaprase, *Shire*) to treat MPS II; and galsulfase (Naglazyme, *Biomarin*) to treat MPS VI. Orphan drug enzyme replacement therapy is now used to treat these and several other diseases routinely. While none has proved to be curative, there have been clear benefits to patients in each case; in some cases, the progression of disease appears to have been slowed or halted (Muenzer et al. 2011).

In such examples, the development of orphan drugs has been the result of collaboration between: an academic basic, translational and clinical researchers; practicing clinicians; patients, patient organizations and advocates; policy makers; regulatory agencies and payors of health services. In some cases, variations have developed in the standard drug development Phase paradigm, which have allowed early access initiatives after a single Phase II, III or combined II, III trial (discussed later). The aim has been to provide access to patients as soon as there is reasonable evidence of a positive risk benefit ratio from efficacy and safety data in a controlled clinical trial.

Factors Necessary for Commercialization of Orphan Drug Candidates

There are multiple factors which must be considered by biopharmaceutical companies in order to consider commercial development of a new drug (new molecular entity) for an orphan disease. Firstly, a description of a well-defined disease entity, with a clearly understood clinical syndrome and characteristics, and a credible mechanism of action for the new drug, with biological demonstration of proof of concept, that is based on a deep and detailed understanding of the basic pathophysiology of the disease.

A carefully studied, and well understood natural history (which is frequently not available) may have to be created in a baseline natural history study, which may also include the collection of appropriate biological specimens (blood, urine etc.) in order to describe metabolites that may be suitable as biomarkers of disease and treatment response. Development of an appropriate natural history may include collaborations between a pharmaceutical company; government research agencies (e.g. NIH, WHO) one or more patient and disease organizations; and, in some cases, emerging pan-disease or group advocacy organizations (such as *Genetic Alliance*, *NORD*, *Orphanet* and *Eurordis*). More recently, companies have been set up specifically to collect and organize natural history (and, potentially, treatment response) data, often using social networking technologies. In some cases, (for example, *Pompe disease*) companies have sponsored extremely detailed retrospective reviews of medical records of patients with the disease. This is in order to construct an adequate natural history profile of the disease, and its patients, to provide a baseline for hypothesis-based comparison to a specific intervention, such as enzyme replacement therapy.

Following definition and purification of the new drug—and scaling up of production in several sequential steps, with extensive pre-clinical development and animal safety and efficacy testing (which often takes many years)—a package of data is developed and submitted to one of the major regulatory agencies along with a proposal for testing in humans.

Clinical Development: Phases and Trials

A clinical development plan is prepared, focusing on key clinical history, examination and clinical laboratory testing features (biomarkers as well as analytes designed to detect any adverse effects of the drug on normal physiology) and imaging parameters. Correlations between all of these variables must be examined and understood as best as possible, and time series (pharmacodynamics and pharmacokinetics) defined. Increasingly, enzyme and protein markers, as well as RNA and DNA analyses are performed, although it has been found that the degree of individual and family variation in DNA mutations (for all but a handful of diseases) defies close clinical correlation. That is, specific DNA mutation data are, for many genes, perhaps surprisingly, poor predictors of clinical outcome for disease related to that gene, when considered alone.

For orphan drug development, a major issue, given the relatively small number of patients affected, is the identification and selection of patients for clinical trials. In the first instance, this is dependent on available medical experts for the disease who can identify patients who have been accurately diagnosed and well characterized. Many patients undergo a protracted and convoluted journey before ultimately reaching what, in the context of modern drug development, should be called a "personal molecular diagnosis". Shire HGT (2013), via their *Rare Disease Impact Report*, present a visualization of the diagnostic odyssey that many patients experience.

The goal of a clinical trial is to formally (logically and mathematically: using probability and statistics) test whether a drug is effective in treating one or more aspects of a disease. For the purposes of such testing, patients enrolled in a trial may also need to be at a specific point in the process of evolution of disease. For instance, a group of patients who have been identified at a young age may not have advanced far enough in some aspect of disease pathophysiology to be capable of showing a large enough response to the drug. At the other extreme, patients whose disease is too far advanced may not be capable of responding either. A clinical trial comprised of patients of either of those two types is doomed to fail, even if the drug actually works in another group of patients of intermediate severity.

Patients must be willing and able to take part in the trial; in some cases (again, more frequently for rare diseases) this may involve frequent long-distance travel or even relocation to a clinical research center. In the case of some rare diseases, competition among trials for patients for testing several different potential new drugs for the same disease (for instance, *Duchene muscular dystrophy*) may impact the companies' ability to fully recruit a trial for a particular treatment.

Historically, the clinical aspect of drug development has been described in four phases.

Phase I

A drug is tested either in healthy volunteers or, more commonly for orphan indications, in a small number of patients with the disease, primarily to establish that there is no major adverse effect from the perspective of patient safety. In this phase, investigators will often consider preliminary assessment of various dose levels or timing and assessment of possible biomarkers, in order to gain insight into appropriate dosing and measurement paradigms for the next phase.

Phase II

A relatively small number of patients is systematically exposed to the drug at various doses, and this may include a placebo group, in order to assess efficacy and an optimal dose, as well as gather further safely information.

Phase III

Often described as the pivotal or registration phase, this involves formal, controlled and systematic testing of a hypothesis of the drug's efficacy, usually focused on one or more primary endpoints, as well as measurement of other clinical parameters, described as secondary endpoints. Safety monitoring continues. Success in two or more Phase III trials, taken with the accumulating database of patients followed from the first two Phases, is often submitted to a regulatory agency for consideration for approval. In the orphan disease area, a single Phase III trial may be submitted (and approved), and elements of the traditional Phase II and III may be consolidated into a single trial.

Phase IV

Often referred to as the post-marketing phase, usually follows registration, and is designed to assess the continuing efficacy and safety of the drug in a more real-world environment. Particularly for drugs for orphan diseases and small patient populations, and increasingly for all new drug approvals, this involves the use of outcome surveys or registries for continued study of the disease and drug response. In the case of orphan drug approvals, the commitment (borne by the company developing the drug) to the maintenance of these registries may extend for 15–20 years.

All of this activity comes at a major cost, and both internal and external (particularly for smaller biotechnology companies developing one, or a few, drugs) financial support may be needed to raise up to US$1B to move from basic research proof of concept, and sufficient animal testing to enable approval by a regulatory agency of an Investigational New Drug application, through to the later phases of completed human trials, sufficient for approval. That amount is over and above the costs of both external (e.g. university laboratories) and internal research to bring the candidate molecule to the point where it is first considered as a possible drug and then selected for commercial development by a pharmaceutical company

Future Developments

Therapeutic products of biotechnology are now moving beyond protein infusions. A variety of other approaches to enhancing, modifying and replacing missing or defective proteins (whether it is an enzyme (catalyst) for chemical reactions or a structural protein component of a tissue) are now under development. Alternative methods of delivery, such as gene therapy employing specially modified viruses to cross cell membranes and reach targets deep in body organs; nanoparticles; and direct delivery of product to body compartments other than the venous system, such as intra-thecal delivery of enzymes in cases where disease affects the central nervous system, are being elaborated. Technical advances in bio-engineering on a large scale are allowing the production of drugs in a cleaner, purer environment, and, in some cases, with higher yields than formerly.

However, hurdles remain, particularly for less common conditions, now often referred to as "ultra-rare" diseases. The financial resources available to support the highly complex processes of commercial drug development are limited. In most economies, the availability of resources for development is driven by the need for a clear return on the investment made, with careful consideration of the risk of such investment, and, therefore, the biotechnology industry and its financial backers are forced to make selections among possible disease and drug targets.

Factors Which may Affect the Decision to Attempt Commercialization of a Product

The factors that can affect commercialization decisions between individual drug and disease targets for commercial development include the following.

Definition of the target disease must be clear, and the number and range of involvement of patients, as well as (in some cases) the geography of affected patients must be understood. For some rare genetic diseases, particularly those whose genes follow an autosomal recessive pattern of inheritance, disease may be localized to a particular place (or set of places), region or even ethnic group.

The **disease pathology** (the basic mechanism that is thought to explain the symptoms, signs and progression of the disease) must be clearly defined. Consideration is given to the ease (or otherwise) of demonstrating proof of concept (usually in animal studies); that is, proof that the specific drug or intervention produces a clear treatment result, understood on the basis of the pathology. Co-localization of the drug and its target in the body must be shown. If a drug is tailored to a specific pathology, but cannot cross cell membranes or barriers (for instance, the blood–brain barrier) then it cannot be expected to work without that barrier first being overcome.

Finally, **the mechanisms of adsorption, distribution, metabolism and excretion** of the drug in the whole patient (and, first, in animals) must be defined and quantitated. This is the field of pharmacokinetics and pharmacodynamics.

There is increasing interest in the use of mechanistic biomarkers as a basis (or surrogate) in some of these processes. A biomarker is some (quantifiable) observation, for instance the chemical concentration of a metabolite in blood, which is considered, and has then been shown, to be a hallmark of some aspect of a disease state. Ideally, it should also respond to any new drug in the same direction, and to a similar magnitude, as the clinical symptoms and signs themselves.

In some rare diseases, there is widespread variation in disease pathophysiology and extent. That is, not all patients are affected in the same way, nor to the same degree. In some cases, this is due to age and the natural history of disease progression: in others, it is variation in the type of mutation within a gene, and interaction of the resulting protein defect with the environment (traditionally referred to by biochemists as the milieu). There are even examples, increasingly revealed by modern gene mutation (DNA) analysis, of unaffected family members carrying the same genetic mutation pattern as others who are affected. Thus, symptoms and signs of the disease must be clearly defined, readily measured, and accurately recorded. For such reasons, a well-defined understanding of the natural history of the disease itself, and its progression—which may also be variable—is key. When natural history is well understood and recorded, the possibility of using that as the "control" in a clinical trial increases and the need for strict randomization of trial participants may be reduced or modified. Such a natural history can also allow an appreciation of the likely duration and intensity of treatment and of possible drug administration (i.e. dose, frequency and route of delivery) decisions.

A well-defined natural history can allow and optimize the absolutely critical choice of appropriate endpoints for a clinical trial. This is critical not only, in the sense that they would succeed (in comparison with controls) in demonstrating the effectiveness of the drug (or even of one drug over another, which is an even higher hurdle) but critical also in the sense that those endpoints accurately predict change in longer-term outcomes from the new drug treatment.

It is an unfortunate truth that, traditionally, quite large numbers of patients are needed for the conduct of clinical trials, particularly Phase III, or registration, trials. The more frequent the disease and the larger the number of well-characterized patients with the disease, the easier it is to develop and execute a clinical development plan. Modern approaches to drug development, and innovative approaches to clinical trial design that specifically consider the needs of low-frequency populations such as those with rare or ultra-rare diseases, are beginning to make better use of all of the information obtained from all of the patients in trials. This permits lower patient number clinical trials. Several regulatory and industry groups currently have working parties focused on the development of novel, smarter, optimized clinical trial and testing designs. In each case, the goal is to gain the most knowledge from the least amount of data in the fewest number of patients.

All of the above considerations mean that those who pursue and who finance the cost of drug development are forced to make difficult decisions about how to structure clinical trial recruitment; they are also forced to make difficult choices, usually in the absence of full, optimal information, between different drugs for orphan disease development. Doing so effectively, and to the optimal benefit of all, requires innovation, flexibility, careful timing and compromise between what may be attractive and what is actually doable.

Implications for Patients and Families

The first step that most patients, families and patient advocacy and support groups can take to promote their disease as a candidate for drug development is to take part in defining their disease, including participating in surveys, registries, natural history studies, including studies that aim to carefully define valid disease and patient subgroups. The second is the formation of strategic alliances, not only with other families, but with other family and patient groups, taking into consideration commonalities such as disease, stage, subtypes, geography, and functional clustering of disease groups and focus. This is particularly important for those whose condition is ultra-rare in frequency. Affinities and alliances to particular funding Foundations, both for rare disease generically, and for some system disease groups should be explored. Relevant, experienced and appropriately qualified scientific and medical advisors should be identified, and an early search (best started by systematic searches of the medical and scientific literature) begun for links to particular industry groups and companies. Disease awareness and political lobbying campaigns and alliances should be considered.

Above all, collaboration between all parties is the key to the successful identification of potential therapies for diseases; successful basic science research leading to proof of concept; the design and conduct of well-powered clinical trials with the best use of the smallest number of patients to demonstrate therapeutic efficacy; and, ultimately to the registration and marketing approval for a drug, as well as in real-world use registries and outcome surveys that are able to demonstrate continuing effectiveness, and form a basis for continued reimbursement by health care payors.

Conclusion

The Orphan Disease Act (1983) in the United States, and subsequent similar legislation worldwide, along with quantum changes in our understanding of genetics, biology and biochemical engineering has opened up treatments previously thought impossible for a wide range of rare diseases. Continuing innovations, particularly in modes of delivery; alternative therapeutic vehicles and delivery systems; in manufacturing process; and in innovative clinical trial designs will further broaden the reach of therapeutics into ultra-rare diseases. Patients and families have important collaborative roles and an increasing voice in this movement, and in the development of efficient drug development and testing strategies.

References

Barton NW, Brady RO. Macrophage-targeted glucocerebrosidase: a therapeutically effectiveenzyme replacement product for Gaucher disease. In: Lauwers A, Scharpè S, editors. Pharmaceutical Enzymes. Marcel Dekker, New York; 1997; 261–83.

Code of Federal Regulations. "Applications for FDA approval to market a new drug", title 21, code of federal regulations, part 314.50, available: http://www.accessdata.fda.gov/scripts/cdrh/cfdocs/cfcfr/CFRSearch.cfm?fr=314.50. (2012a). Accessed: 4/9/2013.

Code of Federal Regulations. "Adequate and well-controlled studies", Title 21: Code of Federal Regulations, Part 314.126, available: http://www.accessdata.fda.gov/scripts/cdrh/cfdocs/cfcfr/CFRSearch.cfm?fr=314.126. (2012b). Accessed 4/9/2013.

Code of Federal Regulations. "Content and format of an application", Title 21, Code of Federal Regulations, Part 314.50H, available: http://www.accessdata.fda.gov/scripts/cdrh/cfdocs/cfcfr/CFRSearch.cfm?fr=314.50. (2012c). Accessed 4/9/2013.

Heemstra HE, de Vrueh RL, van Weely S, Buller HA, Leufkens HG. Predictors of orphan drug approval in the European Union. Eur J Clin Pharmacol. 2008;64(5):545–545.

Heemstra HE, van Weely S, Buller HA, Leufkens HG, de Vrueh RL. Translation of rare disease research into orphan drug development: disease matters. Drug Discov Today. 2009;14(23–24):1166–73.

Muenzer J, Beck M, Eng CM, et al. Long-term, open-labeled extension study of idursulfase in the treatment of Hunter syndrome. Genet Med. 2011;13(2):95–101.

Shire HGT (2013) Rare disease impact report, April 2013, [online], www.rarediseaseimpact.com. Last Accessed 8 Apr 2013.

Talele SS, Xu K, Pariser AR, Braun MM, Farag-El-Massah S, Phillips MI, Thompson BH, Coté TR. Therapies for inborn errors of metabolism: what has the orphan drug act delivered? Pediatrics. 2010;126:101–6.

Part II
Health 2.0

Health 2.0: The Power of the Internet to Raise Awareness of Rare Diseases

Laura Montini

> *Synergy—the bonus that is achieved when things work together harmoniously*
>
> MARK TWAIN

Abstract The internet is enabling people with rare diseases to connect with others with the same condition and to access information and support in ways that would not have been possible in previous decades. Through a series of interviews with key stakeholders, and written in an accessible "patient-friendly" manner, this chapter provides numerous examples of how these vital connections have helped people with rare diseases and how the potential of this technology is only just being realized as a powerful tool to advance research in the area.

Background

In the early days of the internet, some websites were as useful as a book sitting on a shelf on the opposite site of the country. You would not know about it unless someone told you it was there, and even if you found it, much of it might be out of date. Today search engines deliver content from a number of locations on the internet, and the dynamic nature of web pages makes it likely that their information is current. The internet's efficiency is why the collaboratively built and edited *Wikipedia* is one of the top 10 most viewed sites online(Alexa Internet 2013) and why the *Encyclopaedia Britannica* is no longer available in print (Encyclopaedia Britannica 2012). The departure from static pages to ones that users can instantly create, update and delete is characteristic of the shift to Web 2.0, a term that refers to the internet as people use it today.

L. Montini (✉)
Health 2.0, 350 Townsend St, San Francisco, CA 94107, USA
e-mail: Lmmontini@gmail.com

A core tenet of Web 2.0 is that it allows for the harnessing of collective intelligence by letting users generate content and become co-developers of the web. Health 2.0, which spun off of this phenomenon, refers to an ongoing process, enabled by the use of digital technology, that maximizes the knowledge of patients, researchers, healthcare workers and others for the continuous improvement of care.

Emily's Story (SMA Type 1)

One day, just after the calendar turned to the new year of 2013, a photo of a father ice-skating with his seven-year-old daughter appeared online. The shot showed Emily in a winter cap, bundled up in thick blankets, lying in her stroller. Underneath her sat a portable ventilator attached to a long tube that travelled up to Emily's breathing mask. Her father Nate Lee stood behind her, and while keeping a hand on the stroller pushed them forward, beginning another lap around the rink.

Emily is unable to sit or stand up, and she has difficulty breathing. She has Spinal Muscular Atrophy (SMA), a disease that causes muscle weakness over time and affects about four in 100,000 people (National Center for Biotechnology Information 2013). There are four types of SMA, but Emily has type I, the most severe form. A friend who knew Emily and her family well thought that the picture encapsulated the way Nate had cared for Emily since she was an infant. She posted it to the online photo sharing site, *Imgur*, and wrote, "I submit my friend for father of the year. Here he is ice-skating with his daughter who was born with a deadly genetic disease called SMA type 1."

Not long after, Nate heard that there was a picture of him on the internet. It did not take him long to find it, and he was angry. Not at the poster, but at the fact that Emily's picture was on the front page of a highly viewed social media site. His daughter was exposed to comments from all kinds of people who knew nothing about her disease. Though his parental instincts were urging him to spring to action and protect Emily, he did not do anything at first. Instead, he settled down and clicked on the photo.

He could not believe what he saw. The picture of him taking his daughter for her first skate had been viewed nearly one million times. If hundreds of thousands of those who had looked at the photo had not heard of SMA before, now they had. "There's no other single thing has gotten that much attention with such little effort," Nate said. He is right in stating that his story is not typical. Yet every day others have similar experiences, and each are made possible by the ability of the web to spread information like wildfire.

Humans are naturally inclined to share information. Among the ones on the internet, analysts can quantify just how much. Every minute *WordPress* bloggers publish nearly 350 new posts, *Flickr* users add 3,125 new photos, and *Facebook* members share more than 680,000 pieces of content (Domo 2012). Once information becomes available, much of it moves. Sites are equipped with social sharing buttons that transmit web addresses in two clicks or less. The signal from the online peer-to-peer world of healthcare information is not very strong in the midst of non-stop content generation. But as the amount of information on the internet increases, the more patients go there seeking answers. If they cannot find what they are looking for, there is always the option of creating it.

After Nate saw the picture and the many comments below, he did act. His friend who had posted the photo asked if he would lead an SMA question and answer session, called an *IAmA* on Reddit. The discussion was called "IAmA Father of a Child with Spinal Muscular Atrophy." Nate was not new to socializing online, especially when it came to SMA. In 2005 when Emily was diagnosed, he was not finding the information he needed at the rate he needed it. "I felt like I had no control over the knowledge that I was gaining and the decisions that I was making from them," he said.

So Nate created *SMAspace*. The website was his take on the popular online social network, MySpace, but specifically for those closely impacted by SMA. It virtually linked families from across the United States and enabled them to create discussion forums and point each other to education resources. From the start it was a closed group, and to this day Nate still controls membership. He said SMAspace now has more than 2,000 users from about 55 countries. The online conversation on Reddit provided an opportunity that SMAspace, a closed community, did not. Nate saw that he could reach a group that genuinely wanted to know more about the disease. All sorts of people showed up to the discussion. There were the generally curious, the incredibly frank and even the outright critical. Above all, there were good questions. One person asked what the most challenging thing was that Nate had to overcome. Someone wanted to know how he dealt with people who were caught off guard when they saw his daughter in a wheelchair. And a future doctor wondered how to get experience working with children like Emily (Reddit 2013).

A public online forum might not seem like the right setting to talk about personal issues like health; yet many people continue to do it. In an age when 80 percent of American adults are consistently online (The Pew Research Center's Internet and American Life Project 2013), paying bills, taking college courses, and buying groceries, they expect to be able to go online for healthcare information and services as well. The reason this seems like a relatively recent phenomenon is because it is only relatively recently that people started using the internet in large numbers. By the end of the year 2000, just half of American adults were online. As Nate demonstrated, not only doctors, not only patients, and not only caregivers can contribute to the conversation. All of their intelligence is starting to be aggregated online with the use of tools like databases, wikis, virtual communities, and blogs.

Health 2.0

Co-founders of the company *Health 2.0*, Indu Subaiya MD, and Matthew Holt, held the first Health 2.0 Conference in 2007. There they began talking about the term as a progression of four stages: users generating healthcare information, consumers connecting to providers, partnerships reforming delivery, and data ultimately driving decisions (Health 2.0 TV 2013). There are a number of new tools, as well as new uses for old tools, that are making this possible. Patients are publishing information online through blogs and social networks. Many of them have even built these sites themselves. Patient to provider connections are taking place using email, texting and phone communication, though unexpected challenges have placed limits on how those are used. The internet can act as a meeting spot where previously unlikely partnerships between patients and drug companies are formed. Lastly, with the conversion of paper patient records to electronic health records, researchers now have vast amounts of data to study, which will yield findings that impact the way patients are treated. Indu and Matthew described a system that, once each stage is mature, works less like a cycle and more like a network where data flowing in and out of each part constantly enhance the ability of the other to make improvements.

So far there has been progress in each of the four stages, but user-generated healthcare is by far the furthest along. Considered by many to have spurred the Health 2.0 movement, patients and those who care for them have seen the value in passing along knowledge by way of the internet. Over the past several years Nate has witnessed the evolution of these methods. When Emily was a newborn baby in 2005, SMA support group members were circulating information via email. It was better than nothing, but it was not a perfect system. Members got email fatigue as they watched their inboxes fill up. There also was not a way to effectively archive and search past information.

During this time, new ways to socialize on the internet were becoming popular. *MySpace* allowed users to build a profile so that others could associate a name with a person instead of an email address, and message boards brought organization to group discussions. After some searching, Nate was unable to find similar applications that also ensured privacy. So Nate, a jack of many trades, taught himself how to build a private online network for his own rare disease community. And around this time, he was not the only one with the idea.

User-Generated Healthcare

PatientsLikeMe was founded in 2004 by three *Massachusetts Institute of Technology* engineers, Jeff Cole and brothers James and Ben Heywood (PatientsLikeMe 2013). It was born out of their motivation to discover treatments for James' and Ben's brother, Stephen Heywood, who was diagnosed with Amyotrophic Lateral Sclerosis (ALS) when he was 29. ALS, or Lou Gehrig's disease, affects nerve cells in the brain and spinal cord that control voluntary muscle movement. It is usually not diagnosed until patients start to exhibit symptoms at age 50 or older (National Center for Biotechnology Information 2013). Desperate to find a way to extend Stephen's life, James founded the *ALS Therapy Development Institute* in 1999 (ALS Therapy Development Institute 2013). Then in 2006, *PatientsLikeMe* launched its first site, the ALS/Lou Gehrig's disease community. From there *PatientsLikeMe* grew to include other rare conditions as well as more common ones.

When patients first join, they build a profile based on their illnesses, symptoms and treatments, which they can continue to update over time. Not only are they linked up with those who have the same condition, but they can find others who are similar to them, based on characteristics like age, gender and how long they have been living with their disease. During pre-internet days, it was likely that patients could go years or most of a lifetime without meeting anyone like them, even if their disease was not technically a rare one. For example, epilepsy is the fourth most common neurological disorder in the U.S., affecting more than two million people (Epilepsy Foundation 2013). Yet *PatientsLikeMe* polled its own community and found that before joining the site, one in three patients had never known another person with epilepsy (Wicks et al. 2013). Ben Heywood once met a woman in her 20 s who was diagnosed with a rare epilepsy syndrome when she was a child, and she had never encountered anyone else who also lived with it. But upon joining *PatientsLikeMe* she found 11 others who had been diagnosed with that specific syndrome.

The *Pew Research Center's Internet and American Life Project* found that one in five internet users have gone online to search for others who have their illness (Fox 2011). Friends and family members who care for a patient are also likely to go online looking for people in similar situations. "The power of meeting someone like you when you have an illness—or if you're a parent and you're trying to find someone going through it with a child—is incredibly powerful," Ben said. Those who are brought together under these circumstances offer the kind of empathy that can only truly be felt by someone who has shared the same experience. The connections re-emphasize that there are others out there working to improve their lives or the lives of the people they love.

Patients are not the only ones using healthcare social networks to ask questions and share what they know. Physician online communities also help users to find others based on similarities like specialty or area of research. The general structure of online physician communities is similar to patient networks. Users build a personal profile, they can individually message other members, and they can participate in discussion forums. Sites like *Sermo* and *Doximity* are exclusive to physicians. That way members can trust, to an extent, that the information they find on the site is credible.

Doximity Co-founder Nate Gross MD, described the site as a *LinkedIn* for doctors. He said the network includes about 120,000 U.S. physicians. One of the greatest benefits of a platform that lets a number of physicians convene online is the quick access it gives them to multiple opinions. Physician networks allow doctors to create online message boards where they can round up specialists from across the country and ask for advice, particularly about a challenging medical case. Another advantage that comes with doctors being able to gather online is that they can hear about and discuss new treatment possibilities at a much quicker rate than before. "It takes 17 years for medical knowledge coming through, say a clinical trial being published in a journal, to actually spread across the entire United States and affect the way that everyone in the country practices medicine," Nate said. Just two decades ago, most physicians may not have heard about new studies or clinical trials in their field until they physically attended their specialization's medical conference. Now online forums can serve as virtual meeting rooms where doctors can discuss new findings the same day they are published.

Since Health 2.0 applications are designed to harness collective intelligence, the more individuals there are continually pouring knowledge into them, the better they work. This is currently a weakness of online physician networks. The typical physician does not log in and use a social site as frequently as the average *SMAspace* user does. The entire technology industry faces adoption challenges on an ongoing basis. It is difficult to change habits when one way of doing something has been ingrained in a particular society for ages. When it comes to patients the story is different. Disease can suddenly storm into the lives of previously healthy individuals. In this case patients will use all tools available to either navigate their way back to health, or to cope with a new normal. In other cases, patients have lived with illness their entire lives. If an online application can give them the hope of understanding their disease better, they only stand to gain from trying it. On the other hand, doctors have practiced with nondigital systems for decades. Technology that does not fit into their workflow costs them time, which is often time they would otherwise be spending with patients.

Of course, not all physicians are social media or technology-averse, and many see value in spending some time online. One recent survey found that nearly one quarter of physicians use social media each day to scan for medical information (McGowan et al. 2013). European doctors in fact have shown more interest in using physician social networks. *Manhattan Research*'s "Taking the Pulse Europe 2012" report found that 22 percent of doctors used these sites, up from 13 percent in 2011 (Manhattan Research 2013). Whether or not doctors chose to adopt new technology comes down to personal preference as well as the culture that they practice in. It can also come down to their sense of which technologies could potentially do more harm than good.

Patients Connect to Providers

The second stage of the Health 2.0 progression involves patients and doctors regularly using digital tools to communicate with one another. There are numerous ways to connect including telephone, email, video conference and text message. And patients and providers utilize each to some extent. In certain areas of healthcare, doctors have enthusiastically taken to using telecommunications tools. The United States *Department of Veterans Affairs*, a government-run healthcare program for military veterans, provided about 140,000 remote consultations in 2011 (Nebraska State Board of Health Meeting 2012). Telemedicine startups have sprung up across the United States, and many have been met with interest. But doctors in the U.S. face significant barriers to using telemedicine. Currently most have a payment system that only reimburses them for in-office visits. Additionally, in certain states, it is illegal to practice medicine across state lines, which it turns out is a common problem. Physicians are also worried about the possible harm they could do. In the case of phone calls and even video conferencing, doctors do not want to offer the wrong medical advice as a result of not being able to physically see a patient.

Unsurprisingly the majority of patient care still takes place in the office, and this will continue to be the paradigm for years to come. However, some basic but effective doctor-to-patient communications can take place remotely. When it comes to Emily's medical care, her pediatrician encourages her parents to avoid routinely bringing her to into the office so she does not risk getting sick. Instead, her doctor gave Nate his phone number and email address as well as permission to text and page him. This way questions about drug dosages and arrangements for new prescriptions and medical equipment are handled from afar. Digital communication is especially applicable for orphan disease patients when location is a barrier to obtaining the best possible medical advice. Often these patients are unable to find specialists within their local radius who have seen a similar case. But the internet can make even the most physically remote expert accessible. Many rare disease specialists are open to communicating with caregivers, patients and their local providers through email and over the phone. Emily's parents, for example, receive advice from a pulmonologist in Wisconsin and a neurologist in Utah who both have specialized knowledge about SMA.

But even doctors who are open to consulting with patients from a distance proceed with caution. Each medical case is different, and for a physician to be able to provide advice to an individual, they have to have a thorough knowledge of that patient's medical information. Dr. Kathryn Swoboda, an Associate Professor of Neurology and Pediatrics at the *University of Utah School of Medicine*, researches SMA and has been in touch with Nate and many other SMA families. Every time she writes an email to patients or parents, she is cognizant of the fact that it could end up published on the web and offered as advice to other patients whom it was not meant for. "It's a problem for physicians because things can get taken out of context, and when stories get told over and over it can change what the initial intent of the advice was," Kathryn said.

It is extremely frustrating for rare disease patients and their families when their local doctor does not have answers for them. It is what leads them to the internet to seek out medical information and recommendations from their peers. But health professionals remain the most trusted individuals when it comes to getting treatment information. Other medical advice found on the web is largely supplemental (Fox 2011). Because of this trust and a need for convenient access to expert information, there is a role for physician and patient communication on the internet. Given all the of complications that come with bridging the two groups online, though, companies trying to do this have to be thoughtful about their approach. Startups like *Alliance Health* and *MedHelp* have set up an organized structure to get patients and providers in touch. Similar to *PatientsLikeMe*, *Alliance Health* is an online site that hosts condition-specific social networks for patients. *Alliance* launched its first community, *Diabetic Connect* in 2006, which now has about 750,000 registered users.

Each community site includes the same basic groups of members including patients, caregivers, hired community managers, and patient advocates. Though unlike the others, *Diabetic Connect* also includes healthcare professionals. Through a partnership with the renowned research facility, *Joslin Diabetes Center*, an endocrinologist, a few certified diabetes educators, and a registered nurse are members of the group. Joslin staff can comment on posts from *Diabetic Connect* members and respond to frequently asked questions. David Goldsmith, Vice President of Product Partnerships and Development at *Alliance Health*, said that *Alliance* also partnered with the *Mayo Clinic* to create a similar setup for its *Heart Connect* community. The model allows for the proliferation of expert knowledge when it comes to general questions about diseases, drugs and treatment, but it avoids putting a provider in the position of having to offer patient-specific questions.

Partnerships Impact Care Delivery

It is important to note that both *Alliance Health* and *PatientsLikeMe* are for-profit companies. Alliance was originally formed to identify diabetic patients online who were looking for medical supplies, so that supply companies could better target their marketing efforts. Today drug, device and health service companies use the site to advertise to relevant condition-specific communities. *Alliance* states that its overall objective is both to deliver greater value to the healthcare consumer and greater efficiency to the healthcare marketer (LinkedIn 2013).

PatientsLikeMe makes a profit by selling de-identified patient health information from members' profiles to drug, medical equipment and health insurance companies (PatientsLikeMe 2013). The company tells its users that almost every piece of information they submit could be shared with its business partners (PatientsLikeMe Privacy policy 2012). *PatientsLikeMe* says this data helps a pharmaceutical company, for example, to identify the types of patients its drug is ineffective for. The thinking behind it is that the ability to drill down into patient data to examine subgroups, like middle aged women or young adults with a specific comorbidity, paves the way for developing personalized medicines. These business partnerships allow both companies to keep the lights on so they can provide valuable social networking services for the patients who use their sites. *PatientsLikeMe* emphasizes that its model hits a sweet spot between bringing in revenue and improving care. "Every partnership we develop must bring us closer to aligning patient and industry interests," the website states.

PatientsLikeMe has an "openness philosophy", encouraging users to share their health experiences and outcomes in order to enhance the collective disease knowledge base. However patients should be aware of two things before they make decisions about what information they provide and how much they want to publish. One, privacy on the internet is never a sure thing. Even if a member contributes information semi-anonymously under a user name, the more information he shares, the more identifiable he potentially is. *PatientsLikeMe* says that outside data sources can be used in combination with patient information from the site to piece together a person's identity. A stated risk of using the site is that identification could lead to discrimination from an employer or insurer (PatientsLikeMe Privacy policy 2012). And two, members do not have control over how their data is used by a third party. It could be used for drug research, but it could also be used for drug marketing research, an effort patients might not want to be a part of.

In addition to its revenue-generating partnerships, *PatientsLikeMe* has also aligned with nonprofit groups. Several collaborations have specifically been formed to benefit orphan disease research. For example, the company recently partnered with the *AKU Society*, a United Kingdom-based charity that raises awareness and research funds for the rare disease, alkaptonuria (PatientsLikeMe 2013). AKU is a genetic disease that causes early-onset arthritis. The partnership created an AKU patient registry, which like all others on *PatientsLikeMe* can be used by third parties for research.

Pharmaceutical companies and researchers studying the impact of their drugs in orphan disease patients benefit from the availability of these platforms in several ways. A registry gathers patients that are typically hard to locate—specifically because of their rarity—in one place online. With these members keeping detailed and updated accounts of their experiences, researchers learn about the impact of a particular drug at a much faster rate. "The platform allows the system to actually begin to measure the meaningful outcomes and how treatments are affecting patients in the real world, in real time," Ben Heywood said. It also lets them do it much more cheaply than a traditional study, which can cost millions of dollars.

In a way, patient community sites that rely on their patient users and their researcher customers form partnerships between patients and industry. The optimal outcomes from this relationship are better products that improve lives. However, it is understandable why this model makes many people uneasy. The pharmaceutical industry is widely thought of as one of the least trusted segments of the healthcare system. However, the internet provides an opportunity for the industry to change this perception. Just as there is a place for physician to patient communication online, there is also a place for drug company to patient interaction. Currently the drug-related conversations on patient community sites are one-sided. Drug companies can listen to what patients are saying, but they do not reply. *Alliance Health*'s David Goldsmith thinks that is going to change, and in the coming years drug companies will start to converse with patients directly. "The companies that get out in front on this and find ways of actually connecting with the patients more proactively or trying to be far more consumer- or customer-centric, will be the ones that begin to really chart a different course for the industry," he said.

Data Backs Decision Making

Dr. John Mattison is of the same mind that pharmaceutical companies can improve their reputation in the digital age of healthcare. But it's going to take a lot more information than patient anecdotes and self-reported data for them to do it. In the not too distant future, increasing numbers of patients will have access to whole genome sequencing, providing insight into their genetic makeup that can reveal exactly which treatments will work for them and which will not. John (Chief Medical Informatics Officer at *Kaiser Permanente* of Southern California) said that, with this information, pharmaceutical companies can protect patients in advance from taking drugs that they know will not work for them or will give them severe side effects.

The cost of whole genome sequencing has decreased over the past decade, and experts predict that with improvements in technology the procedure will soon fall to $1,000 (Winslow and Wang 2012). This is remarkable given that the cost per whole genome was $100 million in 2001 (National Human Genome Research Institute 2012). "We're entering an era where the genome will be a necessary part of every single health record," John said. Advanced genomics will reveal just how unique every individual truly is, genetically speaking. It will uncover how many different variants of disease exist within broad classifications like diabetes, cystic fibrosis and cancer. It will lead to the diagnosis of diseases that do not have names yet. And it will show just how tailored to the individual treatments must become in order to be effective.

In some ways drug companies are incentivized to invest in orphan disease treatments. In the U.S. the Orphan Drug Act encourages research through grants and reduced taxes. In 2011 the U.S. Food and Drug Administration counted nearly 1,800 projects in development with an orphan disease designation (Long and Works 2013). Also, as generic versions of blockbuster pills are brought to market by competitors, there is economic value in discovering drugs for smaller groups of patients, for which drug companies can charge more money per dose.

However, personalized medicine threatens their current business model. Drug companies make most of their money from patients who do not experience the full intended benefits of their drugs. Technology is getting closer to being able to use genetic typing to figure out who those patients are. If a test shows that in fact, 70 % of patients do not benefit from a particular drug, the manufacturer will be in a tight spot. It will either have to significantly increase its prices or stop making the drug altogether.

This is why pharmaceutical companies will have to get involved in the business of drug designation. John believes it will be part of their job to know which patients will benefit from a drug, and to make sure they receive it. It will also be part of their job to inform patients if they are likely to experience harsh side effects. And it will be their duty to identify the patients who will not benefit from their drug and make sure they do not take it. "Today they're one of the least trusted elements of healthcare, and the first one that gets to that trough is going to be the most trusted and set the standard for the rest of the industry," John said (YouTube 2012).

Personalized medicine is currently only used today in a few areas such as oncology and newborn screening. It will take years before its effects are fully realized. In the meantime, doctors and scientists proceed with rare disease research with the information they have. In most cases, this is limited, but it is certainly growing. One of the ways this is taking place is through a global effort to recognize every known rare disease. *Orphanet* is an organization based in France that makes orphan disease data available to researchers. It was first formed by the French government in 1997 so that officials could make more informed policies on rare diseases. *Orphanet* has since become an international consortium that includes 38 countries. The network includes mostly informaticists and researchers working with rare diseases. However its data, which is available at orphadata.net, is open to everyone, and about one third of its site traffic comes from patients.

Orphanet currently recognizes about 6,000 rare diseases. The organization continually updates its inventory based on its minimum criterion for entry: the disease must occur in two separate clinically homogenous cases. So even if there have only been two known individuals in the world to ever have had the disease, it is entered into the database. Another mission of *Orphanet*'s is to update the *International Classification of Diseases* (ICD) to include rare diseases. ICD is a *World Health Organization* classification system involving codes that correspond with diseases. The ninth and tenth revisions, ICD-9 and ICD-10, are currently used today, and the ICD-11 revision is underway. *Orphanet* Chief Scientific Officer Ana Rath said rare diseases are badly represented in ICD-10, with only about 200 diseases out of the thousands recognized by *Orphanet* included.

It is important to have as many diseases as possible in the ICD. Since it is an international standard it can enable international collaboration. As countries around the world implement electronic health records, there will be more opportunities to bridge information gaps in order to learn more about groups of patients. An individual patient's electronic health record holds information about each condition, treatment and procedure they have experienced since they were born. Many researchers believe that millions of medical history records will change the way clinical trials are conducted. The thought is that each patient with an electronic record is virtually participating in a huge, ongoing clinical trial. So rather than enroll a group of patients in a time-consuming and expensive study to test a hypothesis, scientists can find out if the hypothesis proves true according to existing data. In the future rare diseases can be studied in the same way, even across international borders. But first, they need to be assigned codes in order to be recognized by the system.

Concluding Thoughts

Founder of *O'Reilly Media* Tim O'Reilly delivered the commencement address to the *University of California Berkeley School of Information* class of 2006. In his speech, Tim (known for popularizing the term Web 2.0) related an anecdote that summarized the changes then taking place in computing and still going on today: "Clay Shirky, who studies the effects of the internet on society, was at a conference where he told the story of Thomas Watson. Watson was head of *IBM* from the 1950s until the 1970s, a time when the company produced its first mainframe computers. Watson once famously stated that he saw no need for more than five computers worldwide. Everyone in the audience laughed, thinking of the millions of computers that existed."

Then Clay finished with the thought, "He overstated the number by four." Clay's premise was that the internet joins each of the millions of physical computers into one vast computer, virtually a single brain. And the more that people contribute to the internet, the greater its impact. "The secret of success in the networked era is to create or leverage network effects," Tim remarked (SlideShare 2013).

That was the motivation behind Nate creating *SMAspace*, and it is why the site eventually evolved into an international network. When Emily was born, Nate stated an ambitious mission: that no SMA family in the world would be unable to connect with other families or unable find the information it needed to make decisions. Starting with a simple site, Nate's message was picked up and carried by other families across the U.S., then by more families across dozens of countries. At the very least Health 2.0 makes the world feel a lot smaller. At the most it builds a global network that will lead to better healthcare tailored to each individual.

Acknowledgments Many thanks to the following persons for giving up their time for me to conduct interviews: Nate Lee (*SMAspace*), Ben Heywood (co-founder, *PatientsLikeMe*), Nate Gross MD (co-founder, *Doximity*), Kathryn Swoboda MD (Associate Professor of Neurology and Pediatrics, *University of Utah School of Medicine*), David Goldsmith (Vice President of Product Partnerships and Development, *Alliance Health Networks*), John Mattison MD (Chief Medical Information Officer, *Kaiser Permanente*, Southern California) and Ana Rath (Chief Scientific Officer, *Orphanet*).

References

Alexa Internet. Top Sites.http://www.alexa.com/topsites. Accessed 31 Jan 2013.
ALS Therapy Development Institute. History. http://www.als.net/About-ALS-TDI/ALS-TDI-History.aspx. Accessed 31 Jan 2013.
Domo. How much data is created every minute? http://www.domo.com/blog/2012/06/how-much-data-is-created-every-minute/ (2012). Accessed 31 Jan 2013.
Encyclopaedia Britannica. Change: It's Okay. Really. http://www.britannica.com/blogs/2012/03/change/ (2012). Accessed 31 Jan 2013.
Epilepsy Foundation. About Epilepsy. http://www.epilepsyfoundation.org/aboutepilepsy/. Accessed 31 Jan 2013.
Fox S. Peer-to-Peer bealthcare. Pew internet and American life project. http://pewinternet.org/Reports/2011/P2PHealthcare/Summary-of-Findings.aspx (2011). Accessed 31 Jan 2013.
Health 2.0 TV. Welcome and intro to health 2.0 San Francisco 2010. http://www.health2con.com/tv/welcome-and-intro-to-health-2-0-san-francisco-2010/ (2010). Accessed 31 Jan 2013.
LinkedIn. Alliance health networks. http://www.linkedin.com/company/alliance-health-networks. Accessed 31 Jan 2013.

Long G, Works J. Innovation in the biopharmaceutical pipeline: a multidimensional view. Analysis Group. http://phrma.org/sites/default/files/2435/2013innovationinthebiopharmaceuticalpipeline-analysisgroupfinal.pdf (2013). Accessed 31 Jan 2013.

Manhattan Research. New study finds physician-only social network adoption growing in Europe. http://manhattanresearch.com/News-and-Events/Press-Releases/European-Physician-only-Social-Network-Adoption (2013). Accessed 31 Jan 2013.

McGowan BS, Wasko M, Vartabedian BS, Miller RS, Freiherr DD, Abdolrasulnia M. Understanding the factors that influence the adoption and meaningful use of social media by physicians to share medical information. PubMed. http://www.ncbi.nlm.nih.gov/pubmed/23006336/ (2012). Accessed 31 Jan 2013.

National Center for Biotechnology Information, U.S. National Library of Medicine. Amyotrophic Lateral Sclerosis. http://www.ncbi.nlm.nih.gov/pubmedhealth/PMH0001708/. Accessed 31 Jan 2013.

National Center for Biotechnology Information, U.S. National Library of Medicine. Spinal muscular atrophy. http://www.ncbi.nlm.nih.gov/pubmedhealth/PMH0001991/. Accessed 31 Jan 2013.

National Human Genome Research Institute. DNA sequencing costs. http://www.genome.gov/sequencingcosts/ (2012). Accessed 31 Jan 2013.

Nebraska State Board of Health Meeting. http://dhhs.ne.gov/publichealth/Licensure/Documents/091712bohminutes.pdf (2012). Accessed 31 Jan 2013.

PatientsLikeMe Privacy policy. Alliance health networks. http://www.patientslikeme.com/about/privacy (2012). Accessed 31 Jan 2013.

PatientsLikeMe. About PatientsLikeMe. http://www.patientslikeme.com/about (2013). Accessed 31 Jan 2013.

PatientsLikeMe. Corporate. Alliance health networks. http://www.patientslikeme.com/help/faq/Corporate#m_money (2013). Accessed 31 Jan 2013.

PatientsLikeMe. PatientsLikeMe and AKU society to develop world's first open registry for alkaptonuria patients. http://www.patientslikeme.com/press/20130109/42-patientslikeme-and-aku-society-to-develop-worlds-first-open-registry-for-alkaptonuria-patients (2013). Accessed 31 Jan 2013.

Reddit. IAmA father of a child with spinal muscular atrophy. http://www.reddit.com/r/IAmA/comments/15ua0y/iama_father_of_a_child_with_spinal_muscular/ (2013). Accessed 31 Jan 2013.

SlideShare. Tim O'Reilly's commencement speech at UC Berkeley SIMS. http://www.slideshare.net/GeorgeAppiah/tim-oreillys-commencement-speech-at-uc-berkeley-sims (2013). Accessed 31 Jan 2013.

The Pew Research Center's Internet and American Life Project. Who's Online: Internet User Demographics. http://pewinternet.org/Static-Pages/Trend-Data-(Adults)/Whos-Online.aspx. Accessed 31 Jan 2013.

Wicks P, Keininger D, Massagli M, de la Loge C, Brownstein C, Isojärvi J, Heywood J. Perceived benefits of sharing health data between people with epilepsy on an online platform. Elsevier. http://www.sciencedirect.com/science/article/pii/S1525505011005609 (2012). Accessed 31 Jan 2013.

Winslow R, Wang S. Soon, $1,000 will map your genes. The wall street journal. http://online.wsj.com/article/SB10001424052970204124204577151053537379354.html (2012). Accessed 31 Jan 2013.

YouTube. Code for health: how software is eating the (healthcare) world. http://www.youtube.com/watch?v=tfWJGcfvBXw&list=PLHahN1_6vGt1soUxcBg274jYZNpPC50QW&index=3 (2012). Accessed 31 Jan 2013.

Vignette: Living with NOMID: Michael's Story

Jocelyn Gardner

Michael was born Feb 1, 2004, a full term baby and appeared healthy. All the standard newborn tests were run, including the PKU test. Before leaving the hospital, Michael developed a rash within 24 h of birth, his mother was told it was a normal neo-natal rash and not to worry. He was severely colicky and his mother had been to the doctors many times in the first month, due to his extreme behavior and an increasingly aggressive rash.

By two months of age and many doctors visits later, the family doctor told his mother she was just tired because she was also going through a divorce. He accused her of suffering Munchausen syndrome and suggested she place her kids in foster care to give her a break. Naturally, she did not even consider such a preposterous idea and forged on.

At his second trip to the ER for what appeared to be a severe allergic reaction to an unknown source, his mother was informed that there was no PKU result, investigation showed the sample was received at the Ontario government Lab, however the test results were lost.

Over the next 16 months, hospital trips became routine and by Michael's first birthday, the local hospital began to reject Michael. It would seem they were as afraid of him as their mother, and would just tell her to take him to the Sick Kids Hospital right away.

Michael was suffering with regular fevers, ataxia, he could not bear weight, had an aggressive rash covering his entire body, multiple episodes of toxic joint effusions, eye and ear deficits, high blood pressure, regular breathing problems and vomiting. He was rapidly deteriorating, but there was little being done to diagnose him.

During a visit to a specialist at Sick Kids, a resident reviewing Michael's chart mentioned he was checking him for NOMID. The pediatric rheumatologist had not told Michael's mother about this and when she contacted him he said that that they were certain it could not be NOMID, but were ruling it out along with a list of other potential diagnoses.

Nevertheless, the family immediately looked up NOMID, since most other diseases had been ruled out. The picture was bleak and they desperately wanted to believe the doctor was right, that it could not be NOMID.

Only a week later Michael crashed again and was admitted into Sick Kids for another week. This was his worst episode yet. He was unable to move, lift his head, his fontanel was bulging and his joints were swollen. His strange behavior, inability to focus or engage anyone, inability to balance or even sit up was terrifying. In that week there were hundreds of doctors and students parading through to see this little curiosity, his mother obliging in the desperate hope that someone could help.

He was not given any medication for pain for the first four days, despite dangerously high blood pressure and a toxic synovial infection. He was on a waiting list for an MRI for more than six months at that point and it was never done, even though he was admitted into hospital and displayed obvious neurological symptoms. He was discharged after seven days and not a single doctor kept their word in terms of co-coordinating a multi-discipline meeting, nor was any follow up planned. It was a matter of, "I'm sorry Ma'am, we just don't know."

That admission was the final straw and Michael's mother, my daughter went to the Internet. There was only one website about NOMID on the Internet at the time which had been created by the parent of a child with the condition. She

Vignette: Living with NOMID: Michael's Story

e-mailed the page owner and hours later was contacted and given the name of the lead researcher working on projects concerned with this disease at the National Institutes of Health in Bethesda, Maryland.

The Doctor at the NIH responded to her desperate voice mail and called back the next day. After a lengthy discussion, she felt confident based on the information his mother provided, that NOMID was a probable diagnosis. She asked his mother to have Michael's rheumatologist make a referral, in order to bring Michael to Bethesda, Md. Being a single working parent of two young children, travelling was far from practical, however, desperate to do something for this helpless little boy, she would have done anything.

After the Rheumatologist from Sick Kids spoke to the NIH, he called my daughter and said it was not necessary to go to the States, he would try to get Michael into a study at the hospital but it would take months and require Governmental approvals. When pressed, he admitted there was no guarantee he would be accepted. Knowing from the Doctor in the US it was literally a matter of life or death, my daughter made the only choice she could and travelled to the United States.

At the NIH, Michael underwent extensive testing, including an MRI, spinal tap, numerous blood tests, biopsy, x-rays, hearing evaluation, eye evaluations and bone density to name a few. This was testing that should have been done much sooner. This organization understood the urgency of the condition, and no stone was left unturned.

At the end of a long week of testing, my daughter learned that Michael was a ticking time bomb. He had remarkably high inter-cranial pressure, which had caused a stroke and damaged an area in his brain, and his optic nerves were inflamed and damaged. He was living his entire life in excruciating pain, which had been untreated for all that time. His bone structure and growth was significantly delayed. At 21 months of age he already had early osteoporosis. Not a single test came back within "normal ranges".

Genetic testing confirmed a genetic mutation; in his case this mutation has been reported in less than 20 people worldwide. The NIH agreed to take on Michael as a patient and for him to participate in an experimental treatment, which would be unattainable here in Canada. Michael received his first treatment on October 1, 2005 and made us all believe in miracles.

To this day, when Michael goes to the local hospital, my daughter ends up wasting valuable time explaining his condition and then having the doctors doubt her explanation. They often require her to have the doctors here contact the NIH in order to validate his unusual situation, and for the doctors to get guidance. They will do as directed by the NIH and quickly walk away. No one will commit to his ongoing medical care.

The battle after diagnosis continues, affecting many aspects of Michael's life. When trying to get him into regional day care, hoping to participate in the early intervention programs, it was necessary to disclose his health problems. Suddenly it was necessary to review his case—once his mother supplied all his medical information, it was then refuted because it no space was from a doctor outside

Canada, and the condition is classified as a "rare disease", which they did not understand.

Now an active 9-year-old boy, Michael's disease is under control and he is thriving. His life will always be impacted by his disease and he will require care for the rest of his life; he needs occupational therapy and assistive devices at school due to his fine motor delays. He must make adjustments and sacrifices at times in order to participate in day-to-day life with his peers. Michael is strong, and a great advocate for himself and others. He is well aware of his limits, but will allow nothing hold him back from trying.

Michael continues to travel to the NIH as part of the clinical trial and in order to ensure access to treatment.

We had a small breakthrough last year. When Michael's older brother was in hospital for his appendix, the doctor taking care of him asked questions about Stephen's family history. When he was told that Michael had NOMID he was very interested and asked who Michael's doctor was. When we told him Michael did not have a doctor in Canada he asked if he could take him on as a patient. He explained that he did not know about NOMID but he would like to contact Michael's doctors in the US and wanted to learn about it. As he said, "This is a one in a million opportunity to learn about a rare disease".

Over the past nine years there have been small steps towards the medical profession being open to the possibility of a person having a rare disease, thanks to families and patients speaking out and not sitting back and accepting that there is no help for them.

Living with a rare disease is a challenge; the patient or parent must become a medical student, detective, human rights advocate and politician, in order to navigate the underworld that accompanies the "rare disease" label.

Canadian CAPS Network

Health 2.0 and Information Literacy for Rare and Orphan Diseases

Hannah Spring

> *We read to know we are not alone*
>
> CS LEWIS

Abstract This chapter examines the issues of information literacy for both health professionals and patients in the context of rare and orphan diseases. In particular, the chapter uses a real life case study to consider the information experience of a person in the early stages of a rare condition, both pre and post diagnosis. It also reviews the information experience of the health professionals responsible for their care. The chapter addresses common barriers to information literacy such as lack of information, lack of information skills and difficulties associated with accessing information. Behaviour related to information pre and post diagnosis is also discussed in this chapter, together with the current role of health 2.0 as the provider of both information and emotional support for people with rare diseases. Areas currently being given consideration for future coordination are briefly reviewed as part of the development of information literacy in rare and orphan diseases.

Introduction

The information age has fundamentally changed the way people interact with information and also increased the accessibility of information. Only three decades ago, the doctor was generally viewed as the main information authority in matters of disease and intervention. In hospitals, medical libraries were reserved exclusively for the use of doctors and consultants, whilst other health professionals such as nurses and allied health professionals were not deemed worthy

---

H. Spring (✉)
Health and Life Sciences, York St John University, Lord Mayor's Walk, York YO31 7EX, UK
e-mail: H.Spring@yorksj.ac.uk

of their use. The idea of a patient having access to medical information, or even having the skills to use it was unheard of. The birth of the information society has made a mockery of such values, and in current times individuals are considerably more information rich than they were in previous decades, having greater exposure to health related information (Cegala et al. 2008). Today all disciplines of health professionals have access to health information, are expected to use it for the purposes of evidence based practice, and in fact often complain of suffering from information overload. Furthermore, the general public have access to the same information as the health professionals who provide their care. The issues of information access and use can however be different for those suffering with a rare disease. This chapter uses a case study to highlight the issues of information access and information literacy in the case of an individual in the early stages of a rare condition.

Information and the Experience of People With Rare Diseases: A Case Study

This case study focuses on the real life experience of a 38 year old British woman diagnosed with cluster headache. Although cluster headache does not fit the definition of a rare disease in terms of its prevalence (MacGregor 2011), it is uncommon and particularly difficult to diagnose. It typically affects men in their twenties through to forties, and the average time for achieving a diagnosis is 10 years (MacGregor 2011). It is a member of the primary headaches group and far less common than secondary headaches, which are caused by external factors such as stress or as symptoms of another condition for instance a cold or flu. Cluster headaches are often incorrectly diagnosed and treated as migraine, but in fact need very different treatment, typically including daily injections, or oxygen. The condition is very difficult to achieve complete control over and even with treatment, sufferers are rarely symptom free.

Cluster headache is a debilitating condition which presents as a sudden onset of severe and intense pain, usually on one side of the head and/or face. The pain is usually accompanied by what are termed 'cranial autonomic symptoms' on the same side of the head in which the pain occurs. These include redness and tearing of the eye, a running or blocked nostril, facial sweating and flushing, constriction of the pupil, and extreme restlessness and agitation. The attacks can last from just a few minutes to up to 2 h and can reoccur within a very short space of time, hence the name 'cluster' headache. Some people experience acute attacks which can last between six to eight weeks approximately once per year, whilst others suffer from chronic cluster headache attacks which in extreme cases can happen daily and almost constantly, sometimes up to eight times a day. Although not life threatening, the unpredictability of the condition can have a severe impact on the quality of life of those who suffer from it, in some cases having led to suicide or attempted suicide.

The uncommon nature of cluster headache means that it is not well understood and largely unheard of both by the general public and by many health professionals. The condition is therefore one which is both underdiagnosed and regularly misdiagnosed, and so for those suffering from it, the experience of obtaining a diagnosis is characteristic of and similar to those with a rare disease. There is very little research into cluster headache and very few health professionals with appropriate expertise in the area. In this sense it is an orphan disease and the main funding source for research into cluster headache is through the only charitable organisation that exists for the condition (OUCH (UK) 2011) (Organisation for Understanding Cluster Headaches). The American equivalent of OUCH has recently closed due to lack of volunteers and funding.

Annie's Story

Annie's story began in December of 2010, when late one afternoon she began to experience an aching head and neck, and tightness of the scalp. Throughout the evening, this developed into an unbearable and devastatingly painful headache.

> This was unlike anything I had ever experienced before and I was totally unable to function. I thought maybe I had a brain tumour and felt like I was going to die. The pain was so severe that I was terrified.

Over the course of the next day, unable to go to work, Annie continued to suffer the terrible pain in her head which presented in bouts and seemed ceaselessly to come and go. No amount of pain killers were helping to ease the pain:

> I knew that these were not normal, run of the mill headaches. When you have a bad headache or a migraine, lying still in a darkened room usually helps, but even though I was in excruciating pain, I didn't want to do that. It was strange, but I felt agitated and felt the need to keep moving. I found that going outside and walking in the fresh air would help a little when the attacks came on.

However, 48 hours later, desperate and unable to cope any longer, Annie admitted herself to the local A&E unit. The consultant who saw Annie said it was likely to be a migraine and prescribed some medication for migraine and strong painkillers. He suggested she visit her GP if the pain was no better in 2 days' time, and discharged her. Over the course of the next 2 days, the medication had no effect on Annie's pain and she was still unable to attend work. Following the A&E consultant's instructions, she made an appointment to see her GP who did not seem able to provide a diagnosis, but prescribed a different drug.

> After the appointment, when I went to the chemist to collect my prescription, the pharmacist was very concerned about the drug I had been prescribed. She asked me if I had vertigo and I said no! She said that this drug was very strong and should only ever be prescribed for vertigo and she advised me against taking it if I did not have vertigo. I felt so annoyed with the GP for not explaining what she had prescribed to me, or even that she had diagnosed me with vertigo! It was a total misdiagnosis which could have been

dangerous had I taken the drugs prescribed, and I have never been back to that particular GP since!

Annie continued to take pain killers but they were ineffective and 2 days after the GP appointment, Annie was back in the same frightened place she had been 6 days earlier.

> Although it would ease off at times, it always came back and the pain just would not go away. All I wanted was to make it stop and it seemed like no one was able to help. I was really frightened. Something was obviously very wrong and so I went back to A&E. The consultant I saw on this occasion did not think I had vertigo and I was told I would need to have tests on my eyes. I was also told that if my eye tests were clear, then I may potentially have a brain tumour and the next course of action would be to investigate for this. Although I was terrified, I felt as if something was happening towards getting a diagnosis and this was reassuring in a funny kind of way.

The eye tests were carried out the next day and came up clear. Annie was then instructed to return to her GP to be referred back to the hospital for further investigative tests, this time for a brain tumour. The GP (this time a different one) agreed to do this, but felt Annie may be suffering from neuralgia. A new drug was prescribed which, after a brief Google search, Annie discovered was an anti-depressant drug.

> It turned out that this drug was also used as a pain relieving drug, but the GP had not explained this to me. I just thought that she thought I was depressed, and so I was angry that this had not been mentioned during the consultation. I became very scared again at this point because I honestly thought she didn't believe me. I thought if the GP thought I was depressed, then perhaps she thought it was all psychological and perhaps imagined. It was awful, I felt so helpless and not in control of the situation at all. I felt as if I was at the mercy of a load of supposedly health professionals who didn't understand, weren't listening and didn't believe me… and all the while these attacks of pain were still happening to me.

Annie's experience at her next hospital appointment a few weeks later was better. She underwent general tests under the care of a consultant whom she described as being very good.

> He explained to me really clearly all the tests that would be necessary. He explained that there were three possible lines of investigation and that it was likely that I had either shingles, a brain tumour or cluster headache. He was good, you know? He listened to me and he understood that I didn't have a migraine. He believed me. Even though he wasn't sure at that stage what was wrong with me, it made a difference.

During the time Annie was undergoing tests, she spent some time Googling cluster headache. She discovered that there was very little information about the condition but found some USA based web sites which she felt were of little help and in fact increased her stress levels.

> All the information was really dramatic and scary. It was all about people suffering from cluster headaches killing themselves! It was very depressing and didn't really help me much. I am a practical person and needed good, sound information, not dramatic stories.
>
> The lack of information and a diagnosis was so stressful. As well as all the hospital tests, I had spent a fortune on drugs that were irrelevant. I had spent a lot on travel back

and forth to hospitals for tests. I had also tried physiotherapy, acupuncture, dental and optician consultations, none of which had helped. An ongoing cost like this was not sustainable. I was fearful of how the costs would be covered if my job was lost due to ill health.

Following an MRI scan, further tests and a very long wait, the diagnoses of shingles and brain tumour were eliminated from the medical investigations. Annie was told that the likely diagnosis was cluster headache and was referred back to her GP for treatment, but this was a frustrating experience.

> The GP had never heard of cluster headaches. She just wanted to give me migraine tablets and insisted that's all it was. It was a nightmare. I knew this would not work for me because I had already gone down the migraine medication route at the very beginning of the process when my symptoms started.

It was during this GP consultation however, that Annie was given a general information sheet about conditions causing head pain. Further information resources about head pain were provided at the end of this leaflet, and this gave details of the web site of a UK based organisation called OUCH which Annie then visited.

> This information seemed almost like just an afterthought at the end of the leaflet the GP gave me. It wasn't at all the focus of the main information on the leaflet that's for sure. But when I looked, the UK website was a lot better than some of the American ones I had visited previously. It wasn't full of dramatic stories about suicide and had some really good, matter of fact information about cluster headache. I was able to identify better with what I was reading on this site. There were even some really useful videos featuring experts explaining cluster headache and real life stories from people who suffered with the condition.

It was from reading the information provided by the UK OUCH site that Annie came to learn that cluster headache was sometimes nicknamed 'suicide headache', essentially because the pain experienced by those suffering with the condition and the unpredictable nature of the episodes have been too unbearable for some to live with. Annie telephoned OUCH UK.

> It was amazing. I just kept telling them 'my head hurts, it really hurts', and 'tell me how I can make it stop'. It was the first time that someone acknowledged with complete understanding that, 'yes it does hurt' and 'yes it is dreadful and unbearable'. I was so relieved. It felt good to finally be listened to and believed. I cried a lot.

OUCH explained to Annie that cluster headache is uncommon and that at present there is very little specialist knowledge or research on it. She was advised, however, that there was treatment available and that they could help her, but that she had to have a recognised diagnosis of cluster headache from a neurologist in order to access that help. They gave her details of the nearest hospital with an expert and explained that she should seek a referral from her GP to this particular consultant. With the diagnosis, they would be able to offer her treatment options.

> OUCH helped by giving me back control. They gave me a step-by-step course of action and I felt like, "I know what to do now". I felt empowered because at last I had the information I needed to be able to really do something about it.

On returning to the GP however, Annie experienced further barriers.

> The GP told me that I didn't need a referral. She said I just had a migraine. But this had been going on for months! Who experiences a continuous migraine for months on end?! Don't get me wrong, she was really nice. She wasn't rude or anything, but she just didn't believe me and simply didn't understand. I tried to explain to her about the OUCH web site which ironically she had given me the information about in the first place. I asked her kindly if she would watch the video about cluster headache, but she didn't watch it. It was absolutely awful. I desperately needed the referral to the expert consultant in order to get the help and treatment from OUCH but she just stood in my way. The GP made me feel completely powerless. I felt as if I had lost control again.

Eventually, after considerable effort, Annie succeeded in getting the GP to provide her with a referral; however, this was done as a non-urgent appointment, which resulted in further months of waiting. Unfortunately, when the appointment came through, it was not with the neurologist specialising in cluster headache that Annie needed to see, but the GP was unsympathetic.

During these months of pain and anxiety, Annie had returned to work. Some days she would experience no pain, whilst other days she would experience one attack after another. The pain was impossibly unpredictable.

> When I was diagnosed I was frightened, worried for my job. I didn't know how it would work on a practical level. When I would have an attack of cluster headaches, it was frightening because it hurt so much and I didn't know how to stop it, and I was being denied the treatment I needed to help with this. I didn't know when it would start or stop, and it was really difficult some days at work. I can understand why they call it suicide headache!

Annie worked for the NHS and had informed her employers that she had been diagnosed with cluster headache.

> Work was supportive but didn't understand. Their attitude was just, 'Oh, she gets bad headaches', but this isn't headache you know? It's a recognised condition that causes headaches, but people just didn't get it at all. I had to have an interview with the Human Resources department and their advice to me was that I should not climb ladders. This has never been part of my job description—I have a desk job! Even though everyone was very nice about it, this was the level of understanding I was faced with.

Feeling that all avenues had been exhausted, she spoke to a colleague at work about her need to see the expert neurologist and through contacts at work she was finally able to get an appointment. She feels that she would never have had this opportunity had she not worked for the NHS. 7 months after the initial onset of her cluster headaches, Annie finally was able to see the right specialist.

> I had to jump through various hoops which the Consultant said was routine. It was just playing the game. All the usual tests, which both of us knew I didn't really need had to be done, and I had to have another MRI scan. The consultant said it was likely that I would have to wait a while for the final diagnosis, and I heard nothing for months. It was a real waiting game. I wish they could have just given me the information, you know? Even if I didn't have the actual letter in my hands with the diagnosis, it would have been nice to at least know it was on the way. You shouldn't have to fight so hard.

Eleven months from the onset of Annie's cluster headaches she finally received her official diagnosis from the specialist neurologist that OUCH had referred her to and was able to access the treatment and support she so desperately needed.

> It was the most incredible relief to finally have the official diagnosis. During all those months trying to get a diagnosis, I had been able to seek out the relevant information about cluster headache from OUCH, but now I was able to make an informed decision about available drugs and treatment options. I was able to move on. It wasn't so much, 'what is wrong with me' anymore, but more, 'OK, so now how do I manage this condition'. I needed different information now. OUCH was just invaluable, and as time has gone on, I have learned how to cope on a practical level with my condition. I can't put anything off anymore, you know? In case I have an attack tomorrow.

On reflection, Annie felt that during her experience as a whole the GP was the least useful information source, and OUCH the best.

> You would assume you'd get the right information from the GP, but in actual fact, in the early stages it was a Google search for cluster headache that got me there, and the OUCH web site. But what about people without information access or skills? I can see why some people might feel suicidal.

Annie now uses a support forum that is provided by OUCH for individuals with cluster headache.

> The support forum is really good but it only exists via the OUCH website. They do have face-to-face meet ups around the country, but people with cluster headaches are not great at meeting up because of the headaches! I worry because OUCH doesn't have secure funding. It is a charitable organisation and there is no certainty that it will always survive. I do not see a positive if OUCH were to close. It is the only organisation that exists to provide support and information for people like me.

Discussion

Annie's experience highlights some interesting issues which have a direct impact on information literacy. These are particularly concerned with the apparent lack of information on rare and difficult-to-diagnose conditions, and lack of appropriate professional knowledge and expertise.

Lack of Information in Rare Diseases

Patients rely heavily on health professionals as an information source (Carpenter et al. 2011; Kuehn 2011; Cegala et al. 2008; Sen and Spring 2013). In a study about patient satisfaction regarding online health information seeking amongst cancer patients, Tustin (2010) found that those who were satisfied with the care and quality of communication they received from their physician were less reliant on the internet for information about their condition. Although Annie believed that the health professionals would be the best information resource, this was not her overall experience. Annie's initial difficulty was a lack of information in so far as she was unable to obtain a correct diagnosis. This was exacerbated by poor communication and information transfer between the health professionals and herself. This happened on a number of occasions, initially when Annie was prescribed a drug for vertigo but did not have this explained to her by the GP, and again when she was prescribed a drug for depression. Had the doctor provided clarification about the different uses of the drug, it is likely that Annie would have had less cause for concern and would have understood that the doctor was simply trying Annie with a different kind of painkiller. The consequences of not providing this information led to anger, and heightened fear and anxiety for Annie who felt she was not being believed and that no one could help her.

The approach by health professionals to the diagnosis of health conditions is described by Beale as the difference between horses and zebras, "If you hear hoof beats, think horses, not zebras" meaning that most disease states are common ones, and most patients are typical, i.e. they are 'horses'" (2011, p 312). Routine lines of inquiry are followed initially which is why people suffering from rare conditions often go for considerable lengths of time before a diagnosis is arrived at. Beale identifies that, like Annie, 'zebra' patients tend to bounce from specialist to specialist until all of the familiar possibilities are exhausted, at which point the GP is likely to throw their hands in the air and shift the focus of discussion away from diagnosis and cure to symptom control and pain management. Providing information support to a patient without a diagnosis is however, problematic and this was evident in the case of Annie whose GP, lost for diagnosis, was only able to provide a generic information leaflet about types of headache. Lewis et al. (2010) identify that health professionals have a routine approach to information support and this is a particular concern when patients with rare or difficult-to-diagnose conditions are the information recipient. There is often little known about rare diseases and they are not well understood. The rarity of their nature often means they are not prioritised for research, resulting in a lack of information or an evidence base, and poor media exposure (Carpenter et al. 2011). Therefore, without a diagnosis the giving of incorrect information is likely, but with a rare diagnosis, a lack of information is also likely.

Pre-diagnosis and at the early onset of her cluster headaches, Annie was keen to find out what was wrong with her. Unable to obtain clear information from the health professionals, like many, Annie turned to the internet. Arora et al. (2008) observe that the online experiences of patients seeking information about their condition are often confusing, frustrating and negative. Although Annie herself was a well-educated, information literate person with good access to information sources such as the internet and appropriate skills to search it, she was unable to exploit that resource effectively because without a diagnosis, she did not know what to search for. She was void of keywords. A keyword like 'heachache' is too broad a search term to find appropriate information with. In a survey based research study on information use by patients with prostate cancer (Cegala et al. 2008), a sharp increase in use of the internet for health information was observed after the patients had been diagnosed, whilst pre-diagnosis internet access levels were very low. This suggests that patients are not always able to search effectively for information until they have a clear diagnosis.

This can also be an issue for health professionals. In a study carried out by Tang and Ng (2006, p 1143), they identify that 'everything could be found on the web if only one knew the correct search terms'. They also suggest that the efficiency of a search and usefulness of the information retrieved is often dependent on the knowledge base of the person searching. Therefore patients searching Google are less likely to find a diagnosis for themselves than a 'human expert' (e.g. a doctor) because the doctor is likely to have a better knowledge base to support the selecting of relevant documents. In their research Tang and Ng attempted to use Google to diagnose medical conditions and concluded that Google was more likely to be effective for conditions with unique signs and symptoms that could easily be translated into search terms. They also suggest that 'searches are less likely to be successful on complex diseases with non-specific symptoms or common diseases with rare presentations' (p 1144). From a patient perspective, this was certainly a reality for Annie, who was suffering from an uncommon and complex condition, but which presented on the surface as a common 'headache'. The indexing of terms in medical databases can also present challenges for those searching for literature on rare conditions. Beales (2011) for instance, notes that the rare condition Sicca Syndrome is associated with another condition called Sjogren Syndrome, however, both are distinctly different. The misinterpretation in MeSH makes searches for Sicca more challenging.

In a response to the research by Ng and Tang, a senior house officer shared her experience of using Google to diagnose a patient who was expelling purple urine into her catheter bag (Butcher 2006). This was a phenomenon the SHO had not seen before, and a search using the keywords 'purple urine' in Google produced information on a rare syndrome called Purple Urine Bag Syndrome (PUBS) affecting chronically catheterised or constipated women who have alkaline urine or a urinary tract infection. She agreed with Tang and Ng that Googling can be useful for diagnosing on unusual symptoms, or rarer and eponymous syndromes doctors may not come across on a regular basis, and confesses that she felt it unlikely she would have found the diagnosis through conventional book searches in which common diseases can be easily found, but rarer one less so.

Indeed, it may be easier to identify keywords for medical conditions presenting in a very unique way, and once Annie had been given the term 'cluster headache' as a potential diagnosis in the early stages of her tests, she was able to conduct much more fruitful searches because she had the right keywords. The initial lack of information created high anxiety levels for Annie, but with the right information she was empowered and was able to take action. This is noted by Draucker (1991) and Carpenter (2011), who state that for patients, obtaining information increases a sense of control over their condition and gives them good 'information health' Spring and Sen (2013). For those with rare or difficult-to-diagnose conditions however, the issues of obtaining that information can be more difficult due to the issues highlighted here.

Information Access and Behaviour in Rare Diseases

The study by Tustin (2010) which found patients were less likely to use the internet for information if they were happy with the care they received from their physician reveals some interesting issues around information literacy in healthcare. The first of these is the extent to which people have appropriate skills to access information. Physicians are generally ranked as one of the most preferred and trusted information sources (Carpenter et al. 2011; Kuehn 2011) but when this fails, their patients tend to refer to other sources. Research by Tustin (2010) and Carpenter et al. (2011) observed that younger people and those with higher levels of education have a greater tendency to seek out healthcare information than older people and those with less education, and lack of skills may arguably be related to this. A certain level of information literacy is necessary in order to access web based information and a lack of skills can lead to inequalities in access to information.

From the perspective of the health professional, information literacy skills can also be a concern. Beales (2011) notes that health professionals often do not have the time or skills to search the medical literature. Furthermore, Beales observes that health professionals often do not recognise what they do not know, advising that this can be 'to the detriment of all, including the physician who continues to practice with a blind spot which affects their clinical competence and performance (p 313). Indeed, this was evident in the experience of Annie who was unable to convince her GP of the existence of cluster headaches.

Information skills might be one issue, but choice of information source is another. In common or easily diagnosed conditions the health professional is often a key information source for patients, but in rare or uncommon conditions this is not a common reality. In the case of such conditions, in the absence of literature or specialist physicians, those with the rare condition, their carers, friends and family members are often a more knowledgeable and a heavily exploited resource for health information. This is essentially because this group of individuals possess a lived experience of the condition and through this, have learned about its peculiarities and how to cope with its effects. In time, this group of individuals often become 'information donors' (Spring and Sen 2013) or 'expert patients' (NHS Choices 2013), able to advise others with the condition, and even health professionals in aspects of the disease. The dissemination of experiences and knowledge through charitable support groups and organisations, peer/social networks such as Twitter and Facebook, and personal blogs can therefore be a vital source of information and support for others who also have the condition. This again is reflected in the experience of Annie who found OUCH to be the information lifeline she needed. Indeed, in their study of patients with the rare disease vasculitis, Carpenter et al. (2011) found that both physicians and the internet were considered to be the most credible information sources, and also specialist organisations such as the Vasculitis Foundation.

These communities of support can be accessed through various methods, although the internet is a common approach. Web 2.0 has been revolutionary in connecting people in time and space from all four corners of the globe. Kuehn (2011) observed from her small study that people with health conditions and their friends and family are using web based communities for emotional support. Whilst the idea of support groups is not a new concept, the internet has provided new opportunities for this form of interaction, helping to widen the circle, and creating a 24–7 environment in which people can access information. Kuehn also comments on accounts given by those with rare diseases who have found particular value in being able to access the experiences and first-hand accounts of others with the same condition. The shared knowledge that information mediums such as this provide are noted as being particularly useful for patients with rare diseases who would otherwise have to travel to meet others with the same condition. In the case of rare diseases, the potential for open access sources such as Wikipedia (which have received criticism and accusations of inaccuracy) have greater power in providing a more accurate information source due to the small levels of high expertise often associated with rare diseases.

Inequalities in health related information access remain a concern, with those who are older, in lower socio economic groupings, and with lower education levels being disadvantaged. Certainly, in the case of Annie, her education, skills and access to individuals within the NHS through her employment were key elements in her achievement of a final diagnosis. She admits that without this she would probably still be without the diagnosis, and this highlights the inequalities which exist in access to information.

It should also be acknowledged that support groups and specialist organisations provide a highly valued source of information for those with rare diseases, but are comprised of those who already have a diagnosis. Therefore those without a diagnosis, which in the case of a rare disease can take considerable lengths of time to attain, can continue to face barriers in accessing appropriate information.

Future Development for Information Literacy in Rare Diseases

At present, there are some ideas and pockets of activity linked to tackling aspects associated with information literacy in rare diseases, but they are generally uncoordinated. In the UK, the Department of Health has published a plan for rare diseases (Department of Health 2012) which recommends the use of specialist centres to make diagnoses, improved co-ordination in the care of patients with rare diseases and appropriate training for health professionals in recognising the possibility of a rare disease. It is arguable that better development and support of information literacy is a key aspect of this. In France an equivalent plan exists (de la sante 2006), which acknowledges recognition that there is a lack of information on rare diseases and recommends that centres of reference are established to tackle this. In the USA the National Organisation of Rare Disorders has been established and aims to identify, treat and cure rare disorders through programs of education, advocacy, research and service (Kostrzewski and Baker 2006).

For physicians and health professionals, health and medical librarians have also been highlighted as key information sources. Beales (2011) for instance, suggests that there is an opportunity for librarians to apply broad based search strategies to identify unusual causes of common symptom clusters and publish them for use by clinicians. Elsewhere in the literature, Johnson (2007) has discussed the value in developing a database on all aspects of thalidomide damage including specialist information and care pathways, and this could act as a resource for both thalidomide sufferers and health professionals treating cases.

Conclusion

Using a case study, this chapter has highlighted some of the key issues that exist around information literacy in rare diseases. In particular, barriers to information associated with skills, socio-economic status, age, lack of diagnosis, and lack of information have been highlighted. It remains a reality that whilst in healthcare more generally there is information overload, the opposite remains true for rare diseases. As a consequence of this, Web 2.0 has become an invaluable and a strong source of information support to those living with rare or orphan diseases. For those with a rare condition, informed individuals often take the place of web sites containing more traditional forms of literature and patient information and advice. In current times, whilst healthcare approaches to rare and orphan diseases remain largely uncoordinated, and when the expertise of health professionals is lacking, social networks, blogs, and virtual communities of support prevail as the fundamental information resource for those with rare diseases.

References

Arora NK, Hesse BW, Rimer BK, Viswanath K, Clayman ML, Croyle RT. Frustrated and confused: the American public rates its cancer-related information-seeking experiences. J Gen Intern Med. 2008;23:223–8.

Beales DL. Beyond Horses to Zebras: Sicca syndrome. J Hosp Librarianship. 2011;11:311–24.

Butcher SM. Googling from an SHO point-of-view (rapid response to article Googling for a diagnosis). BMJ. http://www.bmj.com/rapid-response/2011/10/31/googling-sho-point-view(2006).

Carpenter DM, Devellis RF, Hogan SL, Fisher EB, Devellis BM, Jordan JM. Use and perceived credibility of medication information sources for patients with a rare illness: differences by gender. J Health Commun. 2011;16:629–42.

Cegala DJ, Bahnson RR, Clinton SK, David P, Gong MC, Monk JP 3rd, Nag S, Pohar KS. Information seeking and satisfaction with physician-patient communication among prostate cancer survivors. Health Commun. 2008;23:62–9.

De La Sante M. French national plan for rare diseases 2005–2008: ensuring equity in the access to diagnosis, treatment and provision of care. Paris: Ministere delegue a la recherche; 2005, 2006.

Department of Health. UK plan for rare diseases [Online]. London: Department of Health. Available: http://www.dh.gov.uk/health/2012/02/consultation-rare-diseases/ (2012).

Draucker CB. Coping with a difficult-to-diagnose illness: the example of interstitial cystitis. Health Care Women Int. 1991;12:191–8.

Johnson M. Integrating health information: a case study of a health information service for thalidomide survivors. Med Inf Int Med. 2007;32:27–33.

Kostrzewski MS, Baker LM. National organization of rare disorders (NORD) web site. J Consum Health Internet. 2006;10:77–87.

Kuehn BM. Patients go online seeking support, practical advice on health conditions. JAMA. 2011;305:1644–5.

Lewis S, Noyes N, Mackereth S. Knowledge and information needs of young people with epilepsyand their parents: mixed methods systematic review. BMC Pediatr. 2010; 10. Available: http://www.biomedcentral.com/content/pdf/1471-2431-10-103.pdf.

Macgregor A. Cluster headache: video—Dr Anne MacGregor, director of clinical research, city of London migraine clinic [Online]. London: Organisation for the Understanding of Cluster Headache; 2011. Available: http://www.ouchuk.org/html/clusters_video4.asp.

Ouch (UK). Organisation for the understanding of cluster headache (UK) [Online]. London: Organisation for the Understanding of Cluster Headache; 2011.

Sen B, Spring H. Mapping the information-coping trajectory of young people with long term illness: an evidence based approach. J Documentation. 2013;69:5.

Tang H, Ng JHK. Googling for a diagnosis: use of Google as a diagnostic aid: internet based study. BMJ. 2006;333:1143–5.

Tustin N. The role of patient satisfaction in online health information seeking. J Health Commun. 2010;15:3–17.

Further Reading

Aharony N. Web 2.0 in the professional LIS literature: an exploratory analysis. J Librarianship Inf Sci. 2011;43:3–13.

Boot CRL, Meijman FJ. The public and the Internet: multifaceted drives for seeking health information. Health Inf J. 2010;16:145–56.

Case D. Looking for information: a survey of research on information seeking, needs and behaviour. Bingley: Emerald Group Publishing; 2012.

Dahm MR. Tales of time, terms, and patient information-seeking behavior-an exploratory qualitative study. Health Commun. 2012;27:682–9.

Gardner M. Diagnosis using search engines. BMJ. 2007;333:1131.

Giustini D, Wright MD. Twitter: an introduction to microblogging for health librarians. J Canadian Health Librar Assoc. 2009;30:11–7.

Hawn C. Take two aspirin and tweet me in the morning: how Twitter, Facebook and other social medica are reshaping healthcare. Health Aff. 2009;28:361–8.

Jacso P. Google scholar revisited. Online Inf Rev. 2008;32:102–14.

Lasker JN, Sogolow ED, Sharim RR. The Role of an Online Community for People With a Rare Disease: Content Analysis of Messages Posted on a Primary Biliary Cirrhosis. J Med Internet Res. 2005; 7. Available: http://www.ncbi.nlm.nih.gov/pmc/articles/PMC1550634/.

Laurent MLR, Vickers TJ. Seeking health information online: does wikipedia matter? J Am Med Inf Assoc. 2009;16:471–9.

NHS Choices. The expert patients programme. Available: http://www.nhs.uk/NHSEngland/AboutNHSservices/doctors/Pages/expert-patients-programme.aspx (2013).

Quetel J. Rare diseases resource page. Issues Sci Technol Librarianship. 2007;52:2–2.

Rare Disease Day. What is a rare disease? [Online]. Available: http://www.rarediseaseday.org/article/what-is-a-rare-disease (2012).

Social Media and Engaging with Health Providers

Hugh Stephens

> *Good things happen to those who hustle*
>
> ANAIS NIN

Abstract Social media offers considerable opportunities for patients and clinicians to engage and collaborate across geographic borders for clinical care, research and patient support. While social media use in healthcare is currently in its infancy, there is enormous potential for applications dedicated to accessing health information, specialist clinical knowledge and participating in the latest research initiatives. These benefits are amplified for patients and clinicians who are geographically dispersed, such as those working in the field of rare and orphan diseases. Different applications of these tools emerge frequently and technological developments will provide increasingly better platforms for patients and clinicians to collaborate and engage to improve health outcomes.

Background

Although "social media" is a term that has come to prominence only recently, the internet has arguably had "social" functionality since the early 1990s, including bulletin boards, Internet Relay Chat (IRC) public chat rooms and blog platforms such as *LiveJournal* (launched in 1999). The term "social media" has been most commonly applied in recent times to sites such as *Facebook* or *Twitter* and has defined platforms that allow individuals to use the internet to facilitate conversations (Solis 2010). This

---

H. Stephens (✉)
Monash University, 7 David Lane, Windsor, Melbourne, VIC 3181, Australia
e-mail: hugh.stephens@monash.edu

H. Stephens
Mayo Clinic Center for Social Media, Minnesota, USA

includes other social platforms such as web forums or blogs that have been in use for health purposes for many years. Within the definition of social media we include platforms such as *Facebook*, *Twitter*, *YouTube* and *Pinterest*, along with mediums such as blogging, infographics, real-time chat and asynchronous conversation (regardless of platform). Although the popularity of an individual platform inevitably changes over time as the online zeitgeist evolves, health has been a driver of social media activity for many years (Ferguson and Frydman 2004).

The Current Role of Social Media

Social media has been used for healthcare purposes within four key areas:

1. Personal updates from patients to friends and family: using social media to keep friends or family up-to-date with a patient's current status or location.
2. Patients engaging with other non-health-professional stakeholders: for example patient or caregiver communities. The most obvious example of this is online support groups for rare diseases—where there are not enough patients in a local physical community to form a support group.
3. Patients engaging with health professionals, in either a professional or (controversially) personal capacity.
4. Health professionals engaging with other health professionals (such as for professional development)

Within these four areas, the first and second have been the primary drivers of healthcare social media up to this point (Kim and Chung 2007). Multiple studies (Kim and Chung 2001; Gauld and Williams 2009; Weingart et al. 2006) have identified the reasons behind patients using social media platforms, with the main purposes of social media use being:

1. Sharing a story or narrative of care or experience of a condition.
2. Searching for or providing information (including research) about research or experience to improve their own or others' care.
3. Accessing more detailed information after consultation with a healthcare provider.
4. As a coping mechanism in chronic or terminal disease through engaging with other patients, families or stakeholders via an online platform.

For those with rare diseases (or the families, caregivers or friends of those with an rare disease) social media can play a pivotal role in changing patient or stakeholder experiences of a rare condition. Little research has currently been conducted into the use of social media for rare diseases, instead focussing on patients as a whole or specific groups of patients with more common conditions, such as cancers (Ofran et al. 2012). In a 2010 study (Orizio et al. 2010), most (57 %) of the social networks identified were for multiple diseases (e.g. *PatientsLikeMe*) while others such as *Diabetes Sisters* were for only one disease.

As the number and nature of health-oriented social media platforms change and multiply over time, it is likely that there will continue to be more dedicated sites to individual diseases (such as rare diseases) or populations (such as those caring for someone with an rare disease). In contrast, health professionals have been very slow to adopt web 2.0 or social media technologies to engage with patients, predominantly citing concerns about privacy, security, confidentiality and appropriate physician-patient boundaries (Lupiáñez-Villanueva et al. 2009).

Benefits of Social Media Use

Social media can be highly beneficial for all those involved in rare diseases: from the individual patient and their friends or family, through to clinicians who are part of their ongoing care and researchers interested in the disease. The ability for web and social media tools to decrease the "effective distance" between patients and specialist clinicians and researchers is still being developed and new platforms and applications emerge almost daily. The emergence of video consultation platforms and improved internet speeds worldwide support this phenomenon to grow over time (Deloitte Development LLC 2012).

For those who have a rare disease (or are close to someone who does), social media provides a way to instantly tap into a community of others with similar experiences and produce a network who may provide emotional, clinical, knowledge or even financial support. Breaking down these barriers further is the ability for interactions to be anonymous or pseudonymous, ensuring that potential risks such as confidentiality breaches are mitigated and both sides of the communication are comfortable.

Discussion

Initial internet platforms for patients to engage with health providers were for allowing access to patient information, typically as part of a 'patient portal' (Weingart et al. 2006; Bergman et al. 2008; Ginossar 2008; Kuhn 2008). This access included patient data, pathology or laboratory results, prescriptions, appointment history, referral documentation and (later) secure messaging with health providers.

Using Social Media to Access Patient Health Information

These early platforms [the precursors to patient-controlled electronic health records (PCEHRs)] provided patients with the capacity to access and interpret their own healthcare information and have since received hesitant but positive reception from patients (Bergman et al. 2008) alongside concerns of confidentiality. It is anticipated that over time patient access to their own data through such a secure platform will increase in frequency. Public platforms (most notably *PatientsLikeMe*) have emerged that provide patients with the ability to share personal health data (Frost and Massagli 2008). These communities often involve patient groups sharing and exchanging information about disease-specific symptoms, treatment protocols and outcomes. These platforms can provide a very powerful research database for those looking to investigate large numbers of patients or those with rarer diseases (discussed later). One challenge of such communities is the spread of misinformation or non-evidence-based health advice (Chafe et al. 2011; Jacobs and Popick 2012), which physicians identify as a barrier to them recommending the use of these platforms by some patients (Ahmad et al. 2006).

Using Social Media to Access Specialist Knowledge or Information

Social networking platforms also facilitate the sharing of knowledge and information about the latest research findings and developments in patient care (Weingart et al. 2006). This is particularly the case for rare or orphan diseases where patients are often in different locations to healthcare specialists (Tjora et al. 2005), or when patients (or their families) have used social media to connect with and find more information about each other (Kim and Chung 2007). The greatest identified challenge for the use of social media to access knowledge or information is that of authenticity and reliability of information. The internet does not have established protocols or mechanisms to adequately communicate the credibility of a particular source. Many physicians believe that internet-based information *misinforms* patients, and thus require them to take on an additional role as the interpreter of this data on behalf of or in conjunction with the patient (Ahmad et al. 2006). But patients identify the consultation as the time for them to present this information and feel more empowered and informed about managing their care (Sommerhalder et al. 2009), suggesting that this role as an interpreter is likely to continue to grow for many practitioners.

Using Social Media to Join or Facilitate Research

Another common use of social media is to connect patients with the latest research initiatives or clinical trials (Terry 2009). When there are a small group of patients over geographically diverse areas (as is often the case with rare diseases), social media can provide the connection to both recruit and engage patients and researchers in disease or treatment research. Combining platforms designed specifically to connect researchers with patients and patient health record sites-as discussed previously-is a powerful way of providing a valuable service to patients (ability to access their health information easily) with the capacity to rapidly identify patient populations for research, particularly in rarer diseases or disease states. Ultimately tools such as these will improve the care for these patients and increase the ability and ease for researchers to recruit patients with rare diseases.

Issues Resulting From Social Media Use with Health Providers

While the benefits of social media are numerous and diverse, there are still a range of issues that have limited its adoption. As with all healthcare communications, confidentiality and privacy are paramount to ensure that the audience of health information is well defined. Patients and providers alike should ensure that they are aware of the level of access available to others when posting, discussing or providing health information online.

There are additional concerns about the ethics of using online information as part of healthcare and its impact on personal and professional boundaries both on behalf of the physician and patient (Clinton et al. 2010). Information online may assist clinicians in providing care, such as the experiences from patient blogs or diaries; or the latest reactions experienced by the patient and updated through platforms like *Twitter* or *Facebook*. But should a clinician actively search for this information, or only read it when a patient provides them with it? Given that health professionals are typically time-poor, large amounts of information and social media data may prove challenging to integrate into most patients' care.

These potential risks of social media need to be balanced with the benefits that the tools provide for patients and clinicians alike. Ultimately, every patient and clinician is different, and what may work for one clinical relationship will not work for others. It is necessary for patients, their families and clinicians to work together to identify the most appropriate channel(s) in which to discuss care.

The Future Roles of Social Media

Connecting Patients with the Best Providers Regardless of Location

As has been identified, social media has a powerful ability to break down geographic barriers, allowing patients to access specialist care from anywhere. This allows those working with rare diseases greater access to thought and research leaders across the world. Tools such as video conferencing are already coming into everyday practice

(Deloitte Development LLC 2012) but currently have diagnostic and investigative limitations as the 'physical touch', a core part of examination and the patient-physician relationship, cannot be transmitted.

This will undoubtedly change over time, with new technologies ('virtual clinics') and methods developed to maximise the ability of specialists to access patients without requiring significant travel or expense. Social media can provide two-way conversations between clinicians and patients, with the capability to respond to basic questions, connect patient groups or ensure appropriate follow-up form a visit.

Coordinating Complex Care and Multidisciplinary Teams

The increasing use of *Enterprise Social Networks (ESNs)* in the corporate world to allow employees to collaborate in a closed environment paints a promising picture of the future of collaboration in healthcare. Patients with multidisciplinary care teams or complex requirements often suffer from a lack of coordination and collaboration between professionals involved in their treatment, and these tools may provide a centralised platform to engage the clinical team in the care of a patient. This may be independent of the patient, or with them also contributing to the discussion and collaboration. Such coordination could occur on a dedicated platform (i.e. an ESN) or use existing platforms such as *Facebook*, *Sermo* (a physician-only network) or *Twitter*, depending on what platforms the care team actively use. This increased collaboration could result in less task duplication, faster communication between members of a multidisciplinary team (including those in different geographic areas), improved access for researchers to the clinical team and, ultimately, improved patient outcomes.

Conclusion

The use of social media has increased recently, particularly after *Facebook* became a popular platform for engaging with friends, brands and professionals (such as clinicians) alike. The role of social media in physician, patient and researcher relationships is still being developed and will continue to change over time. Those working with or suffering from rare disease will benefit significantly from these platforms, particularly due to their ability to reduce geographic barriers. Social media can be used to increase access to patient information across multiple clinical relationships, to access knowledge or information, and for research purposes. Future developments in internet access, technology changes and more dedicated platforms for using social media in care will continue to allow patients and clinicians alike to innovate, integrate and discover new ways to connect and collaborate.

References

Ahmad F, Hudak PL, Bercovitz K, Hollenberg E, Levinson W. Are physicians ready for patients with Internet-based health information? J Med Internet Res. 2006;8(3):e22. Epub 2006/10/13. doi: 10.2196/jmir.8.3.e22. PubMed PMID: 17032638; PubMed Central PMCID: PMC2018833.

Bergman DAMD, Brown NLP, Wilson S. Teen use of a patient portal: A qualitative study of parent and teen attitudes. Perspectives in health information management. 2008;5(13):1–13. PubMed PMID: 213143991; 18923702.

Chafe R, Born KB, Slutsky AS, Laupacis A. The rise of people power. Nature. 2011;472(7344):410–1.

Clinton BK, Silverman BC, Brendel DH. Patient-targeted googling: the ethics of searching online for patient information. Harvard Rev Psychiatry. 2010;18(2):103–12. doi:10.3109/10673221003683861.

Deloitte Development LLC. Deloitte 2012 Survey of U.S. health care consumers: The performance of the health care system and health care reform. 2012.

Ferguson T, Frydman G. The first generation of e-patients. Bmj. 2004;328(7449):1148–1149. Epub 2004/05/15. doi: 10.1136/bmj.328.7449.1148. PubMed PMID: 15142894; PubMed Central PMCID: PMC411079.

Frost JH, Massagli MP. Social uses of personal health information within PatientsLikeMe, an online patient community: what can happen when patients have access to one another's data. J Med Internet Res. 2008;10(3):e15. Epub 2008/05/28. doi: 10.2196/jmir.1053. PubMed PMID: 18504244; PubMed Central PMCID: PMC2553248.

Gauld R, Williams S. Use of the internet for health information: A study of Australians and New Zealanders. Inform Health Soc Care. 2009;34(3):149–58. doi:10.1080/17538150903102448.

Ginossar T. Online participation: A content analysis of differences in utilization of two online cancer communities by men and women, patients and family members. Health Commun. 2008;23(1):1–12. doi:10.1080/10410230701697100.

Jacobs HS, Popick R. Utilization of internet resources for adolescents coping with chronic conditions. Pediatric Nursing. 2012 2012 July-August:228 + .

Kim S, Chung DS. Characteristics of cancer blog users. J Med Libr Assoc. 2007;95(4):445–450. Epub 2007/11/01. doi: 10.3163/1536-5050.95.4.445. PubMed PMID: 17971894; PubMed Central PMCID: PMC2000789.

Kuhn P. Patient portals. Health Management Technology. 2008 2008/10//:44 + .

Lupiáñez-Villanueva F, Ángel Mayer M, Torrent J. Opportunities and challenges of Web 2.0 within the health care systems: an empirical exploration. Inform Health Soc Care. 2009;34(3):117–26. doi:10.1080/17538150903102265.

Ofran Y, Paltiel O, Pelleg D, Rowe JM, Yom-Tov E. Patterns of information-seeking for cancer on the internet: an analysis of real world data. PLoS One. 2012;7(9):e45921. Epub 2012/10/03. doi: 10.1371/journal.pone.0045921. PubMed PMID: 23029317; PubMed Central PMCID: PMC3448679.

Orizio G, Schulz P, Gasparotti C, Caimi L, Gelatti U. The world of e-patients: A content analysis of online social networks focusing on diseases. Telemed J E Health. 2010;16(10):1060–1066. Epub 2010/11/13. doi: 10.1089/tmj.2010.0085. PubMed PMID: 21070131.

Sommerhalder K, Abraham A, Zufferey MC, Barth J, Abel T. Internet information and medical consultations: experiences from patients' and physicians' perspectives. Patient Educ Couns. 2009;77(2):266–71. doi:10.1016/j.pec.2009.03.028.

Solis B. 2010 7 January 2010. [cited 2012]. Available from: http://www.briansolis.com/2010/01/defining-social-media-the-saga-continues/.

Terry M. Twittering healthcare: social media and medicine. Telemed J E Health. 2009;15(6):507–10. doi:10.1089/tmj.2009.9955.

Tjora A, Tran T, Faxvaag A. Privacy vs usability: a qualitative exploration of patients' experiences with secure Internet communication with their general practitioner. J Med Internet Res. 2005;7(2):e15. doi:10.2196/jmir.7.2.e15. PubMed PMID: 15998606; PubMed Central PMCID: PMC1550647.

Weingart SN, Rind D, Tofias Z, Sands DZ. Who uses the patient internet Portal? The patient-site experience. J Am Medl Inform Assoc. 2006;13(1):91–95. PubMed PMID: 220821966; 16221943.

Vignette: Hope–Overcoming Fabry

Adrian (Ed) Koning

My name is Adrian Francis (Ed) Koning and I have been married to Marlene for nearly 33 years. Together we have three fine young adult sons, two of whom are now married. We also have a three year old granddaughter.

A. (Ed) Koning (✉)
9011 142 St NW, Edmonton, AB T5R 0M6, Canada
e-mail: amkoning@telus.net

I was born on June 12, 1958 and as the youngest of three boys, my upbringing in Edmonton, Alberta, Canada, was typical of the many young immigrant families who came from Europe in the early 1950s. There was a focus on Christian values and any challenge in life was resolved by a strong work ethic without complaint. After graduating high school where I excelled in mathematics and physics, I entered university and studied civil engineering.

In 1979 during my last year of university, I met Marlene. I graduated in the spring of 1980, ready to conquer the world. She joined me as I relocated to Port Alberni, British Columbia on a beautiful Vancouver Island where I took a job as a project engineer in the pulp and paper industry. Our goal from the start was to work hard, save money, have a family and retire financially independent at the age of 55.

After working as a professional engineer and project manager for 17 years in both Canada and the USA, I wanted to relocate to Indonesia in order to get some international experience and make lots of money. However I landed up back in Edmonton instead. I grew up in Edmonton and left nearly 20 years prior as a young man with a great deal of baggage that I had left behind. You know the kind of stuff that comes with growing up in a certain place and then leaving it all behind as a young adult.

I continued to dedicate my energy into my career and within a year got a much better job with more responsibility, money etc. I was working 60 h a week and deep down in my heart knew life was a challenge trying to balance work, family, church, God, time and all the other pressures of life.

Then in January 2001 after thinking I either had the stomach flu or diabetes, I went to my family physician. In a matter of a few days, I was informed that my kidneys were nearly shutdown, that my heart was severely damaged and I could suffer a stroke at any moment! As a result of the kidney biopsy, I was diagnosed with Fabry disease.

Fabry disease is a very rare genetic life-threatening lysosomal storage disorder. There is a genetic defect in the X chromosome which causes an insufficient volume of alpha-galactosidase A (alpha-gal A) enzyme, allowing sugars and fatty acids, globotriaosylceramide (GL-3) to accumulate in the lysosome of each and every cell in the body. For unknown reasons, the GL-3 accumulation can cause kidney failure, cardiomyopathy in the left ventricle often leading to a heart attack, as well as strokes, hearing and balance loss and severe diarrhea. Other symptoms at an early age cause extreme pain in the hands and feet, inability to sweat, cornea swirling in the eyes and spots on the body called angiocharatomas. Average life expectancy for males is between 40-50 years of age. It is considered an ultra orphan disease with a frequency as low as 1 in 117,000 people. As such it is very difficult to diagnose.

I was 43 back then and was told that I would probably not make it to 50. After researching our family medical history, we learned that the longest any of my relatives lived with this disease was 48! Wow what a shock! It seemed that God was telling me, "Oh what a fool you are! Why are you not trusting and focused on me alone. Forget your goal of freedom 55! You probably will not be alive at 50!".

Because it is inherited, all of my family had to be tested. We discovered that my mother and one of my brothers also had Fabry. My mother inherited Fabry from her mother who died at the early age of 32. Moreover, the worst part was discovering that there was no treatment or cure.

I was now very sick, weak and no longer able to work. I had a sample of blood sent to the Mount Sinai School of Medicine in New York City to confirm my Fabry disease and also to participate in a new clinical study that a company called Genzyme was doing to test the effectiveness of Fabrazyme, an enzyme replacement therapy (ERT). I found out I was too sick and did not qualify. After some advocacy and media campaigning, my request to begin compassionate use of Fabrazyme was approved by the government of Canada. ERT is administered every two weeks intravenously and my first infusion was May 28, 2001. There were 3 other Fabry patients who all had their infusions in the same hospital room at the same time. Over the next 7 years we bonded, encouraged and supported each other. We knew we were not alone.

Being diagnosed with Fabry answered many questions that I had about my health challenges, especially as a child. I remembered suffering with extreme pain in my hands and feet when playing sports or games especially during the hot summer months. I could never gain any weight and was tall and very slim, caused by poor food absorption. I could never sweat. I had some red spots on my body that nobody else seemed to have.

Dialysis is another story. I was traumatized by a hemodialysis unit visit and without knowing it chose peritoneal dialysis. A catheter is surgically installed in the peritoneal cavity which is the sack in which all of the organs are located. Dialysis fluid is then exchanged four times a day. Through osmosis the fluid removes contaminates which your kidneys are supposed to remove. We set up a room in our home for PD and it is a challenge to keep everything sterile to avoid a serious infection called peritonitis. After a few months of dialysis and ERT, I returned to work with a new normal. Then in October of 2001, I had a live donor kidney transplant. After a slow 6-month recovery I was able to return to work. I felt resurrected and had been given another chance! I continued to receive ERT on a compassionate basis.

Unfortunately tragedy struck again in August 2003. I suffered a stroke which caused significant hearing and balance loss. As a result I was no longer able to work. In 2007, two weeks before my eldest son's wedding, I suffered what I thought was a heart attack. Then between 2008 and 2009, in order to gain some weight I was put on Total Parenteral Nutrition (TPN), an intravenous feeding program. It was very successful. In addition to this I had some medication changes which helped. I have also suffered from a very serious blood infection, took part in a three month cardio rehabilitation program and had an implantable cardioverter-defibrillator (ICD) installed to help my heart. I learned to enjoy each day.

Living with Fabry has been very challenging for me and my family, although we were also blessed because of it. Over time, we met together and separately with counselors, psychologists, psychiatrists, pastors, friends and support groups. After all, death was staring me in the face. It affected each of us in different ways.

I recall one of the first counselors I met who also happened to have kidney failure. He shared his experiences with me and encouraged me. He taught me that the only thing in my control is my attitude. I remember leaving and thinking that I had to choose a positive attitude and to count my many blessings.

I eventually realized that my focus on securing enough money was a direct result of my lack of self-worth, fear, anxiety and worry, which caused severe anger. I took an anger management course and with the help of many people, overcame much of my fear, worry and anxiety.

Accessing ERT in Canada was, and remains a challenge. Back in 2003 another Fabry patient and I started a patient network which eventually grew into the Canadian Fabry Association. I was elected the first president and our focus was on accessing ERT. In 2004, both ERT treatments, Fabrazyme™ & Replagal™, were approved for use in Canada but not funded by the provincial governments, as the cost of approximately $250,000 per year per patient was deemed not cost effective. It was a matter of educating politicians and other stakeholders throughout Canada. In 2006 an agreement between the manufactures of ERT and both the provincial and federal governments was finally reached. The agreement expired in 2009 and to a limited extent was renewed until 2012. Last year the agreement was extended by another year. The objectives of the CFA remain to provide hope and encouragement, not only to deal with physical aspects of Fabry but the psychological, social and mental trauma associated with living with a life-threatening genetic disease.

In 2005 Marlene and I, together with 3 others founded the Fabry International Network (FIN) in the Netherlands. The mission of FIN is to be a global, independent network of Fabry patient associations whose purpose is to collaborate, communicate and promote best practice to support those affected by Fabry disease. FIN's vision is of a world where every person affected by Fabry disease has the best quality of life possible through early diagnosis, treatment and cure. There are currently over 23 organizations and countries represented.

The future is bright for people living with Fabry. There is an oral treatment currently being tested as well as gene therapy. Gene therapy is a 'cure' to treat at the genetic level, to deliver a correct copy of the a-gal A gene to cells to produce a 'normal' enzyme level. In fact testing on humans has now begun.

Marlene says that I am 5 years past my expiry date. We laugh a lot and love life. I live as best as I can and live day by day. It has now been over 11 years since my kidney transplant. We give thanks for all of our blessings and our life and give thanks to God our Father and the Lord Jesus Christ. We believe that because He was raised from the dead at Easter, we have nothing to fear. Not even Fabry disease or death itself. I have been privileged to have celebrated many life events and hope I can live to see many more.

Many lessons have been learned since 2001, some of which I have already described. Don't sweat the small stuff. Life is a choice! It is hard to be stressed when you're feeling so blessed. And most important, if I can go pee in the morning I know I am going to have a great day!

Empowering the Rare Disease Community: Thirty Years of Progress

Jason R. Barron

> *Today is only one day in all the days that will ever be. But what will happen in all the other days that ever come can depend on what you do today.*
>
> ERNEST HEMINGWAY (For Whom the Bell Tolls)

Abstract This chapter outlines the efforts over the last 30 years of the National Organization for Rare Disorders (NORD) in the United States. The non-profit organization aims to support individuals with rare diseases by advocating and funding research, education, and networking amongst service providers. The chapter introduces NORD, presents its progress and milestones over the past thirty years. We conclude by looking to the future of rare diseases and associated orphan drug development.

Introduction

NORD's vision and guiding principles on which its advocacy initiatives are based are: (a) a national (U.S.) awareness and recognition of the challenges faced by people living with rare diseases and the associated costs to society; (b) a nation where people with rare diseases can secure access to diagnostics and therapies that extend and improve their lives, (c) a social, political, and financial culture of innovation that supports both the basic and translational research necessary to create diagnostic tests and therapies for all rare disorders and (d) a regulatory environment that encourages development and timely approval of safe and effective diagnostics and treatments for patients with rare diseases. NORD is a unique

J. R. Barron (✉)
National Organization for Rare Disorders (NORD), 1779 Massachusetts Avenue NW, Suite 500, Washington DC 20036, USA
e-mail: jbarron@rarediseases.org

federation of voluntary health organizations dedicated to helping people with rare "orphan" diseases and assisting the organizations that serve them. NORD is committed to the identification, treatment, and cure of rare disorders through programs of education, advocacy, research, and service. In the decade before 1983, only 10 new treatments were brought to market solely by industry for diseases that today would be defined as rare. The problem was receiving little attention. Research dollars and expertise were targeted to the development of blockbuster products for common diseases.

Serendipitous Awareness

Leaders of rare-disease patient organizations began to realize that there were certain problems their patients and families shared which were common to all people with rare diseases. As a result, they raised their voices together, calling for national legislation to encourage the development of treatments for rare diseases. A small story in the *LA Times* led to an episode on a popular TV show, *Quincy ME*. Letters began to arrive from people all over the nation who had rare diseases and thought they were alone in their struggles. It became apparent that, while each disease may be rare, these diseases affect millions of Americans when considered together. The ultimate result was the enactment of legislation known as the Orphan Drug Act, and the patient leaders who had worked to bring national recognition to the problem founded NORD as an umbrella organization to represent the rare disease community. Today, NORD provides information, advocacy, research, and patient services to help all patients and families affected by rare diseases.

NORD Programs and Services

NORD serves all stakeholders in the rare disease community, including patients and their families, patient organizations, researchers, medical professionals, and companies developing orphan products. NORD works closely with many government agencies, most notably the National Institutes of Health (NIH) and the Food and Drug Administration (FDA). All NORD programs are focused on one ultimate goal: to improve the lives of individuals and families affected by rare diseases.

Education

NORD provides information about rare diseases through its publications, website, and other educational offerings. One of its most important resources is the

Rare Disease Database, a compendium of 1,200 rare disease reports developed for patients and their families that includes information on symptoms, causes, treatments, clinical trials, and links to other sources of help. NORD also publishes a small collection of physician's booklets on selected rare diseases. Each year, NORD responds to hundreds of thousands of telephone, mail, and email inquiries from individuals, families, teachers, social workers, and medical professionals.

Patient Advocacy and Mentorship Programs

NORD works collaboratively with its growing roster of member organizations, over 200 organizations at present, representing them at several large medical meetings each year and providing opportunities for them to join NORD in advocacy on behalf of their members. NORD also provides mentoring services to assist in the establishment and growth of disease-specific organizations.

Research Grant Program

Through its research program, NORD provides seed money in small grants to academic scientists. If these studies produce promising data, the researchers may go on to obtain government grants or commercial sponsorship that, ultimately, could lead to new diagnostics or treatments for rare diseases. Research grants are awarded to scientists following a competitive proposal process. Awards are made in consultation with NORD's Medical Advisory Committee.

Medical Assistance Programs

NORD partners with pharmaceutical and biotechnology companies to ensure that certain vital medications are available to uninsured or under-insured individuals. NORD's programs have set the standard for fairness, equity, and unbiased eligibility and earned high marks from patient communities, pharmaceutical companies, healthcare professionals, government officials, and the public. NORD also administers co-pay assistance programs for people who can't afford their insurance co-pays; early access programs that, following FDA guidelines, allow patients with serious or life-threatening diseases to access investigational products under certain conditions; and travel assistance programs, since patients with rare diseases often must travel great distances to participate in clinical trials or to see a specialist in their disease.

International Education and Advocacy

Recognizing that rare diseases are a global public health challenge, NORD has entered into a strategic partnership with the European Organization for Rare Diseases (EURORDIS). NORD also has a Memorandum of Understanding with the Japan Patients Association, enacted in January 2013, to connect patients and patient organizations in the U.S. and Japan. In addition, NORD and the Canadian Organization for Rare Disorders (CORD) are exploring ways to work together more closely to benefit rare disease patients and patient organizations. In addition to these formal affiliations, NORD works in partnership with patient advocacy groups around the world on initiatives such as the annual global Rare Disease Day.

Thirty Years of Progress

Patient advocates played a key role in Congressional approval of the Orphan Drug Act. In the decade before 1983, only 10 new products had been solely developed by the pharmaceutical industry for rare diseases. The Orphan Drug Act would provide financial incentives to encourage companies to develop treatments for small patient populations. However, it was stalled in Congress until Abbey Meyers and other representatives of patient organizations formed a coalition to get the legislation approved. It was not an easy task, and it involved learning to work with Congress, the media and, most importantly, each other, but the patient advocates did their job well and the legislation was approved by Congress and signed by President Ronald Reagan on January 4, 1983. Shortly afterward, the patient leaders held a meeting at which they decided to continue their partnership through an organization to represent all Americans affected by rare diseases. That organization was NORD.

Rare Disease Milestones

With rare diseases, progress is often measured in small steps rather than huge leaps. This section presents several milestones (on specific disease fronts) which may ultimately help drive progress for all (see www.rarediseases.org/nord-30th-anniversary).

1979–FDA/NIH Task Force Issues Report

Task force chaired by Marion Finkel, MD, of FDA issues report calling for measures to address the need for more resources to be directed toward drugs "of limited commercial value" (drugs for small patient populations).

1979–1980–House Subcommittee Gathers Evidence

The Subcommittee on Health and the Environment of the House Energy and Commerce Committee, chaired by Representative Henry Waxman, holds hearings on the orphan drug problem.

1979–1980–Patient Advocates Form Ad Hoc Coalition

Leaders of rare disease patient organizations form a coalition to provide advocacy together on behalf of legislation to encourage the development of treatments for people with rare diseases.

1980–Popular TV Show Features the Issue

Actor Jack Klugman and his brother, Maurice, assist the patient advocates in focusing national attention on the problem with an episode of the TV show, Quincy, M.E.

1980–Pre Orphan Drug Act

Only 10 new drugs were developed solely by the pharmaceutical industry for rare diseases in the decade before 1983.

January 4, 1983–Orphan Drug Act Passed

After rare disease patient advocates mobilized support for the Orphan Drug Act, which had been sidelined in Congress, it was approved by the House and Senate in December 1982 and signed by President Ronald Reagan on January 4, 1983. Those same patient leaders then established NORD to continue their collaboration, realizing that "Alone we are rare. Together we are strong."

February 1983–First Orphan Drug Approved

The Food and Drug Administration (FDA) grants first marketing approval to an orphan drug - Panhematin® for acute intermittent porphyria and other acute porphyrias. Desiree Lyon Howe, co-founder and executive director of the American Porphyria Foundation and for many years a NORD board member, participated in research on this orphan drug.

May 4, 1983–NORD Founded

NORD is incorporated to represent the shared interests and goals of all Americans affected by rare diseases. Abbey Meyers, considered the primary consumer advocate responsible for the Orphan Drug Act, is named president.

1984–Orphan Drug Act Amendment

The Orphan Drug Act is amended to define a rare disease as any disease affecting fewer than 200,000 Americans or a disease with a higher prevalence but for which there is no reasonable expectation that a therapy would recover the cost of development.

1985–Orphan Drug Act Amendment

The Orphan Drug Act is amended again to extend marketing exclusivity to both patentable and unpatentable products.

1987–NORD Establishes First-ever Patient Assistance Program

NORD established the first patient assistance program dedicated to helping patients obtain medications they could not afford or that their insurance did not cover.

1988–Orphan Drug Act Amendment

Orphan Drug Act is amended again to require that application for orphan designation be made before the submission of an application for marketing approval, New Drug Application, or Product License Application.

1989–National Commission Issues Report

The National Commission on Orphan Diseases, chaired by Jess Thoene, M.D., board chair and medical advisor to NORD, conducts a major study and issues a report on the experiences of patients and families affected by rare diseases.

1989–NORD Establishes Research Program

NORD establishes a Research Program to be overseen by its medical advisors so that patients and patient organizations may provide grants for the study of diseases with limited or no other source of funding.

1999–European Union adopts Orphan Law

European Union adopts law similar to Orphan Drug Act (Regulation on Orphan Medicinal Products).

2000–NIH Establishes ClinicalTrials.gov

Partly in response to advocacy from NORD and others in the patient community, NIH launches a website (www.clinicaltrials.gov) providing an overview of current clinical trials.

2000–Rare Diseases Act Introduced

Senators Ted Kennedy and Orrin Hatch introduce the Rare Diseases Act, advocated by NORD to enhance federal funding for rare disease research and accelerate the development of treatments. The House later splits this legislation into two separate bills—the Rare Diseases Act and the Orphan Products Development Act.

2002–Rare Diseases Act Signed into Law

The Rare Diseases Act strongly promoted by NORD is signed into law, codifying the NIH Office of Rare Diseases Research and providing for the establishment of the NIH Rare Diseases Clinical Research Network.

2003–NORD Publishes Guide to Rare Disorders

NORD, with Lippincott, Williams & Wilkins, publishes a 600-page textbook written by the world's leading rare disease experts, The NORD Guide to Rare Disorders, for pediatricians and family physicians to encourage earlier diagnosis and treatment.

2005–First Meeting of ICORD

The first meeting of ICORD (the International Council on Rare Diseases and Orphan Products) takes place in Stockholm, with representatives of NIH, FDA, and NORD participating from the U.S. At this meeting, Marlene Haffner, MD, of FDA's Office of Orphan Products Development, commits to developing a joint orphan designation application with her office's counterpart in Europe.

2006–Rare Diseases Clinical Research Network Founded

The National Institutes of Health (NIH) announces the establishment of the Rare Diseases Clinical Research Network with $55 million in funding for rare disease research. This was made possible by the Rare Diseases Act of 2002, advocated by NORD, and through the leadership of the NIH Office of Rare Diseases Research.

2007–FDA and EMEA Partnership

The Food and Drug Administration (FDA) and European Medicines Agency (EMEA) adopt a common application form for sponsors seeking orphan drug designation of medicines in the EU and US.

2008–New NORD Leadership
Abbey Meyers retires after serving for 25 years as NORD's president. She is succeeded by Peter L. Saltonstall.

2008–Compassionate Allowances Program Established
Social Security Commissioner Michael Astrue announces at a NORD patient/family conference that he will establish a "Compassionate Allowances Program" to fast-track the processing of assistance applications from patients with severely disabling diseases.

2008–NIH Establishes Undiagnosed Diseases Program
The Undiagnosed Diseases Program is established at NIH to help patients with baffling diseases obtain a diagnosis through the shared perspectives of teams of rare disease medical experts.

2009–NORD Sponsors first US Rare Disease Day
NORD becomes the national sponsor for Rare Disease Day in the U.S. Launched in Europe by EURORDIS in 2008, Rare Disease Day is observed annually around the world on the last day of February.

2009–NORD and EURORDIS Form Strategic Partnership
NORD and its counterpart in Europe, EURORDIS, sign a Memorandum of Understanding to work together to connect patients and patient organizations in the US and EU for purposes of advocacy, education and awareness.

2010–Rare Disease Office Established in FDA CDER
FDA establishes a new position for which NORD had provided advocacy, Associate Director for Rare Diseases, in the Agency's Center for Drug Evaluation and Research (CDER), Office of New Drugs (OND). Anne Pariser, MD, is named to the position, which for the first time provides a rare disease point of contact within CDER for patients and those developing treatments.

2011–Affordable Care Act Signed
The Affordable Care Act signed by President Obama provides insurance reforms for which NORD lobbied, such as ending annual and lifetime insurance caps and eliminating discrimination based on pre-existing conditions.

2011–First Annual NORD/DIA Conference
NORD and DIA (the non-profit Drug Information Association) co-sponsor the first annual U.S. Conference on Rare Diseases and Orphan Products.

2011–NORD Releases Orphan Drug Approval Survey
NORD releases major new report written by board member Frank Sasinowski documenting ways in which flexibility has been applied by FDA in the review of all non-oncologic orphan drugs approved between 1983 and summer 2010.

2011–NIH Launches NCATS
To promote and advance innovative research, NIH launches the National Center for Advancing Translational Sciences.

2012–FDA Safety and Innovation Act Approved
The FDA Safety and Innovation Act is approved by Congress and signed by President Obama. For two years leading up to this law, NORD meets regularly with senior officials at FDA and legislators and Congressional staff to educate them about patient needs and concerns. NORD calls this new law the most important since the Orphan Drug Act for the rare disease community.

2012–FDA New Drug Approval Hits 16 Year High

The FDA approved 39 new drugs in 2012, the most since 1996. The tally of 39 new drugs and biological products approved by the Food and Drug Administration compares with 30 in 2011 and just 21 in 2010. At least 10 of the drugs had fast track status in 2012, which enabled them to be reviewed more quickly.

2013–30 Years: NORD and The Orphan Drug Act

While many challenges remain, major progress has been made in research, orphan product development, and patient access to needed treatments and services. NORD and the entire rare disease community remain committed to assuring a better future for all who are affected by rare diseases.

Looking Forward: Current Initiatives

The science of rare diseases is advancing with each new day. The NIH is promoting innovation to translate promising early work into clinical trials. Orphan product development is being seen by many as the most exciting frontier in the pharmaceutical and biotechnology industries today. The **Rare Disease Congressional Caucus** was established in 2010 to create a forum for members of Congress to help improve the lives of individuals with rare diseases. Current initiatives include monitoring and commenting on legislation and policies that affect members of the rare disease community.

Reauthorization of the FDA User Fee Programs

The FDA Safety and Innovation Act

For the FDA, the Prescription Drug User Fee Act (PDUFA) is one of the most important pieces of legislation. PDUFA is considered a 'must-pass' bill and is renewed by Congress every 5 years. NORD worked closely with FDA to address some of the special needs of the rare disease community as they relate to PDUFA. Similar to PDUFA, the Medical Device User Fee Act authorizes the FDA to collect user fees from medical device makers to accelerate the review of their products and determine if they are safe and effective. Like PDUFA, MDUFA is reauthorized on a regular basis by Congress and is an important part of ensuring that FDA has the resources it needs to fulfil its mission. During this reauthorization cycle, both bills were combined into the omnibus Food & Drug Administration Safety and Innovation Act or FDASIA. NORD continues its work to engage and advise the FDA to ensure that the patient voice is further incorporated in FDA practice through the timely and mindful implementation of the FDASIA law.

Patient-Centered Outcomes Research Institute

One of the most promising new initiatives to come about from health reform was the consensus that public health is improved by knowing definitively what kind of care works and doesn't work in a given situation. The Patient-Centered Outcomes Research Institute will begin to systematically tackle questions related to standards of clinical care and will have an important impact on healthcare delivery in the United States, including for rare diseases.

Acknowledgments More detail on the content of this chapter can be found on the comprehensive NORD website: www.rarediseases.org.

Part III
Patient Perspectives and Empowerment Issues

The Role of Social Media in Healthcare: Experiences of a Crohn's Disease Patient

Michael Seres

> *A patient comes because she's sleeping 16 h a day, and it takes ten doctors and a coma to diagnose sleeping sickness*
>
> Dr Gregory House, "Fidelity", HOUSE (TV series)

Abstract This chapter outlines the efficacy of using social media in healthcare. Written by a UK-based Crohn's disease patient, the chapter describes how contemporary IT tools and techniques can be used effectively to actively engage patients in their own care. Examples of patient-to-patient interaction via new social media are provided. The concept of the "i-patient" is introduced and the need to empower both patients and healthcare professionals is suggested.

Introduction

I have been a patient for 30 years having been diagnosed with the incurable inflammatory bowel complaint known as Crohn's disease aged 12. Crohn's disease is a long-term condition that causes inflammation of the lining of the digestive system. Inflammation can affect any part of the digestive system, from the mouth to the back passage, but most commonly occurs in the last section of the small intestine (ileum) or the large intestine (colon). Common symptoms of Crohn's disease include: diarrhoea, abdominal pain, fatigue (extreme tiredness), weight loss. Over time, inflammation can damage sections of the digestive system, resulting in additional complications, such as narrowing of the colon (NHS 2011). The UK's *National Institute for Health and Clinical Excellence (NICE)* estimates that around 80,000

M. Seres (✉)
The Grove, RADLETT, Hertfordshire WD7 7NF, UK
e-mail: mseres@michaelseres.com

people in the UK have Crohn's. It is defined as a rare chronic complex long term condition and patients with inflammatory bowel disease are considered "vulnerable patients" in hospital due to their risk of malnutrition and hydration. This is often caused by the fact that patients cannot absorb much of what they eat and drink in their intestines. In rare cases, they cannot often tolerate many foods and require enteral or intravenous feeding.

My patient journey started with joint pains from about 7 years old. Initially doctors assumed that I had brittle bones as I would often break them. After surgeries on both shoulders and knees, I developed a rare condition known as *Ankylosing Spondolytis* (NHS. Ankylosing spondylitis 2012) but my journey did not end there. I underwent over twenty five bowel resections to either remove diseased areas or divide adhesions within the bowel caused by continual operations. I went on to develop intestinal failure and became the 11th person to undergo a rare life-threatening small bowel transplant (NHS. Small bowel transplant 2012) at *The Churchill Hospital* in Oxford, England.

The Role of Social Media

Every one of us will be a patient at some stage in our lives. Whether it is simply a quick discussion with a healthcare professional, or something more complex, our relationship as a patient is inevitably based around the medical professionals that we go and see. The relationship with my doctors and surgeons has been fundamental to my on-going well-being. However, social media and patient-to-patient interaction have also become key components in helping me to manage my health. I have always believed that whatever pain I am in, or whatever medical problem I encounter, I am fortunate enough to have the best medical team to solve the issues. The other challenge I find in dealing with my long term chronic illness is a *mental* one. Being able to cope with the rollercoaster of emotions, the impact your health has on your loved ones, the financial fallout and those middle-of-the-night problems when everyone is asleep and suddenly your feeding pump stops working or your stoma bag leaks. For me, coping with these issues (and many more) is where the value of social media and online communities lie and where they make such a vital contribution.

I started a blog (http://beingapatient.blogspot.com) in the build up to transplant. I had never done anything like this before. With so little known about bowel transplants, it became the ideal platform to let family and friends know what was happening. It meant that my wife did not have to tell the same daily update over and over again and people felt they could keep involved without disturbing us. It also became a way for the transplant team to keep tabs on me and they, in turn, encouraged their medical students to read it to gain a real-time overview of the daily impact a transplant has on the patient and their family.

I have always been a nosey patient. I gain confidence by knowing more, by asking questions and by feeling I am making a contribution to my own well-being. I like to be an empowered and engaged patient (an e-patient). I understand that not everyone wants to be that type of patient. In reality though, we all are: the level of engagement is personalised to an individual. It is very easy to be passive, to automatically follow what you are being told, to not probe too hard or ask the difficult questions. My blog has now been read by over 55,000 unique viewers and is syndicated in 20 sites around the world. The vast majority of readers are patients themselves. It has enabled me to build relationships with inflammatory bowel disease patients, intestinal failure patients and existing and potential bowel transplant patients and their families. From this I was able to create the closed *Facebook* group *Bowel Disease One Global Family*. It is a forum where patients from over 20 countries share experiences, provide practical tips, share the side effects of their medication and support each other. The group is a classic example of how patient-to-patient interaction has enhanced the lives of people who often felt they coped alone (with perhaps only their doctor as their confidante).

Data on the use of online sources to help patients is now easily available. 68 % of UK-based internet users make use of the medium in relation to their health issues; in the USA, this figure rises to 83 %. In both countries, patient-to-patient interaction is one of the biggest growth sectors. Digital and social media offers new and innovative ways for patients and their carers to manage their own illnesses. The key benefits to patients include increased individual engagement in health and a greater understanding of new products and services. Currently, I engage with over 12,000 patients through social media.

Another significant benefit is what has been described in a recent (UK) *National Health Service* (*NHS*) report into digital healthcare as the "gift economy". Social media has enabled a huge rise in volunteer and sharing activity and, through blogs, forums and social media, patients are happy to share information, tell their own stories and volunteer their time. Whilst this is principally for the benefit of other patients, it can also provide a useful information resource for healthcare providers and medical professionals. A 2011 study in Holland concluded that 55 % of patients who shared information online are then likely to take their findings and comments to their medical practitioner (The Health Foundation 2012). The *Social Media Examiner* 2012 report states that *Facebook* is the number one tool used by 92 % of people on social media; it has over 900 million users and health is one of the most popular (Social Media Examiner 2012).

My own journey now includes *Twitter*, *LinkedIn* and the increasingly popular photo sharing site *Pinterest* (21 % of online users' access photo sharing sites) have enabled patients to share pictures of images that have meant something on their bowel disease journey. Using *Pinterest*, I "pinned" a picture of my stoma in order to illustrate that patients have nothing to be ashamed of. Additionally, I hoped it would result in other patients feeling more comfortable with the scars and wounds that are an inevitable consequence of illness. Social media allows barriers to be broken down far more quickly than a face-to-face meeting can do and it empowers patients in a very unique way. Whilst access to patient records is often the primary goal for e-patients, my own view is that the more the health sector can embrace the digital age as a whole the better it will be for Pharmas, medical practitioners and patients.

I passionately believe in encouraging and supporting the development of *open platforms* that allow patients to share stories. The more the patient voice is heard, the more that healthcare institutions will be able to react to requests and concerns which can only have a positive impact on the quality of care received. Whilst social media has been a fantastic learning curve for me, it has also opened my eyes as to how much more of an impact the digital age can have on my own well-being.

Whilst I am lucky that I can communicate with my transplant team via text and email, I wonder how long it will be before other digital advances play a part in my on-going care. Telemedicine is a big growth area and I presume this would also provide a significant cost saving. Currently, I talk to patients daily via video calls on *Skype*—is there any real reason I cannot do this with my medical team? A recent *YouGov* survey found that 29 % of people in the UK would like to see GPs start offering remote consultations via video-link within the next decade (Digital Innovation in Healthcare Working Group 2012).

My Experience as an "I-Patient"

I do not regard myself as an e-patient anymore and I no longer believe that this term is particularly relevant. The term *i-patient* defines me more accurately. *I-patient* to me means an *interactive* patient. I choose to interact in all aspects of my healthcare with social media being the core component. I use text and email daily to interact with my healthcare professionals as well as using social media to interact in so many other ways. Twitter-based *"tweetchats"* (for example, #IBDCHAT, focuses on bowel disease issues for one hour every Sunday) can be used to talk about issues or concerns patients have relating to new medications, hints, tips and so forth. Patients can use this environment to provide vital support to others experiencing similar situations.

On a more strategic level, the *tweetchat* #nhssm (of which I am the patient lead) tackles weekly healthcare topics of interest to patients and healthcare professionals alike. The key component is the impact that those topics have on social media. Whether it is responding to a new government initiative, or discussing the new NHS mobile apps library, this weekly forum provides true interactivity.

Perhaps though my greatest successes in social media have come in mobilising what I choose to term *#patientpower*. The most recent cause has been the campaign to improve hospital food. By taking to social media, a campaign was able to gather momentum that led to television coverage that, in turn, led to action to bring in mandatory food standards for patients in hospitals (The Campaign for Better Hospital Food 2013). At the time of writing, I am part of the team in discussions with the UK government over these standards and we have secured the backing of hospital caterers, the *British Medical Association* and many others. This support could only have been garnered through the use of a proactive *Twitter* and *Facebook* campaign. Being an *i-patient* depends entirely on having a healthcare team with whom to interact. I am able to interact with my surgeon and medical teams continuously; sadly, this is perhaps rare in the current healthcare climate. However, interacting with patients is not new—talking is the simplest form of communication and social media is simply an extension of that.

The Future

It fascinates me that the tools we use in our everyday life have not yet been used properly in health. Nothing that I have talked about in this chapter is new. *Twitter* is now 7 years old and over 200 million tweets are sent daily. *Twitter* now states that tweets about health are up 51 % year-on-year. *Facebook* is now the 4th largest country in the world (if you count its users as a population). *Youtube* has just signed up its one billionth user. It is fair to say that these social media tools are being used constantly, yet take up in health is still slow. This may be due to issues concerning privacy, security, confidentiality, trust and may also extend to inertia or a reticence to change the norm. As an engaged patient, I use all these tools to aide my healthcare continuously. Using a *Google "hangout"* and a closed *Facebook* group are part of my daily life. When I developed a terrible pain and sore on my right eye, I turned to a patient group on *LinkedIn* for help. A patient told me that it sounded like shingles in the eye and I mentioned this to my surgeon; on consultation, it turned out to be exactly that. When my dietician wanted me to try a new transdermal spray, she asked me to find out if any other patients had experience of using this spray. I used my social media network to give her real-time feedback and data. The question is whether this is the future of healthcare. My vision for the future is one where your health is an interactive table of constituent parts which include (a) face to face meetings, (b) in-patient stays, (c) medications and treatment plans, (d) text and email communications, (e) video chats and picture messaging, (f) Tweetchats, (g) virtual community forums and (h) peer-2-peer interactivity.

Conclusion

My condition may be considered to be rare but I suspect that my interactivity may be even rarer. I am unsure as to why the tools and technologies discussed in this chapter are not in wider use within the patient community. I observe in a typical outpatients clinic that nearly all patients are using their mobile devices. When I visit hospital wards, most people enquire about wifi connectivity and regularly upload pictures of their ward and bed to *Facebook*. We communicate constantly via social media in our daily lives sharing what we eat, where we are, a new purchase and so forth. This interactivity needs to be effectively applied to our healthcare and its planning. Such interactivity may already be a reality: it could be argued that the busiest doctor in the world is *Google* (as many patients carry out a search either prior to or following a healthcare consultation).

Many of these tools will require a "mindset" change before they are widely accepted as patient tools. It is my contention that healthcare professionals still regard social media as something to fear rather than embrace but this is definitely changing. Many hospitals now have their own *Facebook* pages (some even have

Twitter feeds) but there is often an inherent fear about what to do when a patient posts something that the hospital may not like. From a patient perspective, honesty and openness is the solution: interactively explain if something cannot be answered online or if it cannot be dealt with immediately. Social media was meant to be used as a communication and engagement tool and as digital health is here to stay, all stakeholders need to embrace it. By 2015, the UK government plans to be paperless by which time every patient will have online access to their own data. How patients choose to engage, share and use that data will change the way healthcare is delivered; interactivity will lead to a rise in *i-patients* and, with time, *i-doctors*.

References

Digital Innovation in Healthcare Working Group. The Digital Dimension of Healthcare. 2012. http://xnet.kp.org/kpinternational/docs/The%20Digital%20Dimension%20of%20Healthcare.pdf. Accessed 26 March 2013.

NHS. Crohn's disease, NHS website. 2011. http://www.nhs.uk/conditions/Crohns-disease/Pages/Introduction.aspx. Accessed 26 March 2013.

NHS. Ankylosing spondylitis, NHS website. 2012. http://www.nhs.uk/conditions/Ankylosing-spondylitis/Pages/Introduction.aspx. Accessed 26 March 2013.

NHS. Small bowel transplant, NHS website. 2012. http://www.nhs.uk/conditions/small-bowel-transplant/Pages/Introductionpage.aspx. Accessed 26 March 2013.

Social Media Examiner. Social media marketing industry report. 2012. http://www.socialmediaexaminer.com/SocialMediaMarketingIndustryReport2012.pdf. Accessed 26 March 2013.

The Health Foundation. Helping people share decision making. 2012. http://www.health.org.uk/public/cms/75/76/313/3448/HelpingPeopleShareDecisionMaking.pdf?realName=rFVU5h.pdf. Accessed 26 March 2013.

The Campaign for Better Hospital Food. 2013. http://www.sustainweb.org/hospitalfood/. Accessed 26 March 2013.

Vignette: The Blessings and Curse of Diagnosis: Myasthenia Gravis

Grainne Pierse

I was just starting my second year of university and heading back into the heat of training with the varsity swim team. I had taken time off from my training over the summer to recover from a bad bout of *clostridium difficile* that I had contracted the previous spring. But it was a new year now and I was excited to get back into my routine.

About three weeks into the school year I started having constant headaches. They were worse whenever I stood or after practices, so swimming was put on hold for a bit. I went to the doctor for them who prescribed migraine medications

G. Pierse (✉)
104-2983 W 4th Avenue, Vancouver, BC V6K 1R5, Canada
e-mail: gpierse5@gmail.com

but that was about all that was done about the problem. A few days later the exhaustion really set in. I was too tired to get out of bed in the morning and going to classes just didn't seem like an option. Then came my droopy eye. The first day I noticed it I thought I was just overreacting and seeing things that weren't really there, but the next day it was much more predominant. It was a struggle to hold the eye open at all. At this point it was time to go back to the doctor. I remember sitting in the waiting room, not really sure where this appointment would take me. Within a few minutes of being in the exam room I knew my doctor was as scared as I was. She sent me to the emergency room straight from her office to see a neurologist.

I don't remember much of that first night in hospital, just a lot of questions being asked and a test here and there. I was admitted to the neurology ward to await an MRI and other tests the following day. The following day I had an MRI and CT scan of my brain. Nothing showed up on either of these tests ruling out any tumours or early signs of multiple sclerosis. I would not have known any of this if it had not been for my brother, a doctor, who explained why every test was being done, what it had shown, and what that meant. After three days in hospital I was discharged with no answers as to what was wrong. I was still exhausted and my droopy eye and the constant headaches had not resolved.

I spent the next few days with my family, with everyone seeming to get progressively more frustrated with me. Nothing was wrong with me so why was I continuing to act like there was? I started making excuses for my slow movement: I was in too much pain from my lumbar puncture to walk much, I hadn't slept well in the hospital and needed to rest. Anything to get out of extra activity.

About a week later I had a follow up with my GP. At this appointment I learned that everyone thought I was severely depressed. My family had contacted my doctor to raise their concerns and she seemed inclined to agree. She tried to prescribe antidepressants but I refused them. I knew I wasn't depressed, I was just in so much pain with the headaches and the exhaustion was never ending.

I went home for thanksgiving about a week later and did my best to smile and act like I was fine. I did however still have the drooping eye and had developed a stutter. To hide this I avoided speaking whenever possible and mostly kept to myself, further convincing my family that I was depressed. This way hard to take. Everyone seemed to think that my symptoms were in my head. I know my family only had my best intentions at heart, but I couldn't help but feel that it was my fault I was sick.

My symptoms eventually faded away. I returned to my swim training, albeit slowly and cautiously, and by Christmas I was almost back to normal. I still had headaches almost every day and the weakness would return at the end of a long week, but I kept this to myself and for the most part I was healthy.

It wasn't until January that I had the first hint of an answer to what had been wrong. It was my final follow up with my neurologist where he told me he thought I had an autoimmune disease called *myasthenia gravis*. It couldn't be tested for while I was asymptomatic but he told me to call him when it came back. I will never forget him saying when my symptoms returned, not if. This small difference in phrasing made me happier than you would think. It was nice to have a glimpse of a diagnosis. Naturally, I went home and learned everything I could about the disease and everything seemed to fit so perfectly.

It was almost a year to the day of the first symptoms appearing that they returned. I returned to my GP for a referral to a neurologist. Despite my being certain I had myasthenia this doctor was not convinced but sent me to the emergency room once again to see a neurologist there. After a few hours in the emergency room, I had only gotten as far as a neuro resident who concluded I was fine to be sent home with a follow up a few days later with another neurologist.

When I saw the neurologist a few days later she was shocked I had been sent home in my condition. By this point I struggled to stand by myself and my breathing was beginning to be laboured. She admitted me to the hospital immediately with a suspected diagnosis of myasthenia. Over the next few days I had the standard battery of myasthenia tests done with all the ones that provided a true diagnosis coming back negative. My symptoms fitted too perfectly though and I responded well to Mestinon, a drug used solely in the treatment of myasthenia.

Once the diagnosis was set I was started on a course of prednisone to get the symptoms under control. Anyone who has ever been on prednisone understands how awful this drug can be, with the chubby cheeks, high anxiety, and insomnia being only a few of the many negative effects. I was also given a course of intravenous immunoglobulins (IVIG) while in hospital. Now, this really is a magic treatment. Within a few days of this treatment I was almost back to normal. The only downside to IVIG is that it wears off. My first course lasted only three weeks and despite the high dose steroids, I was worse than I had been a month earlier.

At this point my steroid dose was increased and I was prescribed IVIG once every four weeks. My doctors hoped I would not need IVIG for longer than three months with the higher doses of steroids, but again attempts to go without caused a relapse, so the treatments were continued. In January I had also started a course of another immunosuppressant called Imuran. This drug, like many immunosuppressants, takes about six months to kick in and a year to become fully active. I was able to stop IVIG around the 6 month mark of this, in June.

June was also the month I had my thymectomy. Thymectomy is the surgical removal of the thymus gland. The thymus aids in development of the immune system in infancy but has atrophied by adulthood, except in some people with myasthenia. A thymectomy is only done if the patient has a thymoma, a tumour of the thymus gland. If they do not, however, the benefits of thymectomy are not clear and it is not always recommended. I went back and forth on the idea of having the surgery for quite some time before finally settling on yes. It was the scariest and most painful thing of my life but I am currently healthy and haven't had any serious flare-ups in the two years since surgery, so I'd like to think it was worth it.

So now you have my story. I would still say that going undiagnosed for a year was the hardest part by far. Yes, it was scary to have this diagnosis that is now with me for the rest of my life, but more than anything it was a relief to finally have answers. Some people ask me now how I have stayed so positive through everything and all I can say is I like being happy. If I was to let this one thing bring me down and define who I was then I would be self-pitying and upset for a very long time. Myasthenia is a part of who I am now but it hasn't changed me. I am still the same person I was before, albeit with a pretty kick-ass scar down my chest and a few extra meds on board. I was even able to get back into swimming four months after my surgery. It started slowly and I trained on a modified program but I did it. Just this past week I was a part of my team's second national title. Yes, I got sick, but I also got better and am a stronger person because of it.

Noah's Hope: Family Experiences of Batten Disease

Tracy VanHoutan

Every Childhood Disease Deserves a Cure

TRACY VANHOUTAN

Abstract This chapter describes one family's experiences of coping with Batten disease, a rare condition which has afflicted two of their three children. The chapter, written by one of the parents, presents a powerful and personalised account of the initial diagnosis and prognosis, the daily consequences of caring for the children, family efforts to find effective treatments and promising research for the future.

Introduction

When I was first asked to write this chapter, I realized I had not written anything for public consumption in many years. But, as I thought about the past few years, I realized that our family's story was worth telling. 5 years ago, life was much different - as many parents of children with rare diseases can tell you. We had a 3 year old son, Noah, and 1 year old twin girls, Laine and Emily. Our lives were busy, but we were happy and felt very blessed to have three healthy young children. But then things began to change, and change quickly. Over the next months, and the subsequent years, we learned more and more about a rare pediatric disease

---

T. VanHoutan (✉)
Noah's Hope Research Fund, 4930 Northcott Ave, Downers Grove, IL 60515, USA
e-mail: tracy@NoahsHope.com

T. VanHoutan
Batten Disease Support and Research Association (BDSRA), 1175 Dublin Road, Columbus, OH 43215, USA

known commonly as *Batten disease*. I hope through my writing of this chapter to help raise awareness of this terrible disease. To learn more about our efforts, please visit us at www.NoahsHope.com and www.BDSRA.org.

Background

Batten disease is named after the British pediatrician who first described it in 1903. Also known as *Spielmeyer-Vogt-Sjogren-Batten* disease, it is the second most common form of a group of genetic disorders called *Neuronal Ceroid Lipofuscinosis* (or NCLs). Although Batten disease is usually regarded as the juvenile form of NCL, it has now become the term to encompass all forms of NCL. The forms of NCL are classified by age of onset and have the same basic cause, progression and outcome. However, the forms of NCL are all genetically different. Over time, affected children suffer mental impairment, worsening seizures, and progressive loss of sight and motor skills. Eventually, children with Batten disease/NCL become blind, bed-ridden and unable to communicate. Presently, it is always fatal. Batten disease is not contagious or, at this time, preventable.

What are the Forms of Batten Disease/NCL?

There are four main types of NCL, including two forms that begin earlier in childhood and a very rare form that strikes adults. The symptoms are similar but they become apparent at different ages and progress at different rates.

Infantile NCL (Santavuori-Haltia disease) onsets between about 6 months and 2 years of age and progresses rapidly. Affected children fail to thrive and have abnormally small heads (microcephaly). Also typical are short, sharp muscle contractions called myoclonic jerks. Initial signs of this disorder include delayed psychomotor development with progressive deterioration, other motor disorders, or seizures. The infantile form has the most rapid progression and children live into their mid-childhood years.

Late Infantile NCL (Jansky-Bielschowsky disease or LINCL-Batten disease) is the most common form of Batten disease (Verity et al. 2010), and typically onsets between ages 2 and 4. The typical early signs are loss of muscle coordination (ataxia) and seizures, along with progressive mental deterioration. This form progresses rapidly and ends in death between ages 8 and 12. This is the disease Noah and Laine are fighting.

Juvenile NCL (Batten disease) typically onsets between the ages of 5 and 8. The typical early signs are progressive vision loss, seizures, ataxia, or clumsiness. This form progresses less rapidly and ends in death in the late teens or early 20s, although some may live into their 30s.

Adult NCL (Kufs Disease or Parry's Disease) generally begins before the age of 40, causes milder symptoms that progress slowly, and does not cause blindness. Although the age of death is variable among affected individuals, this form does shorten life expectancy.

A more precise chart of the forms of Batten disease is shown below (Mole 2013):

Chart: Forms of Batten Disease			
Form	Initials	Gene	Age of Onset
Infantile	INCL	CLN1	6 months–2 years
Late Infantile	LINCL	CLN2	2–4 years
Juvenile	JNCL	CLN3	5–7 years
Adult–Parry disease	ANCL	CLN4	25–40 years
Finnish Late Infantile	fLINCL	CLN5	2–4 years
Variant Late Infantile	vLINCL	CLN6	3–5 years
Turkish Late Infantile	tLINCL	CLN7	2–4 years
Northern Epilepsy	EPMR	CLN8	5–10 years
Variant Juvenile	vJNCL	CLN9	5–7 years
Congenital	CTSD	CLN10	Birth–2 years
Adult		CLN11	Adult onset
(un-named)		CLN12	5–7 years
Adult–(Kufs type B)		CLN13	Adult onset
(un-named)		CLN14	6 months–2 years

How Many People Have These Disorders?

Batten disease/NCL is relatively rare, occurring in an estimated 2–4 of every 100,000 births in the United States. The diseases have been identified worldwide. Although NCLs are classified as rare diseases, they often strike more than one person in families that carry the defective gene. Symptoms of Batten disease/NCLs are linked to a buildup of substances called lipopigments in the body's tissues. These lipopigments are made up of fats and proteins. Their name comes from the technical word *lipo*, which is short for "lipid" or fat, and from the term *pigment*, used because they take on a greenish-yellow color when viewed under an ultraviolet light microscope. The lipopigments build up in cells of the brain and the eye as well as in skin, muscle and many other tissues. Inside the cells, these pigments form deposits with distinctive shapes that can be seen under an electron microscope. Some look like half-moons (or commas) and are called *curvilinear bodies*, others look like fingerprints and are called *fingerprint inclusion bodies* and still others resemble gravel (or sand) and are called *granual osmophilic deposits* (*grods*). These deposits are what doctors look for when they examine a skin sample to diagnose Batten disease. The diseases cause the death of neurons (specific cells found in the brain, retina and central nervous system). The reason for neuron death is still not known.

How are These Disorders Diagnosed?

In the most common form of Batten disease (cln2 or Late Infantile) often the first major symptom is seizures of unknown origin. In addition, children begin to lose motor skills they had once mastered. Progressive loss of mastered language typically follows. In the Juvenile form of Batten disease (cln3 or JNCL), vision loss is often an early sign and Batten disease may be first suspected during an eye exam. An eye doctor can detect a loss of cells within the eye that occurs in the three childhood forms of Batten disease/NCL. However, because such cell loss occurs in other eye diseases, the disorder cannot be diagnosed by this sign alone. Often an eye specialist or other physician who suspects Batten disease/NCL may refer the child to a neurologist, a doctor who specializes in disease of the brain and nervous system. In order to diagnose Batten disease/NCL, the neurologist needs the patient's medical history and information from various laboratory tests. Diagnostic tests used for Batten disease/NCLs include:

- **Skin or tissue sampling**—the doctor can examine a small piece of tissue under an electron microscope. The powerful magnification of the microscope helps the doctor spot typical NCL deposits. These deposits are found in many different tissues, including skin, muscle, conjunctiva, rectal and others. Blood can also be used.
- **Electroencephalogram or EEG**—an EEG uses special patches placed on the scalp to record electrical currents inside the brain. This helps doctors see tell-tale patterns in the brain's electrical activity that suggest that a patient has seizures.
- **Electrical studies of the eyes**—these tests, which include visual-evoked responses (VER) and electro-retinagrams (ERG), can detect various eye problems common in childhood Batten disease/NCLs.
- **Brain scans**—imaging can help doctors look for changes in the brain's appearance. The most commonly used imaging technique is computed tomography (CT), which uses x-rays and a computer to create a sophisticated picture of the brain's tissues and structures. A CT scan may reveal brain areas that are decaying in NCL patients. A second imaging technique that is increasingly common is magnetic resonance imaging, or MRI. MRI uses a combination of magnetic fields and radio waves, instead of radiation, to create a picture of the brain.
- **Enzyme assay**—a recent development in the diagnosis of Batten disease/NCL is the use of enzyme assays that look for specific missing lysosomal enzymes for Infantile and Late Infantile forms only. This is a quick and easy diagnostic test (Sohar et al. 1999).
- **Genetic/DNA Testing**—each form of Batten disease is the result of a different defective gene. Testing for most of the forms is available for diagnosis as well as carrier and prenatal status.

As yet, no specific treatment is known that can halt or reverse the symptoms of Batten disease/NCL. However, seizures can be reduced or controlled with anti-convulsant drugs, and other medical problems can be treated appropriately as they arise. At the same time, physical and occupational therapy may help patients retain function as long as possible. Some reports have described a slowing of the disease in children with LINCL-Batten disease who were treated with various vitamins, supplements and anti-oxidants (Yoon et al. 2011; Le et al. 2012). However, these treatments did not prevent the fatal outcome of the disease. Support and encouragement can help children and families cope with the profound disability and losses caused by NCLs. The largest and oldest institution in the world dedicated to support of families and developing research initiatives is the *Batten Disease Support and Research Association (BDSRA)*. BDSRA enables affected children, adults and families to share common concerns and experiences. Meanwhile, scientists continue to pursue medical research that will someday yield an effective treatment.

Batten Disease: The VanHouton Story

Every parent dreams of the future as they hold their babies in their arms. With every first smile, first step and first somersault, we watched our three children grow stronger and more independent. Noah loved baseball and trains, while his little blonde-haired twin sisters (Laine and Emily) adored coloring, dancing at ballet class, and skipping rope. We used to worry about how we would keep up with such active children and how we would coordinate Little League with soccer practice and dance recitals. We never thought that we would worry if our precious babies would live to participate in those everyday childhood moments.

Our Three Little Stars

In April 2004, my wife Jennifer (Jen) and I were blessed with our first child: Noah. Noah was perfect to us in every way—an energetic and curious little guy who loved trains, running in the park and doing everything a normal, healthy child would do. Noah hit all of his developmental milestones appropriately and was even doing some things early. A year and a half later, we were blessed again with twin girls: Laine and Emily. Life was good—beautiful twin girls and a big brother to look after them. We spent a lot of time back in those days imagining what sort of amazing experiences our children would have together as they grew up.

Early Signs

Noah developed normally as an infant and toddler. He talked and giggled, loved watching trains, wrestling with Daddy and was always on hand to get into a little silly mischief. He loved baseball, and as he grinned under his toddler baseball cap we used to dream about him becoming a pitcher for the Chicago Cubs. But things began to slowly change with Noah. At about age two and a half, we began to notice he didn't socialize with other children in a totally normal way; we noticed some slight speech problems and he began to have slight tremors in his hands and feet. Our first neurologist said that these symptoms were normal for a boy and that he should grow out of them.

On December 18 2007, we were doing some last-minute decorating for Christmas. As I carried some boxes loaded with decorations, I looked to the other side of the room and watched Noah seemingly lose his balance and fall to the floor. As I rushed to his side, he stood up and resumed playing with the puzzle he had been putting together. I returned to my task, only to watch as Noah again lost his balance and collapsed to the floor—only this time he didn't get up. As I picked him up and lay him on the sofa, a horrible realization swept over me. Noah had stopped breathing and was turning gray. As I tried desperately to administer rescue breathing to Noah, Jen called the paramedics. Upon their arrival, Noah began to take shallow breaths on his own and his normal color returned. We had no idea what had happened, but the doctors told us it was probably a seizure. At the time, it was the most frightening day of our lives.

A few weeks later while exiting the car to take the kids sledding at a nearby hill, Noah had another seizure. Paramedics were called and Noah was again transported to the hospital. He was then diagnosed with idiopathic childhood epilepsy. About 9 months after the epilepsy diagnosis, an MRI scan revealed some atrophy of his brain, but no cause. We also noticed Noah's speech began regressing and he had difficulty performing simple activities such as eating with a fork and brushing his teeth.

After several months of different seizure medications and with varying degrees of seizure control, our second neurologist (a "specialist" in hard to diagnose cases) ran a limited screen for genetic diseases. Once the results arrived he informed us that Noah's condition was not genetic. It was at this point he seemed to think there was nothing else we could do and recommended we stop looking. He also resisted putting Noah on the Ketogenic Diet to help control his seizures. That did not really work for us. Soon after this experience with our Neurologist from our childrens' hospital, we traveled to St. Louis to get a second opinion. Not much came of this meeting except that we were told he didn't know what was causing Noah so many problems.

Duke Children's Hospital Sheds Some Light

After several months, a friend of the family put us in touch with Duke Children's Hospital. Within weeks, Jen, Grandma Jacque, and Noah drove 850 miles to Duke University for a two-week evaluation. The team ran every test imaginable to determine the cause of his seizures. By the end of their stay, the team did not have a definitive diagnosis, but several tests were still awaiting results. While at

Duke, Noah started the Ketogenic Diet, a special regimen to control his seizures. We were thrilled when Noah responded very well to the diet. His seizures were reduced dramatically and the fog around him seemed to lift. Noah, Jen, and Grandma Jacque returned home after 2 weeks and Noah was doing great; we were getting our little boy back.

On March 17, 2009, our lives changed in a way no one could imagine. As we were sitting down for our St. Patrick's Day dinner, we received a call from Noah's neurologist at Duke Children's Hospital. Two of the tests which had been sent to separate labs came back confirming the same diagnosis. Noah was diagnosed with Late Infantile Neuronal Ceroid Lipofuscinosis (LINCL), often referred to as Batten disease. Over the next few hours as we researched the disease, our world quickly crumbled. It was never the same again. Over time we learned that Noah would suffer mental impairment, worsening seizures, and progressive loss of sight and motor skills. Eventually Noah would become blind, bed-ridden, tube-fed and unable to communicate. His life would be cut short, probably between the age of 8 and 12, as the disease is presently always fatal. We sat there in disbelief and horror as we tried to come to terms with the fact that Noah's life was likely halfway over.

Noah's News Broke Our Hearts

When we learned that Noah had Batten disease, our world literally fell apart. As we frantically tried to learn as much as we could about this terrible illness, we worked hard to keep everything at home peaceful and uneventful. It was very important to us that the girls would continue their busy lives with as little interruption as possible. They were our shining lights of joy amid our new chaos, and we went to great lengths to keep up their dance classes and play dates. It soon became difficult to continue with business as usual, as Noah experienced more seizures and began having trouble walking and talking. The girls were only 2 years old, but it was clear that they knew something was wrong with Noah. It was also clear that they had no idea how that could be, as Noah was their super-hero big brother who succeeded at everything he touched.

Waiting and Praying

As the weeks passed we finally decided to have our twin girls Laine and Emily tested for Batten disease. We learned that this disease is autosomal recessive, which means that each child of ours had a 25 % chance of having the disease. Laine and Emily had none of the symptoms that Noah had displayed when they were tested for Batten disease at age three. We prayed, and hoped, and wished as we waited for the test results to be returned. The first set showed that Emily did not have Batten disease, but that Laine's test results were inconclusive. As we waited several more weeks for more tests to be run and re-run, our family and friends prayed with us for a miracle. Sadly, it did not come to be. Our world shattered again when we learned that Laine also had Batten disease. It was August 17, 2009, exactly 5 months to the day when Noah was diagnosed.

Our Broken Hearts Broke Again

Laine still had not experienced any symptoms when she was diagnosed. However, like horrifying clockwork, she had her first two seizures within the first month after her diagnosis. Emily found her when she had her first seizure, and told us that, "Lainey has the shakes." The look in her little eyes told us that Emily already knew that her sister was sick.

By the time we knew that Laine also faced Batten disease, the illness had come to live like an unwelcome guest in our house. Yet even as we stayed up nights searching for answers, Noah continued on blissfully unaware. He seemed occasionally frustrated by his loss of control, but blessedly never appeared to be very conscious of his decline. Unlike Noah, six-year-old Laine knew exactly what was happening to her. We will never forget the morning she crawled into our bed, her blue eyes wet with tears and her five-year-old blond hair tickling her pink pajamas. She was so sweet and innocent as she stared up at us and asked, "I can't walk! I can't see! Will I be like Noah too?"

What do you say at a moment like this? What do you say when your little girl can see her future in her sick brother's deteriorating body, and you can't do anything to stop it?

Laine doesn't ask these questions any more. Just months after that pink-pajama moment, she stopped walking and lost most of her ability to speak. But even while she can no longer ask, it's clear to us that she feels trapped inside her own declining body. And that, perhaps, is one of the hardest realities for us to face.

Our Giggling Little Girl

Similar to Noah, Laine began her life as a healthy infant and toddler, successfully progressing through all of her developmental milestones. Our house rang with her giggles as she galloped down the hallway with her twin sister Emily. They were always a pair, whether coloring pictures for Mommy, snuggling during naps or showing off their newest princess dresses. Laine and Emily were together so much that we often merged their names into one when we called them. They twirled in ballet class together, raced across the grass together, and had doll tea parties together. They even shared a room and had trouble sleeping by themselves.

Laine has recently lost her ability to use her adaptive walker and her declining body twitches out of her control. It was a terrible day for Laine when she realized that she could no longer run and keep up with her twin sister Emily. They did manage to walk down the aisle at a friend's wedding in June of 2011—our beautiful twin flower girls. Sadly, that was one of the last walks the girls ever took together. Laine and Emily still share a connection, but it is becoming more remote. Yet as Laine declines, her relationship with Noah grows more special. She lights up around him, and her smile is thrilling to see. We cherish each smile. And while we mourn for the future that could have been, we embrace each day that we have, cherishing our little blue-eyed girl who used to dance.

Current Efforts

As a so called "orphan disease", LINCL-Batten disease receives almost NO FEDERAL FUNDING. As close as we are to real treatment for this heartbreaking disease, none of these experimental approaches can proceed without help and funding. Within the Federal Government, the focal point for research on LINCL-Batten disease and other neurogenetic disorders is the National Institute of Neurological Disorders and Stroke (NINDS). NINDS, part of the National Institutes of Health (NIH), is responsible for supporting and conducting research on the brain and central nervous system. The Batten Disease Support and Research Association (BDSRA) is the largest and oldest organization in the world dedicated to research and support of families affected by Batten disease. More can be found on their website: www.BDSRA.org.

Clinical Trials for Late Infantile Batten Disease

Recently a phase I clinical trial has been completed: Weill Cornell Medical College at Cornell University has completed a phase I safety study using direct intracranial injection of gene therapy vector into the brain. The first clinical study, which involved using AAV serotype II to treat 10 children with LINCL-Batten disease, showed stabilization of the disease in some children (Worgall et al. 2008). At the time of the writing this chapter, the Cornell researchers are recruiting and treating children in another phase I clinical trial using the more aggressive virus Rh10. Early reports from families lead us to believe that this therapy may be slightly slowing progression, but is not stopping the disease. In brief, our family's extensive efforts include:

1. For the past 3 years, we have funded projects at the Center for Advanced Biotechnology and Medicine at Rutgers University for Enzyme Replacement Therapy. The hope is that this proven methodology in other lysosomal storage disorders can move to clinical trial by 2013 (Vuillemenot et al. 2011).
2. In partnership with *Jasper against Batten*, we funded the creation of a Batten Disease mouse colony at King's College London. The work to be done with

these mice will help other scientists around the world to better understand how Batten Disease progresses through the brain and to develop therapeutics that are better targeted.
3. Recruited a new researcher at Rush University in Chicago to look at a new approach in Batten Disease. This researcher is a Parkinson's, Alzheimer's, and Multiple Sclerosis expert who believes he can apply some of his previous work to Batten Disease using an already FDA approved drug to relieve some of the symptoms of Batten disease (Ghosh et al. 2012).
4. Recruited a new researcher at Rice University in Houston to look at how the protein that is responsible for Batten disease becomes inoperable. She is testing a new compound in cell cultures to determine if she can make this protein work correctly again.
5. Funded the upgrade of all computer systems at the national Batten Disease organization (Batten Disease Support and Research Association).
6. Tracy was elected by members of the BDSRA to the national Board of Directors, and is currently the second VP of the organization.
7. In June 2010, Tracy testified before the FDA in Washington DC on the challenges that rare diseases face at the regulatory level, and submitted proposed changes. Final Brownback/Brown report issued to Congress in August of 2011 (VanHoutan 2010a, b).
8. Participated in numerous conference calls with the National Organization for Rare Disorders (NORD) and Rare Disease Legislative Advocates (RDLA).
9. Attended the past four World Lysosomal Disease Network meetings to network with researchers and biotechnology firms, and heighten interest in Batten research.
10. Assisted in the recruitment of seven other Batten Disease foundations to form a small working group dedicated to finding a treatment or cure for Batten Disease.
11. Coordinated with the national Batten organization (BDSRA) and the National Institute for Neurological Disorders and Stroke (NINDS) to convene an international Batten Disease conference at the National Institute of Health (NIH) in November of 2010. The purpose of this conference was to identify gaps in knowledge about Batten disease and to develop collaborative efforts among the attending scientists to develop a concrete plan to transition work from the lab into the clinic. We also expect the NIH will issue a funding request as a result of this conference that could lead to significant funding for Batten Disease research at the federal level.
12. Funding a project at Weill Cornell Medical College to identify a bio marker or surrogate endpoint for Batten Disease that will aid in clinical trial design. This project will measure specific metabolites in the blood or spinal fluid that will allow us to determine if new therapies are effective or not.
13. Made five trips to Washington DC to recruit members of the US House and Senate to join the newly formed "Rare Disease Congressional Caucus". Discussed other issues of the rare disease community including HR3737-"The ULTRA" Act- Advocating for FDA flexibility in using bio-markers in clinical

trial design. The language of the "FAST act" was successfully incorporated into federal law (FDASIA) on July 9, 2012 (Public Law 112–144 2012).
14. Developing an accurate worldwide survey (count) of children affected by Late Infantile Batten disease. We expect to help co-fund a world-wide project soon to expand this work to all forms of Batten disease.
15. In partnership with *Hope-4-Bridget*, co-funded a project at Sanford Children's Research center in South Dakota to develop a test for a high throughput screening of tens of thousands of compounds and pharmaceuticals to determine if they might have an effect in Batten Disease cells.
16. Working with other foundations across the globe on the formation of the Batten Disease International Alliance. Attended the inaugural meeting in March 2012 in London.
17. In September 2011, Tracy was invited to the National Institutes of Health campus in Maryland to participate as a parent advocate for a workshop on cell based therapies for pediatric disorders. This workshop addressed strategies to overcome the barriers to advancing the development and delivery of cell-based therapies for pediatric patients, in particular those with rare and life-threatening diseases. The clinical applications of cellular therapies and regenerative medicine, including the ethical considerations and models of clinical trial design, were examined with intent to optimize overall processes for the future (NIH 2012).

The Future

Enzyme Replacement Therapy

Enzyme replacement therapies are also in different stages of development. As a previously proven therapy option in other lysosomal storage disorders, enzyme replacement therapy represents so-called "low hanging fruit". While some obstacles still need to be overcome, we believe that this may be the first therapy to be widely available to children dying from this devastating disease. Noah's Hope Fund has partnered with the BDSRA to fund several research projects in enzyme replacement therapy. In December 2012, BioMarin, a California Biotech company that focuses or rare diseases, announced at a New York investor conference that they have added a new clinical program to their pipeline. BMN-190 was announced as a pipeline project to treat Late Infantile Batten disease with a synthetically manufactured enzyme that is missing in these children. As I watch the conference via webcast, the company's CEO plays a video we produced to launch Noah's Hope. The video is titled *"Fix You"*. Nearly 3 years of developing this relationship and funding pre-clinical mouse work at Rutgers has paid off. Quite an emotional and bittersweet day as we will soon have a viable treatment for this disease. A big step, but much more work still to be done. And sadly, Laine and Noah have likely progressed too far to be part of the clinical trials slated to begin in Europe in 2013.

Gene Therapy: Current Efforts

Dr. Beverly Davidson at the University of Iowa is working on a gene therapy protocol. She is using virus vectors for gene transfer to the central nervous system. Her work in a mouse model is encouraging, and we are eager to support her work in the future (Chen et al. 2009). Dr. Steven Grey at the University of North Carolina is working on modifying the virus capsid to more effectively deliver therapy. Dr. Ron Crystal of the Weill Cornell Medical College has completed one clinical trial and is currently conducting a second as mentioned previously.

Small Molecule Therapy

Small molecule (drug) therapies are currently being researched. Several pilot programs to repurpose existing drugs are currently underway due to the efforts of our Batten researchers. Our hope is that we can find an already FDA approved drug that will have some beneficial effect.

Researchers are also trying to discover the mechanism by which chronic activation of glia in the brain, especially astrocytes and microglia, leads to damage of the neurons and progressive neuro-degeneration in diseases like Batten disease (Macauley et al. 2011). The overall goal is to utilize knowledge of potentially "druggable" pathways to develop new therapeutics. The normal role of the glia is to cooperate with the neurons to keep the brain operating smoothly. When an injury or change in the brain occurs, the glia mount an inflammation response to fight off the insult and restore the brain to its proper functioning. But in neurodegenerative diseases, the glia are over-activated, producing a state called neuro-inflammation. Neuro-inflammation can lead to nerve cell dysfunction or death, which manifests as dementia. Although neuro-inflammation appears to play a pivotal role in the development and progression of neuro-degeneration, the molecular mechanisms underlying the process and approaches to reduce the neuro-inflammation have received little attention to date from the research community.

Conclusion

As a man, a guy, a father, I have always been inclined towards wanting to fix things. I have actually always been very handy and able to fix just about anything. So it is especially painful and incomprehensible that I am not able to fulfill that role with my own children at this time. In the two years following Noah's diagnosis, we had high hopes that somehow we might be able to save Noah and Laine. We still carry that hope, but the cruel realities of time are working against us. We do still have great hope that we will soon be part of developing treatments and possibly cures for this group of devastating diseases. Much has been accomplished over the past few years due to the efforts of many, but much work remains to be done. Thank you for taking the time to learn a bit more about Batten disease and for letting me share my special children, Noah and Laine with you.

Acknowledgments There are many people who deserve to be thanked here. Batten researchers: thank you for your countless hours reviewing proposals and for patiently teaching me about this disease. Emil Kakkis: for helping me understand the regulatory process better and for inviting me to become a part of significant change. Donors: for your continued support of Noah's Hope which make our research and policy initiatives possible. Mammo and Hoppa: without your time, support and counsel, we would be lost. My wife, Jennifer: your strength amazes me. Thank you…for everything. To the staff of BDSRA: you are the light in a world of darkness. And a special thank you to Lance Johnston, former Executive Director of the BDSRA. Lance is the major reason we have been able to accomplish all that we have. A man with incredible reserves of compassion and knowledge, Lance likely touched more lives than he will ever realize.

References

Chen YH, Chang M, Davidson BL. Molecular signatures of disease brain endothelia provide new sites for CNS-directed enzyme therapy. Nat Med. 2009; 15(10):1215–1218.
Ghosh A, Corbett GT, Gonzalez FJ, Pahan K. 2012. Gemfibrozil and fenofibrate, food and drug Administration-approved lipid-lowering drugs, up-regulate tripeptidyl-peptidase 1 in brain cells via peroxisome proliferator-activated receptor α: implications for late infantile Batten disease therapy. J Biol Chem. 287(46):38922–38935.
Le NM, Parikh S. Late infantile neuronal ceroid lipofuscinosis and dopamine deficiency. J Child Neurol. 2012; 27(2):234–237.
Macauley SL, Pekny M, Sands MS. The role of attenuated astrocyte activation in infantile neuronal ceroid lipofuscinosis. J Neurosci. 2011; 31(43):15575–15585.
Mole S. NCL mutation and patient database. [online]. Available at: <http://www.ucl.ac.uk/ncl/mutation.shtml> (2013). Last Accessed: 2 Jan 2013.
NIH. 2012. Cell therapy for pediatric diseases: a growing frontier production assistance for cellular therapies (PACT) workshop [online]. Available at: <http://www.nhlbi.nih.gov/meetings/workshops/pact.htm>. Last Accessed: 2 Jan 2013.
Public Law 112–144. 112th congress. [online] Available at: <http://www.gpo.gov/fdsys/pkg/PLAW-112publ144/pdf/PLAW-112publ144.pdf> (2012). Last Accessed: 2 Jan 2013.
Sohar I, Sleat DE, Jadot M, Lobel P. Biochemical characterization of a lysosomal protease deficient in classical late infantile neuronal ceroid lipofuscinosis (LINCL) and development of an enzyme-based assay for diagnosis and exclusion of LINCL in human specimens and animal models. J Neurochem. 1999; 73(2):700–711.
VanHoutan T. Accessing the accelerated approval pathway with a rare neurologic disease. [online], Available at: <http://noahshope.com/assets/fda-presentation-on-batten-disease-june2010--v2.pptx≫ 2010a. Last Accessed 2 Jan 2013.
VanHoutan T. Testimony of Tracy VanHoutan, at the FDA public hearing: considerations regarding food and drug administration review and regulation of articles for the treatment of rare diseases. 29 June 2010b. Federal Docket Number: FDA-2010-N-0218.
Verity C, Winstone AM, Stellitano L, Will R, Nicoll A. The epidemiology of progressive intellectual and neurological deterioration in childhood. Arch Dis Child. 2010;95:361–4.
Vuillemenot BR, Katz ML, Coates JR, Kennedy D, Tiger P, Kanazono S, Lobel P, Sohar I, Xu S, Cahayag R, Keve S, Koren E, Bunting S, Tsuruda LS, O'Neill CA. Intrathecal tripeptidyl-peptidase 1 reduces lysosomal storage in a canine model of late infantile neuronal ceroid lipofuscinosis. Mol Genet Metab. 2011; 104(3):325–337.

Worgall S, Sondhi D, Hackett NR, Kosofsky B, Kekatpure MV, Neyzi N, Dyke JP, Ballon D, Heier L, Greenwald BM, Christos P, Mazumdar M, Souweidane MM, Kaplitt MG, Crystal RG. Treatment of late infantile neuronal ceroid lipofuscinosis by CNS administration of a serotype 2 adeno-associated virus expressing CLN2 cDNA. 2008; 19(5):463–474.

Yoon DH, Kwon OY, Mang JY, Jung MJ, Kim do Y, Park YK, Heo TH, Kim SJ. Protective potential of resveratrol against oxidative stress and apoptosis in Batten disease lymphoblast cells. Biochem Biophys Res Commun. 2011; 414(1):49–52.

Using Technology to Share Information: Experiences of Oesophagus Atresia (OA) and Tracheaoesophageal Fistel (TOF)

Caren Kunst

> *A person always doing his or her best becomes a natural leader, just by example*
>
> JOE DIMAGGIO

Abstract This chapter discusses some of the problems faced by parents whose child is diagnosed with the orphan disease Tracheaoesophageal fistula (TOF)/ Esophageal atresia (OA). Effective management of this condition requires parents to share in decisions about care. It is also essential for them to learn to care for their child within the hospital setting and at home. The chapter discusses ways in which they can become active partners in the care team by recording information collected whilst caring for their child. This can then be shared with health professionals in order to ensure that decisions about care are based on the most complete set of information. Parents and carers can also use technology to access information about the condition and treatment and to gain support from people who are dealing with similar problems.

Introduction

The author is a healthcare consultant with a growing interest in digital healthcare. She is also a nurse and an experienced (homecare) mother of a 21 year old son with the rare (orphan) disease Tracheaoesophageal fistula (TOF)/Esophageal atresia (OA). Caren has been an e-coach since 1999, a role that involves providing remote support concerning TOF/OA and its self care management via phone/Skype to over 400 families. She also supports parents all over the world

C. Kunst (✉)
Barmsijs 54, 3435 BP, Nieuwegein, The Netherlands
e-mail: carenwit@planet.nl

using sa such as *Facebook*, *Hyves* (a Dutch social network) and *NING* (an online platform for people to create their own social networks). Her aim is to organize effective and safe self management in home and hospital situations and she empowers parents to take the lead in collecting data. She is interested in working towards a culture change in TOF/OA healthcare by promoting decision making that this is based on real-time data and information. Known as "the Young TOF/OA child patient's advocate" on *Facebook*, she is among the Top five nurses who use Twitter in the Netherlands. In 2011, she won a ticket to *TEDXMaastricht* for her work on *Linked In*. She has also made a movie about TOF/OA self care management and the transition to adult care. Caren Kunst is the builder and co-founder of *EAT*, a federation of European member support groups associated with the rare congenital condition of Oesophageal Atresia (OA) and/or Tracheaoesophageal Fistula (TOF). In 2010, Caren won the *Social Innovation Prize*, awarded by the minister of the Dutch Healthcare Government. This was for writing a handbook entitled, "Selfcaremanagement Oesophagusatresia and Tracheoesophageal fistula" (Goodcaresupport 2010).

Tracheoesophageal Fistula and Esophageal Atresia

> When parents get a baby born with the orphan disease TOF/OA (Wikipedia 2012), they are overwhelmed by emotions. First, just after birth the parents are happy with their child. Suddenly they get very sad when they hear, "your child is very sick with an orphan disease". They are immediately playing a triple-role in the (para) medical team, participating in the (shared) decision making processes.
>
> **Caren Kunst**

Babies born with TOF/OA require intensive neo-natal care prior to corrective surgery, normally within days of birth. Some children have to undergo additional surgical interventions later on in their lives. Respiratory distress and swallowing problems will develop during the first hours after birth. If any of these symptoms are diagnosed, a catheter will be passed into the oesophagus to check for resistance. If resistance is noted, an additional medical examination will be undertaken to confirm a diagnosis of OA. A catheter can be inserted and will show up as white on a regular x-ray film, demonstrating the blind pouch ending. Treatment for OA and TOF is surgery to repair the defect. If OA and TOF is suspected, oral feeding is stopped and intravenous fluids are started. The baby will be positioned to help drain secretions and decrease the chance of aspiration.

OA and TOF is considered to be an orphan disease because it affects a small percentage of the population, occurring in approximately 1 in 3,500 live births (Tracheo-Oesophageal Fistula Support 2007). Whilst many children born with TOF/OA will experience only a few problems, others may have (severe) difficulties with swallowing, even with a repaired oesophagus. They also have problems with digesting food. Some have chronic conditions which include (severe) gastro-oesophageal reflux (where the acidic stomach contents pass back into the lower oesophagus) and they may also have orthopedic, cardiac and/or respiratory problems. Surgery to repair the gap and fistula is rarely an emergency but is normally done within 24 h of diagnosis. Once the baby is well enough for surgery, the oesophagus can usually be repaired; the length of time of hospitalization varies from a few weeks to longer than a year if there are other complications. A small proportion of these TEF/OA children have more medical problems, referred to as the VACTERL (Wikipedia 2013c) (V: Vertebral anomalies; A: Anal atresia; C: Cardiovascular anomalies; TE: Tracheoesophageal fistula; R: Renal (Kidney) and/or radial anomalies; L: Limb defects). The care should be tailored to the unique requirements of each child.

Knowledge, Skills and Data for Better TOF/OA Self Care Management

Most TOF/OA parents are unaware of the requirement to collect clinical data and health information from the time of the child's birth. They are too overwhelmed by their emotions: having a very sick child, wondering how to organize a job in combination with caring for their child, dealing with other children and worrying about the care arrangements of the sick child. Parents also have to cope with the realization that their child has a rare (orphan) disease and they may well be questioning whether they did anything wrong, could it have been prevented and "why us?"

A few days after the birth, parents are more aware of the need to collect clinical data and information about TOF/OA and its implications for their child. Knowledge at that point is a blend of information from the (para)medical team, family and friends, homecare workers and self care management. Parents are ill equipped to take an active and effective part in this process. They often lack knowledge, are under-informed (or informed too late) or are overwhelmed by "information overload" concerning their new situation. Although clinical staff ask parents for permission for surgical treatments, at this emotionally-charged stage, without enough basic information, parents often feel that they are hopelessly reactive—they are not equal partners in the decision-making process.

> They asked me to say YES to the first lifesaving operation, but I was in a "never ending (para)medical data and information-rollercoaster". My wife was an emotional wreck, my child was very sick and I had to decide. How do I decide without any data or information? I'd never heard of TOF/OA…I could only say "PLEASE, do what's needed!
>
> **Father of newly diagnosed TOF/OA baby, relating his experience to Caren Kunst**

Arranging TOF/OA Self Management

The active community of parents with TOF/OA children (located across the world) communicate effectively using social media (such as *LinkedIn*, *Facebook*, *Twitter*, wikis, video sharing sites and *NING*). They are quick learners, highly motivated and can be regarded as "2.0 skilled". They interact with one another and collaborate using the dialogue of new social media (termed *Web 2.0*) to create user-generated content in a virtual community; this is in contrast to more traditional websites where people are limited to the passive viewing of content. The lack of care providers and dedicated IT experts on these TOF/OA platforms is disappointing. They are frequently asked to take part, but they are not always effectively connected. The parents are aware of the need to supply themselves (and other parents) with trusted and validated clinical information about the treatments for TOF/OA but they are also aware that the information collected by care providers and themselves cannot always be shared easily by digital means.

Parents using TOF/OA *Facebook* pages often ask if it is possible to have one shared "partnership" health record (there is currently one entry for the use of formal care providers, one for parents and one for the young adult child). Suggestions include the use of different colors or fonts to clarify and confirm who had posted a comment. Parents commented that even this may not have been necessary if parents/patients had been included at the planning stage. Posting on *Facebook* pages include the desire for such requirements as: management by parents and young adult patients, a "folder of bookmarks", a diary linked to the medical record of care providers, a structured e-learning center, narratives of carers and a link to pertinent information that concerns the patient's health, whether personal or clinical.

The Patient as a Stakeholder: Towards Health 3.0

In order for parents to fully embrace their role as a stakeholder, they need relevant, trusted, open source data. They often voice concerns about the heavy reliance on data for a multitude of purposes. Asking seemingly naive questions can lead to new insights regarding care. Real-time feedback from parents of TOF/OA patients and devices is leading to more accurate insights into conditions and the patient responses to treatment. Integrated data and associated analytics are key to changing the value base of care. Patient information, controlled by patients themselves, may well become more commonplace. Meaningful use of this rich multi-dimensional information is the final driver of the transition to Health 3.0 (the next generation of healthcare). Health literate TOF/OA parents and new business models, underpinned by changes to healthcare funding and service design, are empowering care providers [such as Dutch TOF/OA care providers (Erasmus 2013) in *Erasmusmc Sophia Childrens Hospital* (Erasmus 2013)]. This growing wave of "creative disruption", supported by co-created technology, will be the forerunner of Health 3.0 where cross-disciplinary teams deliver personalized treatment to fully engaged citizens.

Parents using the TOF/OA *Facebook* pages are aware that the global healthcare system is shifting from a fee-per-service to a shared-risk model (mirroring moves from episodic treatments to "joined-up care"). This new model involves both payment for outcomes and a greater degree of risk-sharing among the different parties involved in delivery (including the TOF/OA parents). The impact of cloud computing (Wikipedia 2013a) and global mobility (Rosen 2011) are also shaking up the healthcare- ICT supply market. TOF/OA treatment and diagnosis has traditionally been episodic and isolated, as has IT deployment, with staggered rollouts of different systems, ranging from picture archiving and communication systems (PACS) to electronic health records (EHRs). Interoperability and integration initiatives will involve a considerable amount of open data resources, shifts in mindsets and different IT budgets. The traditional approach to IT deployment (roll out, fix, integrate, update, fix) is also changing.

Some healthcare enterprises are already actively embracing the new capabilities offered by Health 2.0 technologies, while others are being dragged towards more agile technologies in order to accommodate the fast growth of associated phenomena (mobile applications, wirelessly connected devices and social media). TOF/OA health professionals still argue with parents that compliance and security issues are barriers to cloud adoption but these objections grow weaker by the day. Private data clouds can be made compliant (in the U.S.) with the *Health Insurance Portability and Accountability Act (HIPAA)* and the *Health Information Technology for Economic and Clinical Health Act (HITECH)*. Regarding security, privacy and confidentiality; there are strong arguments that data is actually safer in the cloud than on devices (Davies 2013).

Parents as Information Providers

Reading through TOF/OA *Facebook* postings, it is obvious that parents are willing to use technology as part of their role but they need transparency (Kelsey 2013). TOF/OA parents have already suggested the use of video over the internet—parents embrace the opportunity to talk with caregivers without being physically present. When we consider TOF/OA care, technology has already altered the landscape from changes in syringe technology to the development of telemetry and long-distance monitoring.

In discussions with TOF/OA *Facebook* users, parents confirm that it would have been useful to collect data by way of personal IT tools (camera, phone) on the day their child was born. Parents have to remember a lot of details in order to understand what happened before, during and after birth, the current situation and what was happening when symptoms occurred. Parents need to recognize TOF/OA complications and how to deal with them. In the meantime, they also have to learn a lot of skills to care for their child at home and in the hospital. These skills are necessary to enable themselves (and later, their young child) to become an active and responsible partner in the TOF/OA care pathway and to make the best decisions to cope with the long term complications of a TOF/OA. The *Facebook* parents could have made effective use of a pre-structured personal record that was able to collect and contain all (para)medical, care and health-related information. Such a record could take a structured form with space for unstructured notes (for example, diary, pictures and relevant links to the internet).

TOF/OA parents often forget to examine the healthy parts of their child. Normally a child develops socially, mentally and physically, step by step, to a young adult. Healthy children get several screenings, mostly conducted in the primary healthcare system. But when the sick child is in hospital, all screenings are carried out by healthcare professionals, often when the parents are not present. Data about vaccinations, screenings, measuring of the scalp, measurements concerning length and weight and so forth are recorded in a digital or analogue medical file. Most parents are unaware of any gaps in this clinical data. They miss the data and they are usually unaware of how useful this can be at a later stage in the life of the child. The healthcare system and the providers are often not aware of the value of informing parents about the screenings and the importance of collecting clinical data.

The Need for Transparency

Advancing homecare technology is becoming an important part of parental TOF/OA care. With the pace of change and innovation increasing exponentially, TOF/OA parents need to embrace these changes, be connected and use advanced tools. Some parents still encounter care facilities that record information on paper, but their numbers are decreasing. When collecting digital and paper versions of a child's medical data, parents may also wonder:

- What does the law say about a parent's rights to obtain copies of the child's medical records?
- What does the law say about being required to pay for copies and not receiving the original?
- What can be done about mistakes (wrong or missing data) in the medical records?
- Why do most doctors not want patients to have their data?

Some parents report that they have obtained the records with no problems, but more frequently parents are being blocked in their attempts to get them. It is a topic discussed frequently among parents on *Facebook*: that despite having all the requisite paperwork signed, hospitals (in particular) do everything in their power to keep the records. Most doctors believe that they are the owners of medical records. In fact, one study (3,700 doctors in 31 countries) showed that only 31 % of doctors think patients should have full access to their records. Most of the remainder thought it was permissible for patients to have some limited access. 4 % stated that they did not think patients should have any access to their records (Fisher 2013). Transparency means sharing three types of data:

- **Big data**: clinical quality measures, survival rates from international clinical audits of (para)medical specialties (Healthcare IT News 2013)
- **User data**: where the parent TOF/OA is the source of information (for example, the UK's National Health Service (NHS) is rolling out a new "friends-and-family" service to see whether patients recommend services to loved ones)

- **My data**: personal information that empowers the individual to organize their self management.

If parents and care providers gain experience in shared decision making for everyday TOF/OA health problems, both parties will develop and be ready to share decision making when bigger issues are at stake. However, this will not happen without a significant culture shift by patients and care providers. Doctors and parents must accept a new and more engaged role for the individual, where the parent's voice is better heard and the patient's choice is better honored. TOF/OA parents are empowering themselves to achieve a high level of knowledge, skills and experience, as a result of sharing practical experiences, clinical data and homecare information on the internet. *Facebook* parents claim that social media and the internet play an important role in organizing their child's safety, with regards to quality of life and follow-up care. Their needs are challenging the healthcare system to facilitate them.

Starting a Dialogue About Sharing Information: 10 Questions

The dialogue about shared care starts with the realization that while specialists' qualifications and experience are important, they must also support parents and other care providers to record information when they are delivering self care. The dialogue about achieving a partnership to create the best management of care can be started by way of these 10 questions, like:

1. What are the typical questions TOF/OA parents have when they come to this clinic, in the first 24 h (or whatever time period)?
2. Do care providers know how to find the answers to those questions and do they have a genuine interest in talking with the parents within and outside their clinic?
3. Does the care provider at the self management support helpdesk really care about what is happening?
4. Do care providers have any awareness of what patient education/empowerment/shared decision making means?
5. Can the TOF/OA website of the children's hospital clinic be better utilized by extending its reach to diverse specialists in related domains?
6. What type of E-Dialogue tools could be added to the website for the benefit of 2.0 patient education, empowerment etc.?
7. What makes TOF/OA parents an "ideal shared decision team" for the clinic?
8. How could the TOF/OA staff members communicate consistently with new parents in order to facilitate working as a team?
9. Which conditions and underlying causes are being considered?
10. How easy is it for anyone to find the clinic website without knowing any of the physicians or the name of the clinic?

How to Deal with the Healthcare-Culture Phenomenon (Brown 2013)

Parents not only need a wide bandwidth to transfer digital information about complex medical phenomena but also require direct 1-to-1 interactions in order to tackle complex cultural-phenomena. There needs to be genuine communication between parents and the healthcare professionals, even though this may go against cultural norms. Parents have said that "Surgeons are mostly fixers, but less communicators" (Becker 2010). It is important to recognize that sharing is caring—there is little point in developing an impactful collection of data if nobody shares it. When data is used to describe the quality of life and self care it usually relates to (discrete) aspects or attributes of a service or product in healthcare. However, there are still no measurement tools that are suitable for the observations made by TOF/OA parents. Although the information supplied by parents is often difficult to quantify, it should not be considered any less valuable.

TOF/OA Parent's Outcomes Matter: Trying to Improve the Systems with Processes of Patient-Centric TOF/OA Care (Wikipedia 2013b)

Parents often say that TOF/OA care providers still follow very traditional rules, procedures and power-relations. Health professionals have little financial incentive to do things differently and their institutions are often organized in a way that discourages true innovation. Some TOF/OA surgeons, who are also managers of their surgical department, are placed in an ethical dilemma: being a financial manager within the academic setting or competing with other (international) institutions to provide the best TOF/OA care. However, professionals can't provide effective TOF/OA healthcare alone. Patient-centric TOF/OA care is a label that acknowledges the centrality of focus on the TOF/OA patient's needs but which may inadvertently invite a distorted view of professional work. Health outcomes are related to the professionals and their work and the natural history of the relevant TOF/OA disease biology (and genetics); they are also though related to the support given by parents. A conversation about how real life outcomes are produced is fundamental to attracting professional and parental participation in the process of change.

The co-creation processes of designing, introducing, evaluating, and re-designing TOF/OA care (together with TOF/OA parents) is achieving improvements in the rare disease care systems. A culture of trust and respect is called for and surgeons need to embrace competencies required to work with parents and ensure that these are continuously nurtured. The Dutch Chairman of Childsurgeons (Erasmus 2013) has written about the need to change to patient-centric ways of working, particularly with patients who have orphan diseases. Together with Caren Kunst, he won the *Social Innovation prize*, in April 2010 (GSK 2011). 2.0 TOF/OA parents are becoming better informed and have increasing access to the clinical records data and use this to become better carers who are willing to take responsibility for the care of their child's health. TOF/OA care providers should embrace these developments, as patients with TOF/OA who self monitor will probably need to see a doctor less often than those who do not. Adapting to the new technology has not always been straightforward and parents have told their stories of "suffering and eventful success" on *Facebook*. Care providers have to work with them to organize structured self care management. In time, there needs to be a dialogue with a healthcare record provider who can be a true partner in supporting the TOF/OA care team. The requirements of each party are becoming clear in terms of what each person in the shared decision making process needs to do their job. The eventual shared record will provide content for educating patients and professionals. It will have open source content that is accessible for parents and patients, and facilitates collaboration between clinics as appropriate.

Transition to Adult Care

During the time that a young adult patient is transitioning to adult care, they require more support from care providers. It is a big challenge for young adults to get the attention and support that they need to organize their self management. Parents also need to "let go" in order to let their child develop. Care providers in adult care know less about TOF/OA and are not aware of the self management issues of young adults. It can therefore be more difficult to spot problems because young adults often change care providers several times during the process of moving from pediatric care to adult care and the associated medical data may not be passed on. Technology is now offering a way of collecting and effectively organizing information, ensuring it can be passed on when a patient moves to a new care provider.

When a young adult is not sure about the impact of the disease TOF/OA on their life, it can have a negative impact on activities such as studying, looking for work, forming relationships and so forth. A young adult with TOF/OA does not just have physical disabilities; they also experience disabilities in daily activities and have problems participating in activities such as sport, school and social life. Such influences as personality, age, lifestyle, social and cultural background may affect the way a TOF/OA patient's experiences (in both childhood and adulthood). Technology provides a way not only for a young adult to record data, participate in their care and access information, but also a powerful means of connecting with other young people who are facing similar challenges.

Conclusion

This chapter has described how technology can empower parents whose child is diagnosed with the rare disease Tracheoesophageal fistula (TOF)/Esophageal atresia (OA) to become active participants in their child's care. It is clear that the methods of doing this are in their early stages and there are many barriers to be overcome, including some resistance from healthcare providers. It is also clear that the advantages of empowering parents and patients outweigh the challenges involved and the issues discussed in this chapter are also relevant for people living with other rare diseases.

References

Becker S. Chirurg kan veel leren van openheid piloot. Trouw.nl. [online] http://www.trouw.nl/tr/nl/4324/Nieuws/article/detail/1102404/2010/04/13/Chirurg-kan-veel-leren-van-openheid-piloot.dhtml (2010). Last Accessed 26 March 2013.

Brown T. Healing the hospital hierarchy. The New York times (The opinion pages). [online] http://opinionator.blogs.nytimes.com/2013/03/16/healing-the-hospital-hierarchy/ (2013). Last Accessed 26 March 2013.

Davies C. HIMSS13: mobility, cloud, and data equals creative disruption. Ovum website, [online] http://ovum.com/2013/03/11/himss13-mobility-cloud-and-data-equals-creative-disruption/ (2013). Last Accessed 26 March 2013.

Erasmus MC. De Europoort Voor Aangeboren Anatomische Afwijkingen. [online] http://www.erasmusmc.nl/alkh-cs/51026/1324435/pdforatiewijnen (2013). Last Accessed 26 March 2013.

Erasmus MC. Prof. dr. R.M.H. Wijnen. [online] http://www.erasmusmc.nl/kinderheelkunde/teamofpedsurg/chirurgen/3032346/ (2013). Last Accessed 26 March 2013.

Fisher K. Majority of doctors opposed to full access to your own electronic records. Ars Technica, [online] http://arstechnica.com/tech-policy/2013/03/majority-of-doctors-opposed-to-full-access-to-your-own-electronic-records/ (2013). Last Accessed 26 March 2013.

Goodcaresupport. Winning the social innovation prize 2010. [online] https://www.youtube.com/watch?v=QF_1uxR5uiw (2011). Last Accessed 26 March 2013.

GSK. Winnaar sociale innovatie prijs 2010 bekend: good care support qantasie 1. [online] http://www.gsk.nl/nieuws/winnaar_sociale_innovatie_prijs_2010.html (2011). Last Accessed 26 March 2013.

Healthcare IT News. 9 tips for getting started with big data. [online] http://www.healthcareitnews.com/news/9-tips-getting-started-big-data?page=0 (2013). Last Accessed 26 March 2013.

Kelsey T. Transparency in the NHS not only saves lives—it is a fundamental human right. The Guardian, [online] http://www.guardian.co.uk/society/2013/mar/12/nhs-transparency-open-data-initiative (2013). Last Accessed 26 March 2013.

Rosen C. The new meaning of mobility. The New Atlantis, Number 31, Spring 2011. pp. 40–46.

Tracheo-Oesophageal Fistula Support. Tracheo-Oesophageal Fistula support website. [online] http://www.tofs.org.uk/ (2007). Last Accessed 26 March 2013.

Wikipedia. Cloud computing. [online] http://en.wikipedia.org/wiki/Cloud_computing (2013a). Last Accessed 26 March 2013.

Wikipedia. Patient-centered care. [online] http://en.wikipedia.org/wiki/Patient-centered_care (2013b). Last Accessed 26 March 2013.

Wikipedia. Tracheoesophageal fistula. [online] http://en.wikipedia.org/wiki/Tracheoesophageal_fistula (2012). Last Accessed 26 March 2013.

Wikipedia. VACTERL association. [online] http://en.wikipedia.org/wiki/VACTERL_association (2013c). Last Accessed 26 March 2013.

Vignette: MPSIIIA (Sanfilippo)

Roy and Zezee Zeighami

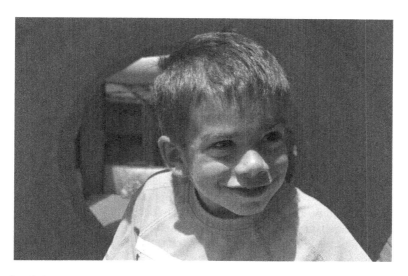

Our family's journey with rare disease began with a chance discovery of some minor bone irregularities in our then three year old son, Reed, at *Dallas Scottish Rite Hospital* in December 2010. I still recall my wife, Zezee, coming home that day and mentioning it to me. Zezee mentioned that the radiologist said that he may have a "storage disease". Zezee said she asked the doctor, "Will it kill him?" and noticed the person pause before they said, "No, whatever it is, it's minor and possibly managed with a change in diet". It wasn't minor.

Roy (✉) · Z. Zeighami
6420 Diamond Drive, McKinney TX 75070, USA
e-mail: zeighami@sf4k.org

We found over the ensuing weeks that Reed was born with MPS IIIA, also known as *Sanfilippo Syndrome*. Children with *Sanfilippo* develop normally up to the age of 3 or 4 and then slowly regress, losing all their skills (talking, walking and even eventually eating). The average lifespan of an affected child is 15. Like all lysosomal storage diseases (LSDs), *Sanfilippo* is caused by a missing enzyme. Unlike some LSDs, *Sanfilippo* is primarily a disease of the central nervous system (CNS). Getting the missing enzyme to the brain is hard, a problem that hasn't yet been solved.

I must say that I hesitated to write this story because all good stories have a start and a finish. My hope is, if I have done my job and we get lucky, we don't yet know how the story ends. It is a fact that society is on the leading edge of an explosion of treatments for genetic disease like *Sanfilippo*. Too many stories that make headlines are happy stories but that is far from the reality that faces most families affected by a rare disease. With 7,000 diseases and only 400 or so treatments, stories of struggle are bound to outnumber stories of triumph.

Our crusade has taken us to Congress, the UN Complex in Geneva, the FDA and to various pharmaceutical companies around the country. The message has been simple: we need more and better treatments and we need them fast.

Our first hope for an approved treatment for *Sanfilippo* is intrathecal ERT (enzyme replacement therapy). *Shire HGT* launched such a program in 2010, just before Reed was diagnosed. On the day of his final diagnosis, we researched online and found a clinical trial with sites in the UK and Amsterdam. Of course, we offered to move to Europe and participate. Unfortunately, the company only allowed involvement from children in participating countries. While the logic made sense, it didn't make it any easier for us to accept. I flew to *LDN World*, the largest annual medical conference, to try and find someone who knew more. I met an official from the company that explained that there was no flexibility and was told that compassionate use would not be an option, as it was far too early in the development of the therapy to consider expanded access (efficacy had not been established, let alone safety). To add insult to injury, I was told that *Shire* had tried to open a US trial site and was rejected. So, I went home to Dallas dejected.

When I got an offer a few weeks later to go to Washington, I decided that I would try to line up a visit with someone at the FDA. I got into see Dr. Tim Cote who was the director of the Office of Orphan Products. I posed the question to him of why the FDA did not allow a trial site and whether they would stand in the way of compassionate use. He wouldn't comment on the trial itself but said that FDA typically did not stand in the way of compassionate use. He proceeded to call the medical director from the drug company and tell him that they would like to see our son gain access to the drug. Honestly, it was surreal. I left the FDA thinking "problem solved" and flew home to Dallas, pondering my next step. I made my mind that I would put Reed on a plane, fly him to North Carolina and discuss compassionate access with Dr. Joe Muenzer. He was supportive and said that he would do it for us. Dr. Muenzer suggested that I let him contact *Shire* and get back to me.

I worked for *Cisco* and had senior executive support for our cause. I relayed the story to Randy Pond, our EVP of operations and penned a letter to be signed by

Vignette: MPSIIIA (Sanfilippo) 175

Randy and our CEO, John Chambers. To this day, it is hard not to cry when you read that letter. It asked that *Shire* allow *Cisco* to move us to Europe or find some way to allow Reed to have access to the drug outside the trial. It was stunning and, I was sure, bound to have an impact. You just can't get a letter from someone like John and not respond. Our response came in the form of a phone call directly to our house phone from the president of the *Shire HGT* division, Dr. Sylvie Gregoire. Dr. Gregoire said that they could not let us into the trial, as the patients had already been selected. She said that the trial would wrap up in early 2012 (which it didn't) and that they would not forget us for the next phase. Further, she said that if—for whatever reason—he didn't fit the protocol, they would consider compassionate use. That was a hope that we clung to.

Perhaps the second worst day since Reed was diagnosed came at a medical conference in Geneva, Switzerland. There we met Professor John Hopwood in person. I'd actually spoken with him by phone soon after Reed was diagnosed. He is a Professor in Australia and a top researcher for *Sanfilippo*. At the conference he dropped a bomb. They had just completed some mouse studies and found that the mouse models no longer responded to therapy if it was begun after the animals showed symptoms. This was very bad news. Of course, my hope had been that it was just a matter of waiting that that Reed would eventually get treated. The hope was that we could (at the very least) stop the disease even if he had lost a bit before that happened. For the first time, I was faced with the reality that perhaps it wouldn't work out. At that meeting, I also met Drs. Doug McCarty and Haiyan Fu. They were developing a gene therapy that would treat the brain and body with a single intravenous injection. A cure sounded good but I couldn't think about it and how long it would take to get that therapy into the clinic.

As we waited for some sort of trial site to open, I spent my time speaking to Pharma companies and advocating for legislation to accelerate the approval of treatments for rare disease. There are multiple approval pathways within the FDA to get a drug approved for a condition. The traditional pathway involves multiple trial phases with a final so called "pivotal" trial that shows clinical efficacy of a drug. Clinical efficacy is a high bar because it requires showing a statistically significant clinical benefit for a patient before the drug is allowed to be sold. While that sounds good, this may be very difficult to do for a small population of heterogeneous patients. Further, many diseases evolve so slowly that proving clinical benefit could take years. One example of that is *Sanfilppo Syndrome type C*, where a measurable decline in patient cognitive ability may take half a decade. Couple that with the amazing rarity of the disease (1 in 1,000,000 births) and that would make developing a therapy difficult indeed. I spent a lot of time working with Dr. Emil Kakkis to get new legislation passed to allow the use of surrogate endpoints for rare disease (they have historically only been used for cancer). The language was incorporated into a bill called ULTRA, which was later re-labeled FAST. In the Summer of 2012, the language from the FAST/ULTRA was incorporated into FDASIA which was a "must pass" bill that sailed through Congress.

In October 2012, I was awarded with one of the first annual *Rare Voice* awards. The biggest honor to me was having the award presented by John Crowley. At that meeting John told me that he was coming to Dallas and mentioned that he would like to come by our house and visit Reed. We learned at the summer 2012 MPS Society conference that, unfortunately for us, *Shire* would seek very young patients for the next phase of their study (so young, in fact, that they probably hadn't even been diagnosed yet).

In light of this fact, I went back to Doug McCarty, the researcher at *Nationwide Hospital* working on gene therapy and asked how I could help push their drug into the clinic. The hospital worked with Al Hawkins, a local entrepreneur, and licensed the technology for both MPS IIIA and IIIB to him. When John made it up to our house, he offered to take a look at the research in flight with me. I showed him the preliminary mouse data from the Dr. McCarty's gene therapy and he was impressed. I also asked if he would mentor Al with a short phone call and, as John does, he went above and beyond. John offered for me and Al to take a trip up to Amicus (Cranbury, NJ) and to make his team available to us. We met with John and many of his senior leaders and explained our plan for the company. *Abeona Therapeutics* was founded in March 2013 and will be focused on the sole goal of delivering systemic AAV9 gene therapy to children with MPS IIIA and IIIB. The outcome is uncertain but, as John said, we will beat nature but will we beat time?

That chapter of the story is yet to be written.

The Empowered Patient in the Health System of the Future

Frank Grossmann, Daniela M. Meier and Therese Stutz Steiger

> *You've got to do your own growing, no matter how tall your father was*
>
> IRISH SAYING

Abstract This chapter outlines the work of a Swiss-based organization which focuses on drug development for rare diseases as well as acting as a support hub for patients and their families. The organization operates as a matchmaker in an innovation network, promoted by its patient empowerment program, which motivates rare disease patients to break out of their isolation and be actively involved in their treatment. The chapter details their approach which proposes the active integration of patients as a full partners into the healthcare process. We include two brief case studies which illustrate how the organization works with patients.

Introduction

There are more than 6,000 diseases which are considered to be *rare*. A disease is considered rare if less than 1 person is affected per 2,000 inhabitants. By comparison, 88 people are affected by Type 1 diabetes and 5 persons in 2,000 suffer from cardiovascular disease. The 6,000 rare disorders are diverse in their symptoms, exist worldwide and nearly 1 in 12 people's lives will be affected (not including their relatives and friends). As there are at least 250 million affected globally, rare diseases as a whole can hardly be considered *rare*.

As each individual rare disease does not always call for huge quantities of drugs, they have not always been of major interest to the big pharmaceutical companies. Large pharmaceutical companies cannot always profitably produce and

---

F. Grossmann (✉) · D. M. Meier · T. S. Steiger
Research Foundation Orphanbiotec, Einsiedlerstrasse 31a, 8820 Wadenswil, Switzerland
e-mail: fg.foundation@orphanbiotec.com

commercialize small amounts of specific drugs across different countries. The problems associated with a rare disease, such as *Moyamoya* (which paralyses children due to stroke and keeps them disabled) are compounded by being ignored by pharmaceutical companies. Silje, 11 years of age from Norway, suffers from Moyamoya. Her mother told us that: *"After Slije was diagnosed with Moyamoya disease, it was a shock for us and a major crisis, as they made it clear, our child was very very sick, and there was no therapy"*. Ignoring such diseases is akin to children being abandoned by their parents. The term *orphan disease* is appropriate for such rare conditions and has been firmly established in recent years. Due to increasing intellectual property being derived from research on these disorders, this new knowledge could be used as a bridge for many other avenues of drug development. The research and best practice from rare diseases could be applied to other diseases and even widespread conditions (*fundamental diseases*).

Despite the plethora of definitions, the same challenge remains. It is envisaged that the global efforts of dedicated individuals and institutions (including patient organisations) would contribute to future research. This would facilitate a better understanding of the rare disease population leading to less suffering for the victims and ultimatley affordable therapies.

What Makes It So Difficult?

The chances of a timely diagnosis for orphan diseases unfortunately remain far too low. Few competence centers (Centers of Excellence) exist and patients must struggle to survive with little or no professional guidance. This scarcity affects many sufferers enormously— patients (and their families) feel extremely isolated due to their unfortunate diagnosis and additionally feel downtrodden and overwhelmed. They become even more isolated from the health system due to their predicament. Physicians are insufficiently trained and, even if a diagnosis is made, the patient and their family are often left to fend for themselves.

Wider society is still largely unaware of orphan diseases, which leads to misunderstanding and discrimination. Patient organizations operate with a narrow remit, often disconnected and independent from one another—knowledge sharing is not always as efficient as it could be. The focus for dedicated organizations is on their own disease and this results in knowledge silos— organizations do not always work together or network effectively to confront the main challenges concerning orphan diseases as a whole. The lack of this wider perspective means that issues are unable to be addressed jointly.

It seems logical that patients and their organizations work together and speak out with one united voice against political and business concerns. Umbrella organizations such as *EURORDIS* (a non-governmental patient-driven alliance of patient organizations in Europe), the *Allianz Chronischer Seltener Erkrankungen* (*ACHSE*) in Germany and the *Canadian Organization for Rare Disorders* (*CORD*) are well known and active in the field. Entities such as *Allianz Seltener*

Krankheiten-Schweiz (*ProRaris*), created as the umbrella patient organization for Switzerland in 2010, are still evolving and have yet to cooperate with others.

Orphanbiotec Foundation

The Orphanbiotec Foundation (*Research Foundation Orphanbiotec*) is a recognized humanitarian foundation (non-profit and tax-exempt) for people with rare diseases. Registered in Zurich, the Foundation's aims and objectives include: funding of research and development of new drugs for rare diseases, expansion of the necessary competence network, interchange of knowledge concerning rare diseases as a result of international partnerships, providing a platform to exchange knowledge and ideas for researchers, raising public awareness and mobilizing the active involvement of patients and their families. The Foundation is working on alternative solutions to finance research and the development of drugs for the medicines of tomorrow, in order to offer affordable and effective medicines.

The Engaged Patient: From Reactive to Proactive

Patients often struggle with being ignored by their healthcare systems, which can result in a passive and resigned attitude. It is no wonder that, over the last 40 years, healthcare systems around the world have nurtured amenable, cooperative patients who willingly receive and accept therapies and medications from their entrusted prescribing physician. This level of trust means that patients are viewed as *customers*, who are discouraged from disagreeing with healthcare professionals or having their own opinions concerning their own treatments. They have no voice and are asked to dutifully swallow their pills with no questions asked.

The Traditional Doctor-Patient Model

This classic top-down nature of the "doctor-patient" relationship does not necessarily work for orphan diseases. Things began to change as healthcare professionals, despite experiencing a massive information overload in general medical knowledge, still could not find appropriate diagnoses and answers for rare conditions. Rare disease patients and their families (customers) had to become more actively involved if they wanted to survive. As patients became more informed, they rapidly connected with others in a similar predicament, a situation which was enabled via the increased global growth of the internet. This enabled them to propose their own solutions and alternative therapies—they gave themselves a voice. Such innovative and necessary patient-led activities have made an important contribution to the increasing "democratization" of healthcare systems.

Costs Versus Need and Expectations

The prevalence of the internet and increasing levels of health literacy mean that patients are able to find out much more about their disease and can readily identify key experts for further advice. Under a traditional health system, the chances of a suitable rare disease treatment option are slim as governments and private institutions struggle with massive cost increases. The focus over the last 40 years has been more on profits—medical feasibility was the holy grail versus medical necessity based on the needs of the patients.

Various players (such as large pharmaceutical cooperations, hospitals and some specialists) have benefited from this legacy and ensured that its evolution would grow to be more complex and expensive. In the meanwhile, some governments have realized that urgent cost cutting is needed. Such a provision was seen in Germany in 2011 when the *Act on the Reform of the Market for Medicinal Products* (*Gesetz zur Neuordnung des Arzneimittelmarktes*—AMNOG), was passed, with the Parliament condemning expensive medicines. In the UK, age limits were set for treating certain diseases to the point of slashing certain therapies altogether (including some rare cancers). Nevertheless, the cost explosion due to the increasing aging population is expected to persist.

Is there still sufficient financial leeway for even more ailing and needy individuals? The few approved drugs after 30 years of legislation in the US (via the *Orphan Drug Act 1983*) are termed *"orphan"* as they only treat a single rare disease and are deemed disproportionately expensive to develop. The costs of the drugs often mean that healthcare providers, even in wealthy countries, refuse to pay for them. The fact remains that rare disease sufferers want and need a proper therapy but lack a resource base (including researchers and financiers) to meet their long term needs.

Switzerland as an Example

Switzerland illustrates a suitable example of the typical problems faced by rare disease patients and their families. The country has one of the most sophisticated health systems in Europe but still lacks a government-supported national action plan for rare diseases. By way of example, families of children who suffer from the disease *Niemann-Pick* (a rare and terminal neurodegenerative disease) currently need to submit a new request each year for reimbursement via social security (due to the fact that health insurers refuse to cover the cost of the life-extending and extremely-costly therapy). This reimbursement approval arrives every year and is often regarded as a "Second Christmas"—conversely, if it does not arrive, it is considered to be a "Death Sentence". This practice applies in Switzerland but is replicated in many other developed countries and demonstrates the potentially devastating ramifications of assessing patients by way of a simple cost-benefit calculation.

Beyond Traditional Approaches

Social Entrepreneurship

As a newcomer to the rare disease field, *Orphanbiotec* opted for a sustainable approach which attempted to challenge the existing structural problems of current systems. It operates under a holistic business model which combines the not-for-profit sector with social entrepreneurship. Not-for-profits (and charities) have traditionally possessed the necessary focus to raise funding to match targeted needs (gaps) which governments and business communities were unable to fulfill. These gaps (much like *orphan* diseases) were overlooked in favor of the more quick, profit-making, commercial interests. Encouraging continuous consumption over the past years warranted companies to assume the responsibility for their business operations. Some not-for-profit organizations have learnt from the commercial sector how to better produce their products and services in the most efficient way in order to help those in need. Thereby, a gradual convergence evolved between both sides: the notion of the responsible or so *Social Entrepreneur*.

Social Entrepreneurship is not an invention of our time as such types of businesses existed hundreds of years ago. On the Hawaiian Islands, the indigenous people always fished and harvested cooperatively by sharing their catch and harvest with one another. In 1860, we have the remarkable efforts of Florence Nightingale, who constantly fought the establishment and created the first professional training school for nurses in the UK (the exemplar for other schools around the world). In 2006, Prof. Muhammad Yunus (a supporter of Orphanbiotec's *Black Nose* campaign) was awarded the Nobel Peace Prize for his pioneering work on microfinance which has changed the lives of more than 100 million poor women worldwide. Such models of cooperative trade and responsible business practices form the basis of Orphanbiotec's work.

Combining Models

The Orphanbiotec business model assumes on the one side that our healthcare system today is currently not able to face the challenge of supplying therapies and cutting costs, and is being pushed to its financial limits. On top of this we have around 6,000 rare diseases awaiting modern therapeutic innovations. The traditional pharmaceutical industry must keep shareholders satisfied with respect to blockbuster drugs (medications with greater than $1 billion revenue annually) and profit expectations. Against such pursuits, in 2009, Orphanbiotec adopted a twin-pronged approach (in 2011, winning the SEIF-*Swiss Social Entrepreneurship Initiative & Foundation Start-Up Award* and Nominee of Swiss W.A. DeVigier Prize 2013):

- a not-for-profit foundation (a spin-off) which is in direct contact with patients, increasing their empowerment
- socially-responsible company which develops affordable therapies and medications for orphan diseases at a lower cost in a sustainable way (depicted in Fig. 1).

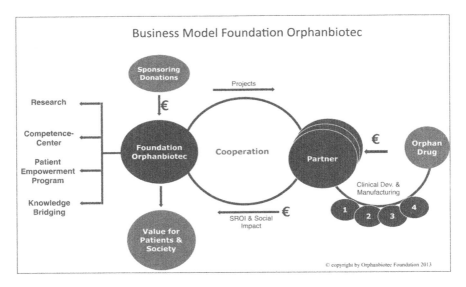

Fig. 1 Orphanbiotec foundation's business model

Comprehensive Action

New ideas and approaches to certain rare diseases are evaluated within the Foundation's competence center and, after a successful proof-of-concept stage, early financing of pre-clinical development is secured with the assistance of professional low-profit partners and responsible investors. The first "for profit" partner has been established and is now developing new and innovative medications with a focus on rare diseases. All for-profit players have limited return expectations and further projects will be financed and developed from the revenues of the initial project.

The collectively-developed medications for rare diseases remain financially-feasible for the payer thanks to this philanthropic approach and engagement. This approach is also economically viable to all partners therefore offering concrete opportunities to achieve mutual goals. The involved for-profit partner acts like a sustainable partner and pays an annual financial benefit back to the Foundation from the revenues of the projects. This is described as a *Social Return on Investment* (and can be seen in Fig. 1). This payment will be used to kick-start new projects creating a virtuous cycle that will lead to more promising avenues for rare diseases.

Holes in the Swiss Cheese

The Orphanbiotec Foundation was established in Switzerland—"the little island in the middle of Europe"—known not only for the highest per capita income and exceptionally high quality of life standards but also for having one of the most expensive healthcare systems in the world. The majority of patients in Switzerland have been over-indulged for a long time and politely await solutions from actors and service providers in the healthcare system. As this wait never ends, even for Swiss rare disease patients, Orphanbiotec decided to empower them.

Switzerland has nearly 500,000 patients affected by a rare disease. Other countries in Europe have earmarked legislation, action plans and some incentive models for rare diseases. In Switzerland, the government has only recently become aware of the *potential* issue of rare diseases due to the action of a sole member of Parliament. Orphanbiotec and its pro-bono staff work with persistence and commitment to navigate a carefully organized "Alpine route" (a trail through difficult terrain, sometimes with no obvious path) to ensure that rare diseases will no longer go unnoticed or remain invisible (like the famous holes of Swiss Emmental Cheese).

Towards Bottom-Up

A patient would traditionally wait for information from the doctor or health care system, but a person with a rare disease cannot always afford to wait. Instead they have to rely on new social media, patient and personal networks and the Internet—without these outlets, they may never receive answers and feedbacks. Many patients on the pathway to a diagnosis and looking for a treatment have gained a wealth of experience compared to conventionally trained doctors—they are often more expert in their condition than their doctor. If a patient is active and empowered, doctors can learn and profit from their experiences. Who though empowers whom? (Lancet 2012).

Patient engagement and integration is a key factor for developing new therapies for rare diseases. Not only does valuable "know-how" flow inwards, but patients also feel more understood and are motivated to work together to find a solution. Foundation Orphanbiotec understands the importance of patient engagement and integration and applies this to the affected patients (and their families) on two levels. On the first level, their experiences and needs are heard within the competence centers and patient advocates are involved in key steps of new drug development projects. Secondly, the Foundation actively connects those affected individuals and encourages collaboration among themselves by way of its patient empowerment program "*Little Orphan Elf*" (see Fig. 2). The Program spans the following aspects:

- self-management program to help gain self-confidence, new abilities and strategies
- how to control and live with symptoms, fear and stress
- motivation training to learn self-efficacy, decision-making and action planning
- day-to-day management and solving problems
- how to maintain and increase quality of life.

In order to encourage patients' participation, Orphanbiotec hosts an approved online forum. The forum is a way for participants to share ideas and experiences—this creates a community of trust and a self-initiative (evidenced by an increasing number of visitors and active participants). The forum's support is accompanied by the "*Little Orphan Elf*" program which relies on the inherent ability of people to trust and engage with each other.

Patient Empowerment

How to Become an Elf

Human nature encourages us to meet face-to-face in order to exchange experiences with one another. For people living with a rare disease, this is often limited due to health conditions. The "Little Orphan Elf" (*Elfen Helfen*® in German) patient empowerment program helps to overcome these hurdles by offering not only free specialized transportation to their social occasions. The program is set up as a casual and supported event that is lively and light-hearted in nature. Those affected by rare diseases can meet up with one another, exchange ideas with experts and motivate each other, irrespective of each other's unique health conditions.

The program also targets close family members and friends whose own lives have been affected by rare diseases and allows them to listen, learn and gain support. "Little Orphan Elf" organizes social occasions where each affected person is assisted by an *Elf* (a person who listens, supports and engages participants). Everyone who joins the program can turn into an *Elf* when they are active and supportive to others. It empowers people from being passively affected to being actively engaged in the healing process of others. Being an Elf is therefore a pivotal part in the healing and support process for all.

The patient empowerment program has been financed through private donations, corporate and social responsibility (CSR) partners. It is expected to include EVIVO-style (www.evivo.ch) patient-to-patient coaching in the future. "Little Orphan Elf's" initiative to start coaching is inspired by successful experiences in increasing the quality of life for other disease areas such as diabetes, obesity, rheumatism and mental illness. Many international studies show that a patient can and must take an active role in their own health management in order to increase their health improvements (The Health Foundation 2011; Osborn et al. 2010; Osborn and Squires 2011) (the core role of the Health Care Management Program, see Fig. 2, the Empowerment of the Patient).

Patients can be Teachers

If the affected patient and their family are motivated and ready to join the "Little Orphan Elf" Program, they must acknowledge that they play an active and direct role and should be involved in their own health improvement. The goal is to motivate

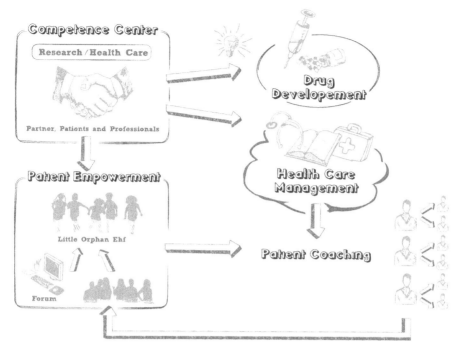

© copyright by Orphanbiotec Foundation 2013

Fig. 2 Health care management and patient empowerment

them to participate and help grow a "Knowledge Management Exchange" with other affected individuals, to gather their experiences and share them. Even though the illnesses are different, they bring similar struggles. It makes little sense to collect and develop 6,000 different ideas and solutions as many patients have the same questions and challenges (Osborn et al. 2010; Osborne et al. 2004; Lorig et al. 2012).

For example, patients with cancer can share tips about dealing with side effects of chemotherapy and people with pain symptoms can talk about what distracts them from the pain. The same holds true for many families whose children are confined to a wheelchair who can share stories of how they cope with the same daily problems (Osborn et al. 2010; Osborne et al. 2004). A motivated and mobilized patient can decide to take the next step to develop into a *Coach* in order to impart their experiences with fellow affected patients. If affected individuals can be directly mobilized and engaged, this increases their chances of experiencing an improvement in their quality of life, due to the improved mental, emotional and physical benefits of engaging in this type of activity (The Health Foundation 2011; Osborn et al. 2010; The Health Foundation Evidence 2011; The Health Foundation Snapshot 2011). The Program, and accompanying initiatives, has no additional expenditures for the healthcare system (which is already struggling with their increasing costs).

Need as a Chance

Based on our own experiences within Switzerland, winning over patients to empower themselves is a new and difficult approach as the level and quality of conventional care is high and patients bestow a lot of faith in the healthcare system. However, this care is not holistic since it is not always focused on patients' personalized needs, their wishes and specific opportunities for improvement. It is not designed to listen to the patients' voice (essential in our view), nor does it ask the family or social network to support their efforts; the emotional situation of the patient is not considered.

We would advocate that patients need to be involved in the process of changing the current way of care delivery by interacting with all stakeholders to get their voice heard. Too often they discover that their fellow sufferers reside in another country or on another continent. When many individuals are engaged a group can be formed and can gain a powerful voice. By making use of the online forum, patients in Europe can network across borders; via the "*Little Orphan Elf*" program, a shift in mindsets can be achieved, empowering patients and families to participate in valuable knowledge sharing.

Everyone Else's Platform

Awareness Campaigns

Empowering affected societies is an important milestone for the Foundation in order to combat the various issues associated with rare and orphan diseases. With the help of several *pro bono* partners, the concept for an awareness campaign was developed (one that was easy to understand and easy to translate across the globe). The "*Black Nose*" campaign (www.black-nose.org) gives those affected by rare diseases a voice and a face and encourages everyone directly affected (and those who are not) the chance to get easily involved. The international campaign serves to impart awareness and knowledge on rare diseases and stimulate broad community involvement.

Bridging and Transmission

The Foundation connects various key stakeholders, does not compete while adding value and remains unique with this approach. National organizations and specialists will be deeply involved due to their know-how and involvement in the projects. In addition to the coaching program, the foundation aims to meet the diverse needs of people with rare diseases. It is hoped that as the program is shown to be successful, it will be rolled out to other countries and further adapted to fit local customs and needs.

Re-empowering those affected pre-supposes that this provides the necessary energy to enable those affected to move out of isolation. Only then can true re-empowerment be addressed. Patients' integration and mobilisation allows them to be autonomous. Providing help to others can be the driving force that can be passed from one person to the next. Fittingly, the teacher can teach others. A resilience study with our reasearch partner will follow and support our coaching program. This will ensure the quality of the program, improve the quality of life for patients and bring cost savings to the healthcare and social systems. We will also involve some experts who can explain how the program works.

Example 1

Engaging with others provides a positive twist to their own struggle

Daniela M Meier

Dr Daniela M Meier is a co-partner of Foundation Orphanbiotec, which implements and develops patient-to-patient coaching.

The Orphanbiotec Foundation is financed by donors and sponsors and is supported by numerous volunteers who work enthusiastically for its various initiatives. What motivates them to get involved and become active? People who were affected by an orphan disease —as a patient or as a relative—would usually confirm that their physicians could not cope with their situation, that even in a hospital, medical specialists were unable to provide adequate care and that none of the patient organisations were an ideal fit for their case. Having struggled through their issues (like a lone fighter) they are keen to share their experience and know-how and prevent others who are affected by an orphan disease from re-living their seemingly endless circle of frustration. They find it fulfilling to support other affected people because this engagement gives meaning and a positive outcome to their own struggle against the disease. This last aspect can be crucial in the long-term healing process of a patient because it empowers the person and can advance them from a "weak or sick" individual to the informed person.

Pancreatic cancer

A businesswoman was diagnosed with pancreatic cancer. She was not only removed from her useful working life during the treatments but her private life was also disrupted when she became dependent on the care of her family members and friends. Her daily life became impersonal, driven by medical treatments and healthcare professionals invading her private space. While recovering from this life-threatening disease, the businesswoman also had to find a way to handle the drastic departure from her prior life.

During the healing process, she struggled to find a way to integrate the disease into her personal life—not just in the sense to accept it physically but also to accept the illness emotionally and to accept how it had changed her. Since such a process may take several years, she felt she needed support that went far beyond the usual therapies offered by the healthcare system which is constructed in such a way to bring patients back to their prior life as quickly as possible. There is little support to deal with collateral difficulties such as pressures from the employer or constraints from the health insurance provider. In the business world in particular, employees are expected to perform and to show strength, a disease (especially an orphan disease that is not commonly understood) becomes a stigma; it requires extra effort to overcome the fear of being fired from one's job under the pretext that one is unable to accomplish the tasks required.

These circumstances could be turned into a positive situation if the businesswoman was able to receive long-term support to accept her new circumstances and re-adjust her life. As a result, she could then reset her priorities regarding a work-life-balance and would be able to re-orientate herself in her professional life. Having overcome the difficulties of her disease, the businesswoman would now be an ideal coach for a patient going through a similar predicament. As a coach, she could meet patients emphathetically and would be able to find the correct words easily (perhaps even daring to make jokes, something which a healthcare professional would not be in a position to do) and suggest "out-of-the-box", non-orthodox, solutions. In order to make this possible, she would need some backing to overcome the fear that she is still regarded as "sick" in the working world. Also, she would require some appropriate training as a patient-to-patient coach so she could then get in contact with other patients. In an environment of like-minded people, she would be even more motivated to get involved. The businesswoman that has been described did indeed manage to integrate that chapter of disease into her life and is now ready to share her experience for the benefit of others via Orphanbiotec.

Example 2

Living with Osteogenesis Imperfecta

Dr Therese Stutz Steiger

I am a patient living with *Osteogenesis Imperfecta* (OI, commonly known as "brittle bone disease" or *Lobstein syndrome*), a congenital disorder. It affected me in childhood the most (I had about 20 bone fractures) as I was largely wheelchair dependent, was always reliant on the help of parents and peers and had limited career choices. During puberty, I had to have large operations mainly in the summer holidays—thanks to those, I am now able to walk. Thanks to the tireless efforts of my parents, I was able to go through the regular school system and eventually study medicine. The dissertation about my own illness aimed to bring together stakeholders in Switzerland and, in 1986, I founded a self-help association—the *Schweizerischen Vereinigung Osteogenesis imperfecta* (*SVOI*), of which I am currently the co-President. We try to set up interdisciplinary OI Centers of Excellence in major Swiss hospitals but the health system, due to its federal structure, unfortunately is not very helpful.

OI biographies typically ebb and flow: phases of relative well-being will be abruptly interrupted when a bone breaks; basically, we have to expect it at any time. Since many of us are walking badly (or not at all) we have to, within varying degrees, live our lives dependent on external aid and support. We are also more often tired than non-sufferers, many of us work only part-time and some also experience depression. The board of *SVOI* offers solidarity to people affected and offers a wealth of knowledge to newly diagnosed patients and their families.

Since 2012, I have been on the board of *ProRaris* which acts as a representative and link between the many small patient organisations. My personal experiences living with this condition, studying and working as a medical doctor and my many years of service at the *Federal Office of Public Health* have convinced me that Orphanbiotec's *Little Orphan Elf* program is important and can form an essential support structure for those directly affected by OI.

The Future is Holistic

Recovering the Health System

We believe that the healthcare system is currently recovering in the infirmary—it will one day recover when we are able to understand what made it ill in the first place. We trust that the current situation of stakeholders organizing the development of therapies with profit maximization in mind (and without the vital input of patients) will be confined to the past. The complex challenges needed for this scenario should be regarded as an opportunity—it is crucial to understand the requirements of the people affected. Rare diseases, with increased and supportive research, can be used as a vehicle for good research practice in relation to medicine and pharmaceutical development.

There is a need for healthcare professionals to recognize that the patient can turn into the best partner for successful treatment—a co-therapist, if this relationship is considered to be a true partnership. As the patient is in a position to make a significant contribution to his or her own quality of health and life, they should therefore be suitably empowered. We need to rethink the system and have the courage to explore new solutions. Old business models should be rethought and changed if deemed to be feasible. Combining sustainable entrepreneurship with a philanthropic approach will lead to innovative solutions and a better future for patients, especially those who suffer from a rare disease, as these patients have been underserved by current healthcare systems.

Conclusions

Orphanbiotec provides innovative support with a holistic approach at a time when new solutions are needed. Patients (as healthcare partners) as well as organizations will profit from leveraging international networks for effective collaboration, thereby providing know-how for all stakeholders (including patient advocacy groups). The bottom-up approach can be used to rethink old habits and conventions, to bring stakeholders together and support sustainable success. A social entrepreneurial approach within continental Europe is still unfamiliar to current healthcare systems, patients, organizations and other partners. We are confident that our new approach also actively pursues and secures funding beyond this continent.

Acknowledgments Special thanks to Daniel Des Roches (Foundation President and Head of the Foundation Board) for his special support, Stephanie Braghero who helped with the translation from its original German content and Jorge Romero for his innovative feedback.

Dedication

This chapter is dedicated to Frank Grossmann's mother, Ilse, who struggled with a rare sarcoma cancer. She enjoyed every moment of her life and personally epitomised a special form of patient empowerment which gave her confidence and the ability to fight the disease. She left us in August 2012.

References

Lancet. Patient empowerment–who empowers whom? Lancet, 2012; 5;379(9827):1677.
Lorig, K., Holman, H., Sobel, D., Laurent, D., González, V. and Minor, M. Living a Healthy Life with Chronic Conditions, 4th Edition, Bull Publishing. 2012.
Osborn, C.Y., Cavanaugh, K., Wallston, K.A. and Rothman, R.L. Self-Efficacy Links Health Literacy and Numeracy to Glycemic Control, Journal of Health Communication: International Perspectives. 2010;15:S2: 146-158.
Osborn, R. and Squires, D. International perspectives on patient engagement: results from the 2011 Commonwealth Fund Survey. J Ambul Care Manage. 2012;35(2):118–218.
Osborne RH, Spinks JM, Wicks IP. Patient education and self-management programs in arthritis. Bone Joint Disord: Prev Control, Med J Aust. 2004;180(5 Suppl):S23–6.
The Health Foundation. Can patients be teachers? 2011. http://www.health.org.uk/public/cms/75/76/313/2809/Can%20patients%20be%20teachers.pdf?realName=br0eQj.pdf, Last Acccessed 1 March 2013.
The Health Foundation. Evidence: Helping people help themselves. 2011. http://www.health.org.uk/public/cms/75/76/313/2434/Helping%20people%20help%20themselves.pdf?realName=8mh12J.pdf, Last Acccessed 1 March 2013.
The Health Foundation. Snapshot: Co-creating Health. 2011. http://www.health.org.uk/public/cms/75/76/313/2339/Co-Creating%20Health%20snapshot%20publication.pdf?realName=UtThFG.pdf, Last Acccessed 1 March 2013.

Vignette: The Journey of a Lifetime

Deb Purcell

Imagine the craziest roller coaster you've ever been on, from the very highest peak to the lowest valley. That roller coaster is what my life has felt like being a mom to a child with MPS II. Incredible highs and terrifying lows.

We began this journey with MPS II (also known as Hunter Syndrome) on February 14, 2006. The appointment should have been routine. All I was looking for was a doctor whose values more closely resembled ours. However, there was nothing routine about the appointment, nor our life since. Not wanting to

---

D. Purcell (✉)
1449 Chamberlain Dr, North Vancouver, BC V7K 1P8, Canada
e-mail: purcelldeb@gmail.com

say anything without confirmation, the doctor only gave us the words, 'possible enlarged liver' and 'possible storage disease.' Two days later, we were asked to bring Trey in for an abdominal ultrasound. A day after that - a Friday- a call came from the pediatrician telling us that Trey did in fact have an enlarged liver and spleen, and that we had an appointment on Monday at BC Children's Hospital. The fear Ryan and I shared was palpable; we could barely make eye contact during those six days between Valentine's Day and the following Monday at Children's Hospital. Eye contact acknowledged fear, which led to tears, and then rationalizations that maybe this wasn't as bad as we imagined. But it was.

That Monday, the day we first heard the word Mucopolysaccharidosis, was the end of my life. It's ironically funny saying that because I'm still here, but I did feel the end of the world, the end of my life, on that day. Speaking of our beautiful 23-month old child, we were told: clawed hands, coarse facial features, stunted growth, possible mental retardation, early death. MPS is a rare, genetic, and progressive disease in which children are missing an enzyme necessary to breakdown cellular waste.

The journey from then until now has been epic. The journey of a lifetime.

The difference between the parent of a child with a progressive disease and the parent of a typical child is that we face death regularly. How can that not be incredible and terrifying?

For months after the diagnosis, the only thing that kept Ryan and I going was Trey, who turned two on March 1, 2006, and Avery, who was three months old. We cried ourselves to sleep, we cried when we woke up in the night, and we cried through the smiles we tried to put on for our kids during the day. I remember walking out of the kitchen or bathroom or bedroom to find Ryan curled up and sobbing in a corner of the house. How do you live, when your son is going to die? That was seven years ago now though, and life has continued.

During the next seven years, I committed countless hours to advocate for Trey. I became an MPS expert in ophthalmology, cardiology, genetics, biochemical disease, plastic surgery, rheumatology, anesthesiology, general surgery, otolaryngology, pulmonology, neurosurgery, and orthopedics.

In order to gain access in Canada to the first ever treatment for MPS II, a weekly four hour intravenous (IV) Enzyme Replacement Therapy called Elaprase that was approved by the FDA in July of 2006, I gained extensive knowledge about the definitions of orphan drug and rare disease and debates surrounding those definitions. I learned about pharmaceutical companies, research and development, tax incentives, healthcare and drug approval processes in different countries, federal and provincial government politics, pharmaceutical company politics, and the impact of media on the government and pharmaceutical companies. I learned how to fight - for my child's life. This fight resulted in Trey becoming the first boy in Canada to receive Elaprase who was not in a clinical trial, on February 5, 2007. It's a date I'll never forget; enzyme - the enzyme Trey so desperately needed - was floating around in his precious body. Elation. Relief. Hope.

I also kept tabs on a developing intrathecal trial at the University of North Carolina (UNC) that would inject the enzyme Trey is missing into his central nervous system via lumbar puncture or port-a-cath. It could treat the progressive brain disease in MPS II that IV Elaprase could not. You see, MPS II has a range of severities. About 80% of children with MPS II end up with cognitive decline, starting between the ages of 18 months-6 years, and ending with death between 7–20 years of age. Imagine watching all the skills your child learned - talking, walking, running, eating, smiling, calling you 'mom'—disappearing. It's one of the worst kinds of torture I can imagine. Kids with typical cognition live a heck of a lot longer (20–60 years) and can go to University, travel, work, get married etc. Hope.

While I waited to find out which type of MPS II Trey had, I kept busy. I prayed. I had another baby. With the help of many, I raised over a quarter of a million dollars for research. In 2010, I was part of a pilot program to train parents to do the weekly Elaprase infusions at home, and I've been doing Trey's infusions ever since. Every six months we flew to North Carolina to have Trey's IQ tested to hopefully give us a sooner rather than later indication of which type of MPS II Trey had. Trey had multiple surgeries during this time and even more appointments. I spoke with every newly diagnosed MPS II parent who called and I responded to every email. If there was a chance to raise awareness or funds for Trey's disease, I took it. I couldn't not do everything possible to save my son and to help other families who were going through the same as we did.

I waited, living in limbo, for four and a half years. I tried my very hardest to appreciate every moment and to stay grounded, and there were moments, but looking back, those moments were cut too short by my mind. With every success and failure of Trey's—learning to ride his bike, learning to put in his own hearing aids, fighting with his brother—just watching him sleep or eat, I was haunted by life and death. Did the fact that he just focused and finished that puzzle mean his brain is okay and he is going to live? Is the reason he can't cut easily with scissors because he has joint stiffness or because his brain is affected and he is going to die? Is there going to be a day when I'm never going to see that smile again? Will Avery remember this moment, when his brother was okay? That little thought at the back of my mind would almost always sneak in. Rare were the moments that I could appreciate them for what they were. My entire world was consumed by MPS.

In August of 2010 I found out once and for all that Trey's brain was affected. That was the second steepest hill on the ride, following diagnosis. After picking myself up from that low, I got back on board with hope and began climbing the hill again: the intrathecal clinical trial was open and there was still a chance to save Trey's life.

How to describe waiting to get into the trial... I did not sleep much. I did not eat much. I was in constant fight or flight mode, much the same as I had been since Trey's diagnosis. I planned how often to call the doctors and nurses involved in the trial, to give them the impression that I would fight tooth and nail for my son, but not call them so much that I would irritate them. I also called them to get updates on the trial because the criteria changed constantly. At one point, a few weeks after being told Trey was next in line to be assessed for the trial (to save his life)—celebration!—I was told that because the criteria had changed, there was no longer a line, and they would let me know if they were interested in Trey. Panic. I researched hydrocephalus incessantly because an intracranial pressure above thirty would exclude Trey from the trial. I wondered if Trey's meltdowns were due to cognitive decline—which I had now adjusted to and which the trial I counted on to fix—or hydrocephalus, which meant death. When an IQ below 50 was part of the exclusion criteria, I hired daily tutors and wouldn't let our kids watch TV, to make sure Trey's IQ would qualify. And this is just the 'book-friendly' version.

Trey did end up qualifying for the trial on August 19, 2011, another date I'll never forget, and he began monthly intrathecal doses at UNC that October. I will never again feel such relief. Never.

Although the trial has had its own ups and downs, and not minor ones (imagine emergency surgery to remove an infected port that if not removed soon enough could lead to meningitis and death, being away from your three and five year old children ten days per month, monthly general anesthetics and lumbar punctures), it is the first time since Trey's diagnosis that I have been able to fully enjoy moments for what they are. I can breathe again. I am not sure if it's because I am no longer fighting death as intensely (I still have smaller battles such as refusing to pin Trey down kicking and screaming for tests I don't see as necessary or advocating for doctors not to give Trey drugs I don't think he needs), because I no longer have to wait, or because Trey is not dying and losing everything in front of my eyes, but I finally feel free and alive again. It is wonderful. Trey may not be a typical child, but I'm not attached to typical. Typical is overrated. I'm attached to life.

Having been on this roller coaster ride many times now, I know another big downward spiral will come, whether with MPS or something else. Smaller curves continue to be thrown at us, like Sadie, our daughter, thinking I don't love her because I go away so much and Avery, our other son, feeling left out because he doesn't get as much attention as Trey (1-on-1 trips to UNC with Ryan and I, speech therapy, education assistants, occupational therapy etc.), but I have realized that this roller coaster is life. In search for some way to make the lows less terrifying and intense, I have found yoga and learned to meditate. So, while I am at the top for the moment, I am taking the opportunity to find steadiness, calm, and mental clarity, so that the next time the ride takes a big dip, I can continue to breath and hopefully enjoy the ride. So far, so good.

Personalized Medicine: A Cautionary Tale or Instructional Epic

Dorothy Weinstein

> *It is more important to know what sort of person has a disease than to know what sort of disease a person has*
>
> HIPPOCRATES

Abstract With the twenty-first century just beginning, the practice of medicine would forever change as the successful sequencing of the human genome was completed. This blockbuster discovery reset the landscape of medicine and the provision of health care. Personalized medicine once based on best practices and characterized by a constructive and caring doctor/patient relationship now includes a vast frontier of treatment options based on an individual's genetic blueprint. Patient-centered outcomes research and renewed interest in patient empowerment are also components to personalized medicine where genetics leads the way. Personalized medicine will always remain a dynamic practice as science moves forward. For it to succeed, personalized medicine must be crafted in the best interests of the patient within the context of our current health policy.

Introduction

This chapter discusses the historical underpinnings of the concept of personalized medicine and the health policies that govern or contribute to its practice. The term was coined in the early part of the twenty-first century when successful sequencing of the human genome provided the knowledge base on which to design individualized treatment plans. Genomic advances have led to breakthroughs in diagnosing rare diseases and in designing treatment approaches to better suit the medical conditions and the individual needs of patients.

D. Weinstein (✉)
9705 Carmel Court, Bethesda, MD 20817, USA
e-mail: dorothyweinstein@juno.com

Personalized medicine has broadened to include more than genomic information. Patient engagement, and the usefulness and appropriateness of evidence based medicine, narrative based medicine, and anticipatory medicine are important considerations in the management of patients with rare diseases. Finally, this chapter focuses on how public policy can help to foster the practice of personalized medicine and includes an overview of health care policy as it relates to patients with rare diseases.

Personalized Medicine: The Prologue

Personalized Medicine: we all have an idea of what that means. Try defining it yourself, and then look it up. Depending on your source, you will discover a vast array of credible definitions. With rapid advances in biomedical science, the definition continues to shift as the latest scientific knowledge is quickly put into practice. The concept of personalized medicine was a natural result of early efforts to sequence the human genome. The Human Genome Project began formally as a U.S. government initiative in October 1990 with the National Institutes of Health (NIH) as the lead agency. The Department of Energy (DOE), seeking data on protecting the genome from potential gene-mutating effects of radiation, was an early partner. The project immediately welcomed international collaborators and was eventually renamed the International Human Genome Sequencing Consortium. The mission of the Consortium was an inclusive worldwide effort aimed at understanding molecular heritage. The consortium's dual inseparable goals were to learn the order of the DNA in the human genome and then to locate the actual genes within the DNA structure. The NIH collaborative project required that all sequencing information be "freely and publicly available within 24 h of its assembly." This founding principle guaranteed broad access for any and all scientists—in academia and in industry.[1]

Private enterprise jumped into the sequencing effort around 1992. Craig Venter, a key NIH scientist, left the federal agency and established the nonprofit Institute for Genomic Research. In 1998, Venter founded Celera Genomics, a direct competitor to the public collaborative gene sequencing effort. Venter believed he had invented a superior technique that would more efficiently sequence the human genome. This technique, "shot-gun sequencing," was instrumental in Celera's success and in catching up to the collaborative public effort. In 2001, almost in a dead heat, the NIH and Celera independently announced the completion of their human genome sequencing. Each had sequenced 90 percent of all three billion base pairs in the human genome (Chial 2008). Today, researchers remain at work identifying

---

[1] National Library of Medicine. Genetics home reference: human genome sequencing project. Lister Hill National Center for Biomedical Communications, U.S. Library of Medicine, NIH, DHHS, US. gov, http://ghr.nlm.nih.g, 26 Nov 2012.

individual genes and the biological underpinnings of what controls each strand of DNA. The latest research findings estimate that any given human genome contains up to 30,000 genes.

The Human Genome Project began as a mind-boggling adventure in biomedical research whose mission seemed sky-high and even unattainable. As the research grew more promising and the discoveries continued, a new era in medicine was heralded, and personalized medicine was born. As a consequence of the human genome research, early definitions of personalized medicine were founded on the ability to customize treatment based solely on an individual's genetic make-up (Ginsburg and Willard 2009). Sources all over the world adopted this concept, ranging from the popular common language of Wikipedia to the U.S. Government's more formal definition. Although Wikipedia's definition admittedly is less sophisticated than others, the idea of personalized medicine, regardless of source, is founded on the premise of care based on an individual patient's genomic history.

In its familiar visual depiction as a double-stranded helix, the genome is of remarkable complexity. As we all learned in science class, a gene is the basic physical and functional unit of heredity. Genes are made up of DNA or deoxyribonucleic acid. Human DNA consists of about three billion bases or protein combinations, and more than 99 % of DNA is the same in all people. The remaining DNA (less than one percent) is what makes each of us unique, singular, and different.[2] The knowledge we gained (and continue to gain) from sequencing the human genome is one of the greatest advances in the history of science. Among the desired applications is the identification of specific genes that cause disease and the ability to use this genetic information to design personal health care treatment plans. We now know that rare diseases may be due to errors in one or more genes and in their interactions. In order for personalized medicine to work, knowledge of individual genetic information and the latest research must be combined with the interests and involvement of the actual patient. Personalized medicine is now defined in much more robust and comprehensive terms and includes a broader scope of knowledge about the patient, not only in terms of laboratory data but also in terms of patient preferences.

A report by the Duke University Center for Personalized Medicine defines the practice as "a rapidly advancing field of health care that promises greater precision and effectiveness than traditional medicine because it is informed by each person's unique clinical, social, genetic, genomic, and environmental information." The report continues by stating that, personalized medicine, "takes an integrated, coordinated, evidence-based approach to individualizing patient care across the continuum from health to disease".[3] The Duke report incorporates humanizing elements

---

[2] National Library of Medicine. Genetics Home Reference: What is DNA? Lister Hill National Center for Biomedical Communications, U.S. Library of Medicine, NIH, DHHS, US.govhttp://ghr.nlm.nih.g, 26 Nov 2012.

[3] Duke Medicine; Duke University Health Systems; Duke University Center for Personalized Medicine, copyright 2004-2010; www/dukepersonalizedmedicine.org. Accessed 2012.

which are as important to treatment design as genomic information. Personalized medicine has evolved from a straightforward genomic treatment approach to one that incorporates a diverse set of tools to uniquely customize a treatment plan. Patients with rare conditions are among the populations to benefit from this.

Rare Disease: A Twenty-First Century Rewrite

Rare disease, by definition, is bounded by the dual complications of pinpointing accurate diagnosis and providing successful treatment. With the tools of genomics, personalized medicine can now truly work to capture what is in the best interest of the patient with a rare disease. The National Organization of Rare Disorders (NORD) estimates that there are approximately 30 million Americans who live with one or more of the documented 7,000 rare diseases. In health policy, a rare disease or condition is defined as one affecting fewer than 200,000 people.[4] The President's Council on Advisors on Science and Technology suggests classification of "individuals into subpopulations that differ in their susceptibility to a particular disease or their response to a specific treatment".[5] Applying this idea, rare diseases can now be classified or thought of in different ways that may provide more or different information for treatment design. Personalized medicine is important for all patient populations with rare conditions. With research providing molecular information on a more granular level, those suffering with rare conditions may be the greatest beneficiaries.

We know now that individuals may react differently to medications based on their genetic makeup. Physicians can predict how a patient will respond to a certain drug and then, customize the treatment regimen using drugs and dosages most suitable for that person. In fact, the Food and Drug Administration now provides genetic information on labels of more than 100 drugs.[6] Application of the information garnered from genome sequencing has not only been a critical element in therapeutics but, equally important, in diagnostics. In fact, the two go hand in hand, since accurate diagnosis is important to designing a successful therapeutic plan. Biomarkers, a wide-ranging group of indicators, are the instrument for collecting key biological information on which to base diagnoses and treatment design.

Biomarkers can objectively measure and evaluate normal to pathogenic biological processes, disease susceptibility, exposure to dangerous toxins or diseases, and responses to therapeutic interventions. They come in many different forms including enzymes, vitamins, lipids, antibodies, sugars, metabolites, microorganisms, amino

---

[4] National Organization for Rare Disorders, copyright 2013, www.rarediseases.org.

[5] President's Council on Advisors on Science and Technology Report to the President on Propelling Innovation in Drug Discovery, Development, and Evaluation.

[6] Research! America. Genomics Research: Transforming Health and Powering the Bioeconomy.

acids, and breakdown products from proteins.[7] The list is long and evolving. New biomarkers are being discovered all the time. Biomarkers have enhanced the ability to diagnose rare disease, many times shoring up medical conjecture with absolute findings. While not diagnostically perfect, a biomarker's ability to make more accurate diagnoses is a critical tool in crafting a treatment approach. Historically, rare diseases have often defied diagnosis. With greater diagnostic certainty, personalized medicine can now be put into play for the patient with a rare condition. Genomic research has identified specific sites where disorders originate opening a new frontier of gene therapy.

A report from the Institute of Medicine states that, "Greater knowledge of how diseases work at the genetic and molecular level has allowed researchers to pursue new targets for therapy and better predict how certain biopharmaceuticals will affect specific groups of people" (Field and Boat 2010). Patients with rare diseases can take advantage of individually designed biopharmaceutical treatment. The Tufts Center for the Study of Drug Development found that 94 % of surveyed biopharmaceutical companies are investing in personalized medicine, and 100 % of those companies surveyed are using biomarkers in the research of new compounds. While this research requires large up-front investments in new tools and training, the research has developed targeted treatment options for patients with rare diseases. Most recently, new personalized genomic therapies are now being used to treat some forms of lung cancer, melanoma, and cystic fibrosis.[8] So, as pharmaceutical companies develop the products to personalize care on a genomic level, providers and patients have the opportunity and challenge to design treatment plans based not only on genomics, but also on the patient's choices.

Patient Engagement: Primer or Self-Help Guide

Along with genomics, treatment plans are now personalized with a variety of meritorious approaches worthy of consideration. In this section, I discuss some of the most current developments in patient treatment design including the use of such tools as evidence based, narrative based, and anticipatory medicine. Evidence based medicine (EBM) is a much debated approach in today's health care delivery. Often misunderstood, its proper application can help build a solid treatment plan. Evidence based medicine was defined by Canadian physician and early proponent, David Sackett who stated that, "Evidence based medicine is the conscientious, explicit, and judicious use of current best evidence in making decisions about the

---

[7] Debenham P. *Biomarkers and Gene Tests—A Personal Insight*, p. 2 Paul Debenham. http://webarchive.nationalarchives.gov.uk/20061009085702/hgc.gov.uk/client/content.asp?contentid=677, August 5, 2007.

[8] Scott R. Bringing Metcalfe's Law to Genomic Medicine. Harvard Personalized Medicine Conference. Harvard Personalized Medicine Conference, Showcases Progress in the Field. Randy Scott: Bringing Metcalfe's Law to Genomic Medicine.

care of individual patients." He went on to add, "The practice of evidence based medicine means integrating individual clinical expertise with the best available external clinical evidence from systematic research" (BMJ 1996). If interpreted and applied correctly, available research can be a useful tool in personalizing the care of the patient with a rare disease.

To better understand the use of evidence based medicine, comparative effectiveness research must be understood. Comparative effectiveness research (CER) is defined as the conduct and synthesis of systematic research; comparing different interventions and strategies to prevent, diagnose, treat and monitor health conditions.[9] While cost is a vital consideration in health care decision-making, comparative effectiveness research does not compare the costs of treatments.

Critics have voiced their concerns that comparative effectiveness research is only useful for the "average" patient with conditions that are within a common spectrum, but not applicable to patients with rare conditions. Yet, the success of genomic medicine depends on quality comparative effectiveness research. Such research is essential in order to compare clinical outcomes from genome-based approaches to traditional non-genome-based approaches. Whether the disease or condition is common or rare, comparative effectiveness research is a useful tool[10] and the basis on which evidence based medicine is practiced. One of the greatest obstacles to the meaningful use of personalized medicine is the lack of adequately designed studies assessing clinical utility (Garber and Tunis 2009). Well-designed research comparing outcomes is necessary for informing strong evidence based medicine.

One reason medicine is a fascinating profession is precisely because it is not just a quantitative science. Where science meets art is in our humanity. As the twenty-first century sweeps us up in technological wonderment, medicine continues to grapple with the incorporation of human nature into the rubric of health care treatment design. The age-old question of what defines us has no absolute answer and remains a subject of continued debate. What makes us human, alive, sentient beings is surely a topic that ancient Greeks discussed in their symposia and that mankind will always debate. The very foundation of humanity is in our own uniqueness: our own narrative, our own story, our own voice. Ira Gershwin brilliantly summed this up in his song, *Let's Call the Whole Thing Off*, with the lyrics, "You say potato and I say potato." Our nuances and subtleties, the "poetry" (if you will) of our lives makes us unique. We live within our own timelines filled with continuing stories that lay a historical foundation. The very science of medicine is confronted with the very humanity of its application.

There is a rising movement in health care delivery called narrative based medicine (NBM), an effort to "re-humanize" medicine. Drs. Johna and Rahman state, "The human capacity to understand the meaning and significance of stories is

---

[9] Federal Coordinating Council on Comparative Effectiveness Research, www.hhs.gov/recovery/overview/index.html.

[10] Khoury M. et al. Comparative effectiveness Research and Genomic Medicine: An Evolving Partnership.

being recognized as critical for effective medical practice. Both patients and physicians find some comfort in storytelling". The term "narrative based medicine," was coined specifically to distinguish itself from evidence based medicine (Johna et al. 2011), not as a replacement for the term.

According to Drs. Kalitzkus and Matthiessen, "the development of narrative based medicine has to be understood in the context of patient-centered approaches," i.e., bringing the patient as a subject back into medicine. Crediting the work done on this subject to Dr. Von Weizsacker, they continue by pointing out that narrative based medicine is not simply a description of the patient's morbidity, "it is the description of the life of the illness in that specific human being" (Kalitzku and Matthiessen 2009). So, in simple terms, treatment decisions should also include information about the patient: desires, preferences, lifestyle choices, environment, support systems, and more.

In health care planning, patients often fall into the comfortable rut of relying on facts and figures and what can be objectively measured to support decision-making. Numbers and percentages give us a sense of confidence and certainty and provide a strong foundation for justifying our decisions. However, these formulas and calculations, as argued above, are only one component in the complicated algorithm of health care planning and viewed alone, can leave a patient with a false sense of confidence and certainty.

How can a patient know in advance how he or she will feel? A disconnect commonly exists between the way healthy people view themselves and the way people with medical conditions view themselves. Patients and physicians confront profound complexities and uncertainties in weighing risk and benefit. Nuanced considerations and individualized judgment is paramount to good decision-making. Expert recommendations that are sweeping and generic should raise immediate red flags to any patient.[11] The poetry of personalized medicine is based on the vital dimensions of life, and these dimensions are not easy to quantify. Genomic research is now working to identify specific sites where disorders originate. The possibility of gene therapy being used to delete genetic errors and insert corrections is on the horizon for mainstream treatment. What was once fodder for science fiction is now a real possibility.

Anticipatory medicine is another growing area enabled by our genomic capabilities. This area focuses on anticipating the health of the patient and interventions to prevent or treat the anticipated disease. In theory, early intervention, amelioration, or treatment of a looming disease would be a remarkable way to eliminate illness, suffering, and death. In practice, questions quickly arise about how to manage circumstances when genomic testing points to the uncertainty of disease susceptibility and becomes even more challenging when the patient is a child. What is the purpose of information if there is no treatment and if there is treatment, how are decisions made when there is susceptibility to a disease but not

---

[11] Gorman M. Building Patient Preference into Research Agendas. The National Working Group of Evidence–Based Medicine. Advancing the Evidence of Experience: Practical Issues for Patient/Consumer Inclusion.

a definite certainty?[12] In an ideal world, anticipatory medicine would be an amazing way to improve quality of life and increase longevity but needs to be thoughtfully implemented. With an overview of possible decision-enhancing approaches combined with the reality of genomic information, let's turn to our fictional patients.

Perspective: Leading Characters

Although these scenarios below may appear contrived, their purpose is to provide a rubric in which to view the challenges of personalized medicine. Both Patient A and Patient B have been diagnosed with a rare neurological genetic condition, XYZ syndrome. The course of the illness is unpredictable and can result in debilitating pain, impaired motor control, and the eventual use of a wheelchair. While it is not a fatal illness, per se, it usually shortens lifespan and significantly diminishes quality of life.

New on the market is Treatment Bold, which includes a bone marrow transplant and immunosuppressive medication for a lifetime. Treatment Bold has a 55 % chance of an outright cure. However, to even gamble for the cure, a transplant involves lengthy investigations into a good marrow match, a risky procedure, months of isolation in a specialized quarantine hospital room waiting to see if the transplant is successful, and a lifetime of medication. Treatment Status Quo, until recently, has been the only treatment available. Treatment Status Quo involves monthly infusions of a biologic which helps to dramatically control pain and to slow down the debilitating neurological consequences. However, in 45 % of the cases, patients plateau on the biologic, and then begin to see any benefits diminishing and symptoms reverting.

With this information in mind, let's meet Patient A and Patient B.

Patient A is a 59-year-old white male, married with two children in college. He is an engineer with an aerospace company. His 56-year-old wife, a high school English teacher, retired last year and is about to publish her first novel. Patient A has been planning to work until 60, in order to build retirement income with the intention of realizing a long-sought shared dream of traveling worldwide with his wife.

Patient B is a 24-year-old African American female who is finishing her high school education by earning a GED. She abruptly left high school 8 years ago when she realized she was pregnant. After delivering a healthy baby boy, she was left unsupported by the father. She then took an inventory job at a local chain pharmacy while she lived at home with her parents who helped to take care of the infant. For the past 3 years, she has lived independently with her son and was promoted to day manager at the pharmacy. She is excited and proud about her

---

[12] Hoffman E. Anticipatory and Pre-emptive Medicine-A Strategic Initiative DRAFT Children's National Medical Center.

upcoming high school completion certificate and plans to start training at the local community college to be a radiological technician. She lives in a small apartment with her son and is able to share childcare services with other neighbors for after-school needs. Until the diagnosis of XYZ syndrome, Patient B finally believed she had gotten her life plans back on track since her unexpected pregnancy.

With the same rare illness, Patient A and Patient B face some very real choices. The two treatments (i.e., Treatment Bold and Treatment Status Quo) are now part of the evidence based medicine treatment cannon for XYZ syndrome. But to truly personalize their care, many factors must be considered in their treatment decision-making algorithm, e.g., genetics, age, gender, race, lifestyle preferences, economics, and environment.

Patient A may choose the less aggressive biologic treatment that addresses pain management and mobility. This treatment approach would allow travel in the upcoming years, with the hope that he is one of the lucky 55 % who respond positively to Treatment Status Quo. Patient B may chose Treatment Bold which would entail going back to living with her parents in order to receive care for her son and support while she endures through the long treatment course. She believes she can finish her GED before treatment and can defer future plans to go back to school. Of course, treatment is also contingent on finding a bone marrow match.

Sit back and think about these decisions. Scramble the genetics, gender, race, age, lifestyle, support systems, treatment options, employment, financial situations, and piece together what the final puzzles could look like. No one picture is right or wrong. In designing the treatment cocktail, physicians must work hand-in-hand with the patient to take best practices, including available research, and mix with the patient's profile, lifestyle preferences, desires—the patient's narrative. The ability to make a personalized plan must be a coordinated mutual decision by provider and patient. Access to immediate information has to be tempered by prudent judgment in the decision-making process. With the use and sophistication of electronic medical records and such advances as telemedicine, doctors can be informed immediately and gain access to experts worldwide at the click of a mouse. The landscape of today's medicine is dynamic and needs to be examined continually as we are able to adapt the latest technological advances to our own human natures. Geneticists hunting for the cause of a mysterious illness more frequently than not come upon an incidental finding that might signal a risk for another disease, perhaps cancer, Alzheimer's, or Parkinson's. Incidental findings are part and parcel of doing genetic testing. Some sequencing mistakes cause diseases, some do not—leaving scientists with a vast unknown gray area (Rochman 2012).

Historically, rare diseases have often defied diagnosis. Now, with genomic diagnostic tests, health care providers are afforded more precise information in order to diagnose a condition. One day such tests will probably be easily bought over-the-counter in any pharmacy or grocery store, much like the reliable home pregnancy test. With greater diagnostic certainty, personalized medicine can truly now be put into play for the patient with a rare condition. Many biopharmaceuticals are designed to target the genetic malfunction of the individual patient's disease. Referred to as "designer drugs," these compounds are crafted based on the exact measurements or data obtained from the patient's genome. However, governments and society as a whole, need to develop policy that is cautionary and beneficial as health care policy is created.

Policy Bestseller: From Bench to Bedside

Health care policy is a constant source of debate worldwide. My discussion will focus on policy in the United States since that has been my primary field of interest. I will discuss broad policy areas: research priorities and the role of serendipity, incentives to foster innovation, the Patient Protection and Affordable Care Act of 2010, advocacy, and a sampling of government initiatives specific to rare disease.

Research: Priority and Serendipity

Biomedical research, whether publicly or privately funded, is the foundation to the health and well-being of any population of people. Since World War II, the United States has built a medical research establishment that is unparalleled worldwide and has brought forward spectacular advances in health care. Government, academia, and industry make up this strong research alliance. Sustained research is a vital element in continuing to address the unmet needs of those with rare conditions. Governments must maintain robust biomedical research budgets and set strong and reasonable research agendas (Rochman 2012). Public policy must acknowledge the vital importance to the larger prosperity of any nation by making biomedical research a top budget priority. Freedom to experiment, whether by public or private funding, is a crucial element in the advancement of science. Of course, I will add my belief that research also must be subject to policy that incorporates appropriate and strong regulation, evaluation, accountability, oversight, and transparency in the honest pursuit of knowledge.

Research is always a gamble. Sometimes advances come quickly and naturally, but more often, success in discovery and invention is a long and winding path. Research may take serendipitous turns. Many examples of fortuitous discoveries have resulted from what seemed like futile attempts. With wit and wisdom, former National Institutes of Health director and winner of the Nobel Prize in Medicine, Harold Varmus, said, "Medical research is still more a game of pool than billiards. You score points regardless of which pocket the ball goes into" (Kornberg 2012). Author Kornberg states in his article entitled "Of Serendipity and Science":

> Investigations that seemed totally irrelevant to any practical objective have yielded most of the major discoveries of medicine–X-rays were discovered by a physicist observing discharges in vacuum tubes, penicillin came from enzyme studies of bacterial lysis, and the polio vaccine came from learning how to grow cells in culture. Cisplatin, a widely

used drug in cancer chemotherapy, came about from studying whether electric fields affect the growth of bacteria and observing inhibition due to the unexpected electrolysis of the platinum electrodes. Once again, genetic engineering and recombinant DNA depended on reagents developed in exploring DNA biochemistry. All these discoveries have come from the pursuit of curiosity about questions in physics, chemistry and biology, apparently unrelated at the outset to a specific medical or practical problem (Kornberg 2012).

A new approved drug often turns out to have limited or relatively insignificant use for the particular medical condition for which it was initially researched. More important uses are discovered only after the drug is in use and doctors begin testing it for other conditions. This was the case for the therapeutic protein interferon alfa-2a, sold under the brand name called Roferon-A. First approved for a rare blood condition called hairy cell leukemia, Roferon-A was later found to be effective for a number of other far more prevalent conditions, some chronic forms of hepatitis and leukemia (Conko et al. 2009). Over 24 % of all drugs currently on the market were discovered with the aid of serendipity and thus, may never have been discovered without the curiosity, observation, wisdom and tenacity of the researchers. Additionally, drug repositioning or a new use for an existing drug molecule to treat disease is a wonderful coup to treatment and is a great way to help companies increase profits. The reintroduction of thalidomide into clinical use for multiple myeloma is an example in point (Hargrave et al. 2012).

Given the relatively low odds of success and the high costs of drug development, pharmaceutical and biotechnology companies usually focus on potential therapies with the highest likelihood of generating a good financial return. This risk/reward gamble has meant that potential therapies for rare diseases, including therapies for life-threatening conditions, have often languished in the early development pipeline. Moreover, conventional approaches to drug development are often not feasible for rare diseases, which offer not only small markets but also small populations for participation in clinical trials.[13] Treatments found effective, but costly, for rare diseases have frequently resulted in later application to other more common conditions affecting larger treatment populations. Return on investment would logically be greater with use by larger populations of patients and, therefore, provide incentives for continued research.

Incentives to Foster Innovation

Legislators continue to search for incentives to foster innovation in both the public and private sectors. Rare disease research is daunting; the truth is that scientists usually do not have any way to know or measure the targeted treatment

---

[13] Committee on Accelerating Rare Diseases Research and Orphan Product Development, Institute of Medicine. Development of New Therapeutic Drugs and Biologics for Rare Diseases. Washington, DC: The National Academies Press 2011, p. 47.

population. Stimulating innovation in health care is an important component to finding treatments for rare conditions. Pharmaceutical companies have to decide the relative risk of going forward with research to bring a profitable product to market and to recoup the estimated $4–11 billion invested to bring a new drug to market (Herper 2012). Without incentives, companies are very reluctant to undertake research, gambling on the odds of profitability.

Central to the debate on fostering innovation is intellectual property (IP) law or a synthesis of laws that combine intellectual property with regulations, approval processes, taxation, and funding vehicles (National Health Council 2009). What falls under the rubric of intellectual property is often confusing and legally esoteric but includes patents, copyrights, trademarks, and trade secrets. The area most relevant to health care is patent law, in general, and more specifically, patent filing rules and patent exclusivity. Legislation usually combines intellectual property provisions with Food and Drug Administration regulatory issues, taxation, federal research initiatives, and funding of medical training. These issues are complicated and hard to tackle given all the stakeholders lobbying for their constituencies. Stakeholders include patients, consumers, providers (doctors, nurses, physician assistants, etc.), insurance companies, pharmaceutical companies, device manufacturers, the federal government and families. Although not comprehensive, the paragraph below provides a very brief overview of major legislation for drugs and biologics. This history is intended to set some groundwork for the rest of this section on innovation.

In the United States, the "Biologics Control Act of 1902" was the first national regulation to control either pharmaceuticals or biologics and the first premarket approval statute in history. It was passed in response to the deaths of a high number of children from tetanus contamination of smallpox vaccines and diphtheria antitoxins. The Pure Food and Drug Act of 1906 governs the FDA's regulation and approval of drugs and was amended in 1938 to include pre-market control over new drugs to ensure safety and controls over manufacturing establishments. In 1984, the Hatch-Waxman Act modified the Food and Drug Act again by providing abbreviated pathways for the approval of subsequent versions of drugs and provided a period of data exclusivity for innovator drugs and first generic drugs (National Health Council 2009).

With this short background, I would like to turn to areas affecting the rare disease population. The Orphan Drug Act of 1983 is one of the great success stories in health care innovation. This landmark piece of legislation was designed to facilitate the development and commercialization of drugs to treat rare diseases, termed "orphan" drugs. By definition, an orphan drug treats a small population (under 200,000 people) and thus, is largely unprofitable.[14] The mission of the Food and Drug Administration Office (FDA) of Orphan Products Development (OOPD) is to advance the evaluation and development of products (drugs,

---

[14] FDA, DHHS, PART 316 ORPHAN DRUGS-Table of Contents, Subpart C Designation of an Orphan Drug sec. 316.2.

biologics, devices, or medical foods) that demonstrate promise for the diagnosis and/or treatment of rare diseases.[15]

The legislation was designed to motivate companies to conduct research to bring a pharmaceutical to market for a rare or "orphan" condition and provides a number of incentives: a lighter regulatory burden for developing new orphan drugs, a seven-year monopoly, and a ninety-percent tax credit for the cost of clinical trials. Orphan drug designation does not indicate that the therapeutic agent is safe, effective, or legal to manufacture and market in the United States; those areas are handled through other offices in the Food and Drug Administration. Instead, the designation means only that the sponsor qualifies for certain benefits from the federal government. As a side note on the legislation, the Act also established an Office of Rare Diseases within the Food and Drug Administration.

The legislation has been a remarkable success. The Food and Drug Administration has approved more than 300 orphan drugs, with 1,100 more under development. One of the first developed under the law was AZT, the early AIDS treatment. Two years later, Congress expanded the law to include biological and chemical drugs, which helped spur the biotechnology industry (Green 2012). The Humanitarian Use Device or HUD program was established in 1990 with passage of the Safe Medical Devices Act and creates an alternative pathway for getting market approval for medical devices that may help people with rare diseases affecting fewer than 4,000 individuals in the United States per year (Green 2012).

Patient Protection and Affordable Care Act

The Act, passed in 2009, commonly dubbed Obama Care, is the most comprehensive overarching health care legislation passed since Medicare. The legislation is mammoth, and some parts are difficult to understand and will become even more daunting when regulations are finalized and when implementation is fully realized. Proponents of the legislation are delighted by the promise of universal access to care for all Americans, increased research and evaluation, parity with premium costs between genders, coverage of preventive and wellness services, cost assistance with premiums, and the eventual implementation of electronic medical records systems by all providers and institutions.

Critics say that the legislation flies in the face of personalized medicine with a standardized, one-size-fits-all approach. Among the concerns is that comparative effectiveness research, funded to the tune of $1 billion, will establish a cookbook of standards on which physicians will inform their evidence based medical practices. In trying to lower or stabilize health care costs, patient's individuality will vanish into larger epidemiological groupings. The ability to serve all Americans is a concern since there is great uncertainty about the capabilities of the current

---

[15] See footnote 13.

labor force and the pipeline of providers being trained. Finally, electronic medical records have troublesome issues related to the tremendous requirements in privacy/confidentiality and the enormous costs of implementation.

Advocacy

Organizations representing all types of patient populations or health policy interests are based in Washington DC or have an office in the city. Advocacy is the lifeblood of many of these organizations which work or "lobby" to design, change, and/or implement policy to benefit their stakeholders. Many of these groups work together on common causes. As a case in point, the National Health Council, located in the heart of downtown Washington DC, brings together all segments of the health community to provide a united voice for the more than 133 million people with chronic diseases and disabilities and their family caregivers. One relevant example of legislation they are working to pass is MODDERN (Modernizing our Drug and Diagnostics Evaluation and Regulatory Network) Cures Act. This proposed legislation is designed to change health policy with a three-pronged approach: speed up the development of new and better treatments for conditions with few or no medical options; increase the number of tools that can predict which patients will receive the most benefit from particular medicines; and give patients quicker access to new diagnostic tests once approved by the FDA.[16]

The MODDERN Cures Act addresses the many promising treatments that get left behind in the drug development process. Not uncommonly, drugs that address unmet medical needs, particularly for rare diseases, take longer to develop, and even if the product reaches market, the company's patent may have expired or be close to expiring.[17] While the course of the MODDERN Cures Act is unclear at publication time, it remains a useful example of creative legislative approaches to fostering innovation in medical care. Other legislation that the National Health Council is spearheading includes clarification of the FDA approval process for biomarkers. As discussed earlier, biomarkers aid in diagnosis, allow better target treatments, reduce costs for determining safety and efficacy, and complement efforts to personalize medicine. Without certainty in the approval process, companies have a difficult time attracting investors to back their biomarker research (National Health Council 2009).

Similarly, the Council is also looking to establish a regulatory pathway for biologics. A biologic is often used for the treatment of rare and serious diseases and defined as a biological product such as a virus, therapeutic serum, toxin, antitoxin,

---

[16] National Health Council. Medicare Advisory: NCATS Initiative and MODDERN Cures Act: Complementary Programs to Advance Discovery of New Treatments and Cures, www.puttingpatientsfirst.net/moddern/.

[17] See footnote 8.

vaccine, blood, blood component or applicable to the prevention, treatment or cure of a disease or condition. As biologics come off their patents, the market is ripe for generic biologics but no such regulatory pathway exists for generics (Field and Boat 2010).

Medical Education

Medical education is evolving, including not only the quantitative learning that is vital to being a good physician but also the subjective or human angle to medicine. Many medical schools have medical humanities components to their programs. The area of medical humanities challenges the student to understand how the hard facts and gray areas have to be navigated to capture each individual patient. In an article in the *New England Journal of Medicine*, David Watts says that poems and stories—even just a few a week—can show students the richness of human relationships. In other words, imaginative literature can reignite the compassionate spark that spurred students toward the healing arts in the first place (Watts 2012). In connecting the dots from facts and figures of treatment approaches with patient preferences, the humanities and arts provide profound insight and perspective into the human condition, the gamut of emotions from joy to suffering, personhood, and our responsibility to each other and our environment. Literature, on a basic level, offers a valuable historical perspective on medical practice and on a more sophisticated level, provides insight into our own personal narratives.

Sampling of Government Initiatives

I believe that the Orphan Drug Act opened the door to vast programs on rare disease throughout the federal government. Examples of such entities are briefly described in this section. Established in 2012 at the National Institutes of Health, the National Center for Advancing Translational Sciences (NCATS) works to develop innovations to reduce, remove or bypass costly and time-consuming regulations that impede efforts to speed the delivery of new drugs, diagnostics and medical devices to patients, i.e., translating the basic science into clinical use. Within this center, the TRND (Therapeutics for Rare and Neglected Diseases) program is designed to advance the entire field of drug development by encouraging scientific and technological innovations aimed at improving success rates in the crucial preclinical stage of development.[18] When research moves from the laboratory to the clinical setting, it is commonly described as moving from "bench to bedside."

---

[18] National Center for Advancing Translational Sciences, NIH, DHHS. www.ncatsnih.gov.

In other efforts, the National Institutes of Health collaborates with industry to spur therapeutic innovation. The program, Discovering New Therapeutic Uses for Existing Molecules, intends to make promising molecular compounds that are not being actively developed available to academic researchers in order to evaluate new potential therapeutic uses. The National Institutes of Health is partnering with some of the giant pharmaceutical companies to give researchers access to compounds that had been part of clinical studies but were not proven effective for the specific use for which they were originally developed.[19]

Twenty-nine years after passage of the Orphan Drug Act, in 2011, Congress passed the Creating HOPE Act for Pediatric Cancer Research. The law relies on the same model as the Orphan Drug Act, offering pharmaceutical companies various regulatory and marketing incentives to devise new treatments for rare cancers that strike children, but do not appear to be sufficiently profitable for drug makers to pursue on their own (NCI Bulletin 2012). I want to acknowledge the exponentially increasing health care costs we face in America and the truly remarkable biomedical research establishment this country has also built. The push and pull of cost and discovery is real, but policy must sustain the innovation ecosystem.

Personalized Medicine: The Epilogue

I have tried to outline the major issues facing the opportunities and limits of personalized medicine and the importance and relevance of the concept and practice in health care today. As a collective whole, rare diseases have always challenged medical practice and remain a significant unmet health care need. The burgeoning era of genomics has opened up a rapidly expanding area focusing on the molecular basis of rare diseases, targets for therapeutic interventions, and development of therapies based on these advances.[20]

It is indeed an exciting time. Scientists have unraveled the genetic basis for a number of rare, single-gene conditions, and research is focusing on identifying the basis for many multiple gene diseases. The ability to sequence the human genome defines personalized medicine but is only one character in the whole story. The synthesis of genomic diagnostics with individualized genomic therapeutics has spawned a new age of health care delivery, and there is no turning back. We stand in awe and marvel at how medicine has advanced and what it can do to save lives and improve quality of life. But science alone is just one weapon used to combat disease and suffering. Our uniqueness as breathing, living beings with varying pleasures, subjective thoughts, opinions, lifestyle differences and preferences must also be guiding factors in our health care decisions.

---

[19] NIH, DHHS. NIH Launches Collaborative Program with Industry and Researchers to Spur Therapeutic Development. NIH NEWS. May 3, 2012.

[20] See footnote 17.

I am always fascinated when I get my car fixed these days. Mechanics hook the car's electronic system up to fancy computerized diagnostic machinery which can very accurately find out what is wrong. The car may need a new fuse, more brake fluid, or a transmission overhaul, and the decision to fix the car is binary, either "yes" or "no." But for humans, our health care decisions cannot be binary and are based on multiple factors influenced by who we are and what we want. And while facts and figures are important, decisions cannot be solely based on quantitative data. We live rich and complicated lives, feel emotion, connect to others, and think abstract thoughts. Many independent variables function into the equation of personalized medicine, and each variable needs to be considered for its appropriateness for any given patient.

Providers must develop and nurture skills of observation, analysis, empathy, and self-reflection—skills that are essential for humane medical care—and to learn to practice medicine within civic and social contexts, acknowledging how the cultural environment interacts with the individual experience of illness and the practice of medicine. Health care decision making will not become easier for future generations, but society can help to address this challenge by training our providers, not only in the science of medicine but also in the humanity of its patients. Atul Gwande, a surgeon at Harvard's Brigham and Women's Hospital and one of contemporary medicine's prolific authors and thinkers, wrote:

> We look for medicine to be an orderly field of knowledge and procedure. But it is not. It is an imperfect science, an enterprise of constantly changing knowledge, uncertain information, fallible individuals, and at the same time lives on the line. There is science in what we do, yes, but also habit, intuition, and sometimes plain old guessing. The gap between what we know and what we aim for persists. And this gap complicates everything we do (Gwande 2007).

Patients must become and remain empowered and informed advocates for their own best interests. To the extent possible, they should challenge themselves to understand as much as feasible about their disease and any and all options, including those on the horizon. Patients should feel comfortable with their medical care provider and be able to engage in dialogue that may require asking some tough questions. At the risk of sounding trivial and trite, patients must find their voice, tell their stories, and make their preferences clear by understanding what they truly want. Policy must address the most urgent needs in health care including a robust and well-funded biomedical research agenda, seamless translation of meritorious research into clinical practice, and incentives to foster innovation in traditionally unprofitable areas. The challenge is how best to leverage our scientific and technological advances in ways that will deliver improved health care outcomes to patients with rare diseases. As our knowledge grows, so too, should our policy.

References

BMJ. Evidence based medicine: what it is and what it isn't, Editorials, BMJ, 1996;312:71–72 (13 Jan).
Chial H. DNA sequencing technologies key to the human genome project. Nat. Educ. 2008;1(1).
Conko G, Miller H. Obamacare threatens personalized medicine. Forbes, Forbes Commentary 2009.

Field M, Boat T, editors. Development of new therapeutic drugs and biologics for rare diseases: rare diseases and orphan products, accelerating research and development, Institute of Medicine (US) committee on accelerating rare diseases research and orphan product development. Washington DC: National Academies Press; 2010.

Garber A, Tunis S. Does comparative-effectiveness research threaten personalized medicine? Perspective, New England J Med. 2009;360:1925–27.

Ginsburg G, Willard H. Genomic and personalized medicine: foundations and applications. Transl Res. 2009;154(6):277–87 (Epub 2009 Oct 1).

Green J. Jack Klugman's Secret, Lifesaving Legacy. Wonkblog, Washington Post online, Dec 25, 2012.

Gwande A. Complications: a surgeon's notes on imperfect science 2007;7.

Hargrave T, et al. Serendipity in Anticancer Drug Discovery. World J Clin Oncol. 2012;3(1):1–6.

Herper M. The truly staggering cost of inventing new drugs. Forbes 2012.

Johna S, Rahman S. Humanity before science: narrative medicine, clinical practice, and medical education. Perm J. 2011;15(4):92–4.

Kalitzku V, Matthiessen P. Narrative based medicine: potential, pitfalls, and practice. Permanente J. 2009;13(1):80–6.

Kornberg A. Of serendipity and science. World J Clin Oncol. 2012;3(1):1–6 (Published online 2012 Jan 10).

National Health Council. Akin Gump Strauss Hauer and Feld LLP. Intellectual Property Law and Biosimilars: Summary and Analysis of Federal Legislation 2009;1–2, 26, 10.

NCI Bulletin. Congressional Caucus Hosts Childhood Cancer Summit 2012;9:19 (Oct 2).

Rochman B. The DNA Dilemma: a test that could change your life. Time Mag. 2012;42–5 (Dec 24).

Watts D. Cure for the common cold. NEJM 2012;367:1184–5 (Sep 27, 2012).

Part IV
Closing Gaps: Promising Research and Future Considerations

Managing Communication for People with Amyotrophic Lateral Sclerosis: The Role of the Brain-Computer Interface

Gaye Lightbody, Brendan Allison and Paul McCullagh

Leave no stone unturned
EURIPIDES, Heraclidae

Abstract Amyotrophic Lateral Sclerosis (ALS) is a neurodegenerative neurological condition categorized as an orphan disease and at present the primary treatment is managing symptoms. It leads to severe paralysis, resulting in the need for the patient to use assistive technologies to support them in their daily activities. When the condition is severe, mainstream technologies may no longer offer the support required, due to the need for reliable residual movement. Brain computer interfaces (BCI) have the potential to become a powerful assistive technology for some individuals with the most severe of neuromuscular disorders. With only 'thought' as an input medium the user could harness control and communication. Undoubtedly, the availability of such technology could have a major positive impact on the life of a patient with ALS, supporting their inclusion in the world and contact with people around them. However, despite decades of research and development, BCIs are still not commonplace. Many recent advances have been

---

G. Lightbody (✉)
School of Computing and Mathematics, University of Ulster, Jordanstown campus,
Shore Road, Newtownabbey, Northern Ireland, Antrim, BT37 0QB, UK
e-mail: g.lightbody@ulster.ac.uk

B. Allison
Visiting Scholar, Cognitive Science Dept, UC San Diego, 9500 Gilman Drive,
La Jolla, CA 92093, USA
e-mail: ballison@ucsd.edu

P. McCullagh
Computer Science Research Institute, School of Computing and Mathematics,
University of Ulster, Jordanstown campus, Shore Road, Newtownabbey,
Northern Ireland, Antrim, BT37 0QB, UK
e-mail: pj.mccullagh@ulster.ac.uk

made but some factors still prevent widespread deployment of BCI. This chapter will introduce the background of BCI and provide a short discussion about the problems associated with BCI technology, balanced with thoughts about its potential, challenges and hopes for the future.

Introduction

There are many conditions or injuries that can lead to paralysis, such as stroke, brain trauma or Multiple Sclerosis, with the level of severity varying widely. One such lesser known condition is Amyotrophic Lateral Sclerosis (ALS). It is a neurodegenerative neurological condition that leads to severe paralysis and is often referred to as Motor Neuron Disease (MND) or sometimes Lou Gehrig's disease, after the celebrated baseball player who contracted it (Gehrig 2013) and brought it to the consciousness of the general public. It is a rare orphan disease that affects in the region of 1–2 in 100,000 people (GARD 2013; ALSA 2013a, b), although statistics vary regarding its incidence. In the UK it is estimated that 5,000 people have the ALS form of MND, with around 2 people in 100,000 developing it each year (Patient 2013).

ALS is an aggressive disease that progresses rapidly in most cases. It affects the nerves for establishing movement. These nerves, known as motor neurons link the brain to the spinal cord and onto the peripheral nerves for function control. Degeneration of these nerves leads to a decline in voluntary muscular movement of the limb and trunk (ALSA 2013a; Orphanet 2013). For some sufferers, eye muscles may be affected but not in all cases (Birbaumer and Cohen 2007). It is non-contagious and the cause is unknown but it is estimated that about 5–10 % of those with the condition have an inherited form of the disease. Variations of ALS exist. 'Spinal ALS' has initial symptoms which start with muscle weakness in the arms and legs, leading to paralysis in these regions which then progresses to the neck and head. The symptoms of 'Bulbar ALS' start within the neck and mouth regions and then progresses to other parts of the body. Progressive Muscular Atrophy (PMA) and Primary Lateral Sclerosis (PLS) are less severe forms with a better prognosis.

ALS is classified as an orphan disease. There is limited scope for drug intervention (Orphanet 2013), although Riluzole was approved in 1995 by the Food and Drug Administration (FDA). This drug has been shown to slow the progress of ALS, leading to a possible modest increase in survival time (ALSA 2013b). At present the "primary treatment is managing ALS symptoms" (Ensrud 2005). The expected lifespan of an individual diagnosed with ALS varies, with up to 20 % of people living beyond 5 years and 10 % of people living beyond 10 years (ALSA 2013a). Life expectancy is strongly linked to the patient's choice to accept (or decline) life supporting treatment such as artificial respiration when the paralysis has become severe enough to prevent breathing (Nijboer and Broermann 2010).

Against this outcome, the fear of losing the ability for interaction is a key concern for the individual, as highlighted by Blain-Moraes et al. (2012). The authors stated that, "the existence of the human self hinges on successful interaction with others ... those who cannot engage in communicative interaction are, consequently, at risk of not being accorded personhood by others."

Nijboer and Broermann (2010) discussed the difficulty that ALS patients have in deciding whether or not to write a living will, detailing their wishes to accept or decline life-prolonging treatment. They highlighted the importance of the patients' expected quality of life in making these decisions and stressed that often there is not sufficient information made available to the patients in terms of life-sustaining treatment and communication technologies. Hayashi and Oppenheimer (2003) reported that 24 % of the patients in their study survived 10 years past respiratory decline due to artificial ventilation. At this extreme of the disease, paralysis will have extended to the point that the person will be in a Locked-In State (LIS). With cognitive function often remaining intact the healthy brain is effectively trapped inside the immobile body. A range of assistive technologies, from eye trackers (Calvo et al. 2008) to sip switches (Jones et al. 2008) can offer a reasonable form of communication, but only when residual muscular movement exists. Jean-Dominique Bauby, a French journalist struck down by a massive stroke in 1995 that left him with only residual movement of his left eye, authored his memoir, 'The Diving Bell and the Butterfly` (Bauby 1998), using only eye blinks in response to a repeatedly recited alphabet by his carer. In this book he gave an insight into the life of a person with LIS. Bauby wrote "Other than my eye, two things aren't paralyzed, my imagination and my memory." The book has since been interpreted as a major movie to critical acclaim (Thomas 2008).

As the severity of the condition of an ALS sufferer progresses, other assistive technologies that rely on some level of residual movement will no longer offer the pathway for communication. Those at the most severe levels of paralysis are considered to be in a Completely Locked-In State (CLIS). Can interfacing directly with the brain through the use of recording brain signals and using computers "bridge the gap between the inner and outer world"? This was the question asked by Nijboer and Broermann (2010), who applied such technology to try to help those in a LIS or CLIS state. Such a mechanism is commonly termed a Brain Computer Interface (BCI). It is important to note however, that "BCIs are not treatment for the disease; they do not affect a person's health or the progression of ALS in any way. They are an assistive technology that can potentially make a significant difference in the quality of life for people with ALS" (Ourand 2004).

BCI has been studied in electrophysiology laboratories for over 30 years (Wolpaw and Wolpaw 2012; Allison et al. 2013) and significant progress has been made regarding accuracy, speed, robustness and mobility. The questions we pose in this chapter are: can BCI offer the mechanism to enhance social inclusion and empowerment through communication and control? And if so, what hindrances are there in making it readily available for those that would benefit most? In order to first understand the role and potential impact of BCIs, it is important to present some background information on how relevant information within the brain can be harnessed to enable communication without the necessity of motor movement.

Brain Power

A BCI is a device that can potentially facilitate communication without the need for voice articulation or peripheral movement. A common misconception of brain computer interfacing is that the computer is literally reading the mind of the subject. This next section aims to explain some of the underpinning anatomy and physiology of the brain that enables interaction to occur when all other assistive technology fails. How is it possible to exploit the power of the brain?

Regions within the brain are associated with and responsible for cognitive, sensory and motor actions, as shown in Fig. 1. Application of BCI technology relies upon the understanding of the human brain function and has used this knowledge to develop ways for a person to convey information. It is not simply a case of random thoughts being translated into actions. There are two main mechanisms that may be used to achieve communication; the first picks up responses from planned external stimuli and in the second, the user is trained to perform predefined mental tasks to convey their wishes.

BCIs have four general components (Allison et al. 2013; see Fig. 2). First, a device must measure the brain's electrical activity[1] and extract selected brain signals. Typically, the electrical activity is measured using electrodes placed on the

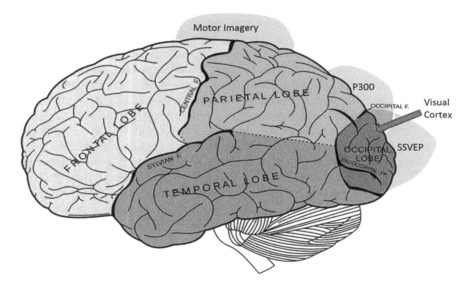

Fig. 1 Overview of the regions of the brain used for BCI. *SSVEP* BCIs rely on activity over the visual areas in the occipital lobe. P300 BCIs use these areas as well, along with parietal electrodes. Motor imagery BCIs rely largely on electrodes around the *left* and *right* central fissure, bridging the frontal and parietal lobes, which contain the brain's primary motor and sensory areas

---

[1] BCIs can also use other physiological properties such as blood flow (Andersson et al. 2010), but these are less common and not considered in this chapter.

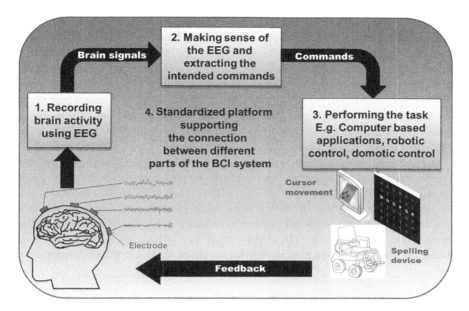

Fig. 2 Main components of a non-invasive EEG based BCI system (Adapted from Allison 2011; Wolpaw and Wolpaw 2012)

surface of the scalp and connected to an amplifier. This process has been in clinical use for many years in the electrophysiology department and is referred to as the electroencephalogram (EEG). Particular electrode configurations are used for different types of BCI and there may be variations between users to achieve the best outcome. Second, a signal processing system must translate these measures of brain activity into a message or command. In turn, these meanings can be used to interact with external devices, offering a medium for communication, expression and control. Classic examples of BCI applications are spellers (Sugiarto et al. 2009; Guger et al. 2012a, b), robotic control (McFarland and Wolpaw 2008) and domotic control (Babiloni et al. 2009; McCullagh et al. 2011; Ware 2010). More recently BCI has found application in gaming (Nijholt 2009) and for self-expression (Miranda et al. 2011; Münßinger et al. 2010). Finally, a platform is necessary to manage the interactions between these different components and the user (Wolpaw et al. 2002; Pfurtscheller et al. 2008; Allison 2011; Brunner et al. 2013).

Recording the EEG

Most BCIs rely on the scalp recorded EEG to measure brain activity (Mason et al. 2007). However, some groups work with invasive BCIs that rely on subdural sensors implanted on or in the brain (Velliste et al. 2008; Hochberg et al. 2012). There has been substantial discussion about the practical and ethical issues involved in the decision to use invasive or noninvasive BCIs (Birbaumer 2006; Millán and Carmena 2010; Nijboer et al. 2011; Allison et al. 2013). Both directions merit further study, since different approaches may suit different individuals, based on their needs, preferences, spared abilities and other factors. Any potential BCI user, especially someone who might rely on the BCI as a primary means of communication, should be fully informed about the risks, challenges and potential limitations of any BCI they might use. The remaining sections of this chapter focus on noninvasive EEG based BCI, which is the modality that could have practical application outside the dedicated research laboratory.

It is not surprising that surveyed users repeatedly comment on the discomfort of gel electrodes and the length of time required for preparation and clean up (Blain-Moraes et al. 2012; Huggins et al. 2011). In a typical set up the electrodes may be placed on the user's scalp with the use of an electrode cap to guide the location of the electrodes, with gel needed to enhance the connection with the scalp and improve signal quality. These systems do not provide an aesthetic and user-friendly solution for home use but they are essential for research and development.

Conveying and Extracting the Information

While the exact definition of a "BCI" has become somewhat fuzzy in the last few years, amidst efforts to expand the term, most BCI research groups focus on real-time systems that allow people to send information via direct measures of brain activity (Allison 2011; Zander and Kothe 2011). BCI can be separated into two categories: one in which the user receives some visual or auditory stimulus which in turn invokes a response in their EEG; the other is based on intended actions of the user and requires no external stimulus. The following paragraphs provide a brief overview of the more typical BCI paradigms.

The visual cortex positioned within the occipital lobe receives and processes information from the eyes. This region can be stimulated to evoke a response distinctive enough to be captured within the subject's EEG recording and classified by a computer algorithm. Garcia-Molina provides details about how the responses can be harnessed into a useful application (Molina and Mihajlovic 2010). The brain's sensory components can be stimulated to give a response, referred to as an evoked potential. A well-known mechanism for evoking such a response in the subject's EEG is the Steady State Visual Evoked Potential (SSVEP) (Zhu et al. 2010). The subject views a visual stimulus that oscillates at a particular frequency. The resulting response in the EEG is detectable using electrodes placed over the occipital region (Fig. 1) and can be distinguished from the background electrical activity.

Stimulus mechanisms include Light Emitting Diodes (LEDs) (Fig. 3a) and reversible checkerboard icons (Fig. 3b). By using a variation of stimulus frequencies it can be possible to differentiate between the responses in the EEG and therefore between the LEDs being observed by the user. If each LED relates to some defined context, then the translated information from the EEG can be then used

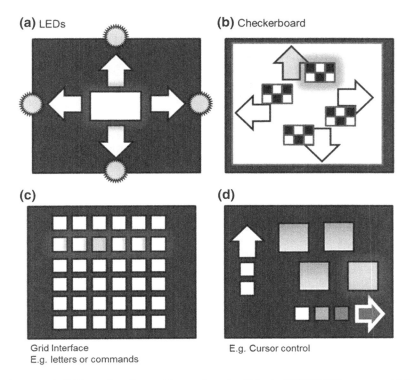

Fig. 3 Example BCI user interface and interactions. (**a–b**) In the SSVEP BCI, users would focus on one of the four LEDs or revisable icons, which would each oscillate at a different frequency. When the BCI detects this frequency over the occipital areas, it recognizes which is the user's target. (**c**) In the P300 BCI, users count each time the target item flashes. This will produce a P300 that does not occur when the other items flash. The BCI detects this P300 and thereby identifies which item is the target. (**d**) In the ERD/ERS BCI, users might imagine hand of foot movements to drive a cursor in two directions via EEG activity over the brain's primary movement and touch areas

to control a computer mediated application. Although in theory a broad range of frequencies can be used, ranging from about 5–45 Hz, as the number of frequencies increases, so does the difficulty in distinguishing between the desired outcomes. As such, a typical SSVEP based BCI system may rely on 4 flashing LEDs, enabling a four way navigation mechanism ideal for movement around a computer screen, as illustrated by Fig. 3a. With this mechanism a range of applications can be supported and examples of spellers and user interface control have been reported (Allison et al. 2010, 2012; Guger et al. 2012a; Wolpaw and Wolpaw 2012). The operation for the reversible checkerboard icons is similar. Zhu et al. (2010) provides an overview of SSVEP stimulus mechanisms.

A different type of response in reaction to an 'unexpected' visual or auditory cue may be elicited using the 'oddball paradigm', whereby a rare target event is interleaved with many non-target events (Fig. 3c). A resulting event within the

EEG gets the P300 label from the location of the evoked response wave appearing in the region of 300 ms post stimulus onset. It is most strongly captured using electrodes over the occipital and parietal areas (Fig. 1). P300 BCIs require very little training and are reliable in field settings for most users (Sellers et al. 2010; Guger et al. 2012b; Mak et al. 2012).

By imagining a movement it is possible to initiate activity within the sensorimotor cortex within the brain (Fig. 1). The process involves the modulation of a motor component of the EEG known as the 'mu-rhythm'. The paradigm is often referred to as Event-Related Desynchronization (ERD) and Event-Related Synchronization (ERS). By extracting this intention from the EEG a user can use trained imagined activities as control mechanisms, without the need for external stimuli. Typical examples would be imagining left-hand, right-hand or foot movement although other paradigms do exist (relate to Fig. 3d). In all cases, sophisticated signal processing is required to extract these intended movement components and the software needs to be matched to the user. With non-invasive electrodes, fine dexterity is not feasible, and typically only a 2 or 3 way decision is possible. With suitable technology such as intelligent robotic wheelchairs or tailored user interfaces, control can be achieved (Graimann et al. 2010; Millán et al. 2010; de Laar et al. 2013).

Example Applications

The last decade has seen the first commercialization of BCI based products (Allison et al. 2013). Unsurprisingly, much of the commercialization has been targeted towards the healthy user. In particular BCI is appealing in the areas of gaming and entertainment, facilitating another channel for communication between the user and the game. Often signals such as facial electromyography (EMG) are used in addition to the EEG. The resulting systems need fewer electrodes since only simple information needs to be conveyed. As a result the average price is lower, opening up a wider target consumer market. Some of these consumer devices retail for under $100. MindWave Mobile from Neurosky (MindWave 2013) was developed for iOS and Android platforms. The system uses attention, meditation and eye blink, combined with raw EEG to gain the control. Applications that can be controlled using the system include MindPlay for controlling video. The Emotiv EPOC (about $299) from Emotiv Systems (Emotiv 2013a) is a more advanced and complex device with a greater range of capabilities. It uses 14 electrodes with 2 reference electrodes and is targeted both to the consumer and research markets (Emotiv 2013b). It should be stressed that such devices listed above are not tailored as yet to the assistive technology market and most rely on some muscular movement (EMG). Nevertheless, this highlights the potential for such technology and the range of applications, and this will help fuel opportunities for further BCI research.

A pure BCI system which uses only brainwaves has been commercialized by g.tec Medical Engineering Company (IntendiX 2013) and has been on the market now for a couple of years. It is the first commercial BCI system for home use, and

(a) Shows the IntendiX Spelling Matrix at the beginning of a "copy-spelling" run.

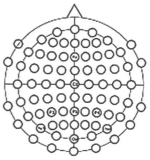

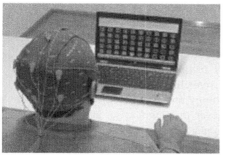

(b) Presents a photograph of a subject using the system in free spelling mode

(c) Shows the electrode montage used for both dry and gel electrodes. The ground is the left mastoid and the reference is on the right mastoid

Fig. 4 The intendiX system, from Guger et al. 2012b

relies on the P300 paradigm. In the speller version of the system (see Fig. 4), the user sees 50 tiled letters and numbers on a grid with 5 rows and ten columns on a computer screen (Guger et al. 2009, 2012b). The rows and columns flash in a random sequence. The user is asked to focus on the letter of interest. As the row and column containing that letter flash, a P300 response is elicited in the user's EEG. However, when other rows or columns flash, the P300 does not occur. IntendiX, like most P300 BCIs, requires averages of at least 3 flashes for adequate signal quality. In many respects, this is a BCI version of the system used by Bauby to write his memoirs. g.tec has developed new modules for IntendiX, providing new applications with new target items and corresponding commands, enabling control over domestic devices (Intendix 2013b), a platform for creative expression (Intendix 2013c) and a system for gaming (Indendix 2013d).

BCI opened up a communication channel using spelling applications to enable verbalization. Other examples of BCI application include control of devices within the home. From a central system the user's commands can be sent through a central hub in the home onto devices which can then be controlled (McCullagh et al. 2011). It could be the switching on or off of a light, the opening of a door or possibly the control of a multimedia entertainment center and TV control. BCI has also been demonsted as an avenue for creative expression using music (Miranda et al.) and art (Münßinger et al. 2010).

Case Study: BCI as an Assistive Technology for ALS

Javier has always enjoyed sitting on his patio, enjoying the warm Catalan sun. Some2011times, he would invite friends and family and play music well into the night. Shortly after he was diagnosed with ALS, Javier learned to use assistive technologies based on eye, finger, and tongue activity. These provided him with decent control over various activities of daily living

DLs), including a music player and smart home controls to open the door to his patio and control the lights. As his control declined, Javier found these systems increasingly fatiguing. His doctor recommended a BCI being developed through The Guttmann Institute, a local rehabilitation hospital. A local graduate student in engineering set him up with a BCI that could control all of the same functions as other assistive technologies. The "hybrid" BCI system could also let him use other assistive technologies when he is not tired.

Javier is impressed with the technology and with the efforts of the local student. He has strong support whenever problems arise, and his BCI can control a wide range of customized applications. Javier appreciates that his experience is not typical—most people do not have such technical support. However, Javier also has major problems with the electrode cap. He prefers to feel independent and dislikes asking a nurse or his wife to help. He finds the gel messy and uncomfortable, and does not enjoy having his hair washed afterward. He is also concerned about ongoing support—will this student be available forever?

Elke has come to accept that her career as a painter is over. Her ALS has made it impossible to paint and over the course of a few years, has left her with only limited control of facial muscles. She learns of a BrainPainting system designed to allow people to paint using a P300 BCI. She tries it, and after several hours, produces an original painting. She is moved to tears, stating (via a P300 BCI speller) that she "feels like an artist again" and recognizes her own style.

Over time, she produces many more BrainPaintings, gaining some attention from local and national media. Since she is very close to a local university, she has help from a local graduate student for technical issues. She can also afford an extra 2,000 Euros for a dry electrode cap, and she does not especially mind wearing it. Her main concern is the speed. She can spell and paint, but much more slowly than before. The system is not perfectly accurate and mistakes are frustrating. She also dislikes the appearance. She feels that it makes her look odd and highlights her dependence on an assistive technology. The cables are especially annoying to her. She would prefer a system with electrodes embedded in a hair beret or earrings. The manufacturer has replied that they are working on it.

Are We There Yet?

Today's BCIs systems still have many limitations (Babiloni et al. 2009; Sellers et al. 2010; Allison et al. 2013). BCI technology is at the point of migration from the laboratory setting to the domestic environment, which entails many challenges and limitations. But the technology has been at this point for some time. What is stopping BCI being deployed widespread? Over five years ago there was a significant impetus within the research community and funding bodies such as the European Union Framework Programme financed large projects (Brain 2013; Brainable 2013; TOBI 2013; Future BNCI 2013) to enhance BCI technology, with the vision of moving it out of the laboratory and into the domestic setting. At that time some of the key scientific advances noted to help achieve this were:

- A convenient setup: The objective was to develop an inexpensive and straightforward EEG acquisition system that can be easily mounted on the head without expert supervision. The importance of aesthetics is highlighted.
- Individualized BCI: The creation of a flexible, reliable BCI system that can automatically identify and optimize important BCI parameters with minimal hassle to the user.
- Application suite: The creation of a straightforward, easy to use link between the BCI system and exemplar applications. Broadening the availability of applications and enabling a modular system consistent with use for a diverse group of applications from multimedia to domotic control.
- Evaluation: To involve target users to inform the development and provide valuable evaluation results for comparison with healthy user trials.

There have been significant movements forwards in these areas but the problems are complex and many goals are still relevant. The key aspects are summarized below.

A Convenient Setup

Several recent reports on users' opinions of BCI have reported negativity towards gel EEG caps and electrodes (Blain-Moraes et al. 2012; Ekandem et al. 2012; Huggins et al. 2011). This is not a surprise and research to develop dry (Gargiulo et al. 2010) and water based electrodes (Mihajlovic et al. 2012) has been undertaken. Volosyak et al. (2010) reported similar results when comparing gel based electrodes with water based electrodes, which rely on simple tap water to moisturize the electrodes. Sufficiently high quality EEGs were achieved even after 4 h of continuous usage. Guger et al. (2012b) found that a dry electrode system was effective for nearly all users, who attained accuracy comparable to state of the art gel-based electrode systems. The ability for non-technical application of electrodes with no clean up time will open up BCI as a more practical day to day technology. It also offers the promise of an easily made customizable cap tailored to the user, creating greater comfort and aesthetics.

The complexity and set up is a major hindrance in the widespread adoption of BCI technology. Blain-Moraes et al. (2012) evaluated the opinions of group of target users with ALS. They found that those with technical knowledge expressed confidence in "their ability to learn to autonomously use and operate BCI". However, those uncomfortable with technology found the complexity of the BCI systems overwhelming, with one user stating, "how can this be made accessible to the computer illiterate or technology illiterate…?" Without existing technology competence the naïve user felt that the BCI operation was beyond their capacity for independent use.

The shrinking of the technology is an absolute must for enhancing battery life and system portability. Brunner et al. (2011b) provide some insight in the current status of software and hardware development for BCI.

Individualized BCI

There is significant complexity involved in providing successful BCI for the individual. Some key issues recur in the literature, namely, technical complexity, the need for strong carer and family support, the need for training, on-going technical support and the BCI accuracy disparity between users. The latter is also highlighted by Allison (Allison et al. 2010a), stating that there is no "universal BCI".

One of the main disparities in BCI technology is the difference in user efficacy between individual users and groups of users. The concept of a generalized BCI setup that can be deployed to the masses actually only has potential impact on certain groups of users. For those in greatest need of a method of non-muscular communication and control the resulting system is not readily available. Commercial BCI technology (Intendix 2013a; Emotiv 2013a) demonstrates potential success for non-disabled users (Guger et al. 2009, 2012b) but whether it will be suitable for the broad spectrum of users remains to be seen. The general BCI literature discusses the measures and support needed for long term domestic BCI use and highlights many challenges (Sellers et al. 2010). Nevertheless, for those who could operate such easily available and refined BCI systems, the technology could prove to be beneficial. Different users with or without brain injury may have a broad spectrum of BCI capabilities, referred to as BCI literacy (Allison et al. 2010). In reality the development of any assistive technology requires a combination of both clinical and technical expertise, tailoring the technology and applications to the user. The Brain Communication Foundation (Brain Communication 2013) is a non-profit organization with the aim of developing BCI for users in which there is a need for such a tailored approach, unsuitable for a more commercial product.

Automated Configuration

How do we determine a personalized BCI for the user? There are a range of potential solutions but matching the person and their needs to a possible solution is a multifaceted task. It relies on the ability to enhance the BCI system from the best position of the electrodes, to tailoring the algorithms, and providing the applications, services and support needed.

How do we determine without long trials with a user what form of BCI will be best suited to their needs and characteristics? Mak et al. (2012) are looking for key parameters that might show a user's feasibility for use with the P300-BCI. If they can determine the EEG features that best correlate with P300 performance they can use this to not only determine suitable candidates for long-term P300 BCI operation, but they can also monitor performance online, an important aspect for remote technical support.

A realistic overview of what the near future could achieve is considered with some suggestions of what such a system would entail and with what caveats it would operate under. FutureBNCI (2013) was established within a cluster of thirteen European funded BCI projects with the joint aim to promote and guide BCI research, development and application. They provided an insight into the future perspectives (Allison 2011; Allison et al. 2013) with a clear overview of combining BCI modalities resulting in hybrid systems. Such technology could combine different BCI mechanisms (Allison et al. 2012; Brunner et al. 2011a; Pfurtscheller et al. 2010) or combine BCI with other input modalities such as eye-trackers. A fluid approach to what is best for the user which steps across the technical boundaries is now the goal. Millán et al. (2010) discussed the need for increasing the level of automation within the system to compensate for low accuracies and creating more context-aware systems that improve with use (Allison et al. 2012; Wolpaw and Wolpaw 2012).

Application Suite

There remains a strong need for tools that can tailor each BCI to each user and there are many aspects of a BCI that could be customized: the sensor system (such as different electrode types and montages); many details of pattern classification (such as which electrodes and frequencies are used for control); the type of brain activity used for control (such as P300 or SSVEP); the application being controlled (such as a speller or internet browser); the interface (such as different displays and feedback methods); and other details. A reliable "BCI Wizard" is needed, a software platform that would walk each user through a series of tests and questions to help identify which options would be best for that user. There has been substantial progress toward such a goal, with many improved open-source software platforms that require less expert help than previous versions (Brunner et al. 2013). The first home commercial BCI platform, called intendiX (IntendiX 2012), also uses software that is aimed at non-expert users. These software tools have reduced the burden on users and their carers, but a practical BCI that can provide a wide range of assistive technology solutions for end users requires further research and development.

BCI in Use: Beyond Evaluation

Nam et al. (2010) report that BCI's lack of acceptance could be a consequence of a lack of understanding of the usability of BCI systems. Finding the right opportunities to make BCI usable and accessible offer the potential to turn BCIs into practical assistive technologies that can help users interact with their family and carers, as well as home-based technologies including assistive devices, home appliances, or computer and internet technologies, A key challenge to this is to minimize the work in deploying BCI systems successfully for users and their supporters.

There is a growing need for a "BCI service provider", referring to people or companies that can provide expert support. These service providers should ideally be certified through an entity consisting of appropriately qualified and experienced experts to avoid misrepresentations of providers' capabilities. The providers may often need to travel to users' homes as well as provide remote support and should be familiar with the challenges unique to any patient populations they might encounter. But what needs to be done to achieve this vision and at a feasible cost?

Discussion: Can BCI Provide a Solution for OD:LIS?

At the beginning of the chapter we asked the questions:

> Can BCI offer the mechanism to enhance social inclusion and empowerment through communication and control? And if so, what hindrances are there in making it readily available for those that would benefit most?

There are many obstacles to the uptake of a BCI system. One of the major issues when collecting the EEG is artefacts caused by movements such as eye blinks, facial twitching and jaw clenches. BCI systems require the user to suppress such movements which may be involuntary due to their underlying condition. Ourand (2004) gives an informative overview of BCI for the ALS Association. She makes a key point that "when the individual is 'concentrating' so fiercely on regulating brain activity and limiting muscle movements, it can severely impact non-verbal pragmatic language and interactions with those in the immediate environment". She also highlights the commitment in time and energy that users may need to invest in training for BCI use before a usable system is achieved. There is certainly a disparity in efficacy between users and indeed user groups. Studies involving target users systematically report lower accuracies than healthy users (Mulvenna et al. 2012).

Kübler et al. (2005) investigated a number of variations of BCI systems working with healthy users and 7 users with ALS (pre locked in state). They comment on the disparity between the user groups, highlighting the need for longer training sessions for the patients with ALS. Healthy users were able to achieve a level of control over a small number of sessions but the patients needed 20 sessions to achieve a 70 % accuracy using BCI with imagined movement (ERD/ERS).

Such disparity can be expected for many varied reasons such as the underlying neurological condition of the user (brain injury for example), the difficulty in controlling involuntary movement, the ability to maintain visual focus on the objects of interest on a screen or maintaining concentration on the imagined movement. Ourand (2004) adds that BCI solutions using visual stimuli such as P300 and SSVEP can be "very fatiguing, and are often not useful in the presence of certain visual impairments, since some approaches require visual prompts from the screen. Because some people who are locked-in have limited vision, research in auditory interfaces is an important focus for the BCI community." Much more investigation needs to be done with target users, to determine how best to tailor BCI for their diverse needs.

All this said, can BCI provide a feasible assistive technological solution for people with ALS? There have been some long term studies with users but the numbers are limited. Sellers et al. (2010) report on their involvement with a user with ALS in which the user had two and a half years of independent use of a P300 based BCI. There were issues, common to most BCI systems, such as difficulty of use, high technical complexity, functionality and lack of user personalization. These factors made it difficult for long term support but the research team has endeavored to make some improvements. Neuper et al. (2003) provide an example of a BCI system for spelling (using imagined movement (ERD/ERS BCI)) established within the patient's home (clinical) setting and training was performed over several months. Technical assistance was also provided on-line. An average spelling accuracy of 70 % was achieved.

So there are some examples of successful BCI use out of the laboratory and with target users. But it is a complex task to achieve and is dependent on many factors. Potential users need to be screened to determine their feasibility for BCI use, as highlighted by Vaughan et al. (2006). As already discussed, there are many forms of BCI and within each of these types, a range of characteristics can be tailored for the specific user. BCI literacy varies from person to person for even those without any underlying neurological condition (Allison et al. 2010). For example, some people may not demonstrate a strong response in their EEG in response to a flashing light (for SSVEP-BCI), yet they may be able to use the Intendix Speller (Intendix 2013a) which uses a different type of visual stimulus (P300 BCI). Others may find it difficult using visual stimuli of any form due to sight issues and may need to use imagined movement (ERD/ERS BCI) or even auditory BCI (Nijboer et al. 2008; Birbaumer et al. 2012).

The Wadsworth Centre for Brain Computer Interfaces (Wadsworth 2013) are at the stage of development whereby they can offer a research version of the system for home use for specified users who have undergone initial suitability investigations. They report that they have used domestic BCI within their homes for several months or more. "One has now been using the system up to eight hours per day for two and a half years." They provide guidelines as to who might be eligible to be involved in their BCI trials:

- Severely paralyzed by any of a variety of neuromuscular disorders such as ALS, cerebral palsy, muscular dystrophy, multiple sclerosis, and high-level spinal cord injury.
- Too disabled to use conventional assistive communication technology such as systems that use muscle activity or eye movements.
- In adequately stable physical condition, with stable physical and social environments, and with caregiver(s) who have basic computer skills.
- Able to see and to understand instructions.
- Able to use the BCI system as determined in a screening evaluation.
- In a geographical location and an environment that allows the Wadsworth BCI group to provide ongoing technical support.

They report a common set of problems for the widespread deployment of BCI, namely, the substantial level of technical support required. The cost the system hardware is reported to be in the region of 5,000 dollars. However, it is expected that this financial value is not the true cost, due to the large support overhead. The goal of the Wadsworth center is to reduce this overhead by means of simplification of the BCI system. They also plan to develop more applications and deploy the system on a Windows platform. FDA approval is to be sought to enable widespread dissemination of the system beyond the research capacity. They point to the Brain Communication Foundation as a possible source for funding.

Positive stories have reached the media. The BrainGate (2012, 2013) neural interface was reported recently in the news, depicting a video of one of the two end users involved in their trials. The woman, who is paralysed, had implanted electrodes and through the BCI paradigm that uses imagined movement she was able to control a robotic arm. Although positive, the researchers made the important point that the user had trained long and hard to be able to gain the control.

Conclusions

BCIs offer a possible mechanism for communication and interaction with external devices using solely non-muscular interaction. Since BCIs do not require movement, they may provide a potential medium for interaction for the most severely paralyzed people, who have little or no reliable control of voluntary movements. They offer a potential assistive technology for people with neuromuscular disorders which, when the conditions are severe can lead to a locked in state for the patient. At this point mainstream assistive communication devices may not be helpful as they rely on residual motor movements. BCI could provide a feasible technology to reinstate a level of interaction and control to the user.

Until recently, most BCI research efforts focused on helping such users, with relatively little focus on other user demographics. This has begun to change, as various improvements to BCIs and underlying technologies have drawn attention to other user groups who might also benefit, leading to a potential diverse user group. For example, people who have lost an arm or the ability to control an arm, might use BCIs to control a device to restore function (Velliste et al. 2008; Hochberg et al. 2012; Mattia et al. 2013). It has sparked interest within gaming (Allison 2007; Nijholt 2009; Tangermann et al. 2009), and may offer an avenue for bio-feedback rehabilitation for conditions such as autism (Zhu et al. 2011). BCIs might also help facilitate stroke recovery (Gomez-Rodriguez et al. 2011; Ortner et al. 2012; Mattia et al. 2013).

In terms of an assistive technology, the opportunity to gain control and express oneself without movement is at one end of the diverse spectrum of the user characteristics, with the gaming community at the other extreme. Between these two extremes, BCI could act as an alternative assistive device that alleviates the stress on the user by switching technology when one becomes tiresome or ineffective. People with mild to moderate disabilities might use a BCI as a supplementary or complementary communication system when other assistive technologies are unavailable or impractical. Indeed, one of the most active BCI research areas involves hybrid BCIs, which combine BCIs with other communication devices to provide users with a suite of communication options (Pfurtscheller et al. 2010; Brunner et al. 2011a; Allison et al. 2012a; Müller-Putz et al. 2012).

"The solutions have the potential to improve productivity and extend communication for education, vocation, recreation and leisure activities. It is indisputable that when BCI technology becomes a routine, everyday symptom management device, individuals will likely experience increased independence and improved quality of life." (Ourand 2004)

"Despite the wealth of interest and solid work in this field, it has to be said that overall the field is still in the research and development phase. Although clinical trials of devices are on the near horizon, the field has more work to accomplish before the technology is readily available and is a proven intervention for people with ALS. With the generosity, dedication and involvement of people with ALS and their families, the clinical studies to test the practicality and effectiveness of services will help immeasurably to move the field forward." (Ourand 2004)

All this is only possible with the unwavering support network of carers, family and friends.

References

Allison B, Graimann B, Gräser A. Why use a BCI if you are healthy. ACE Workshop-Brain-Computer Interfaces and Games, 2007. pp. 7–11.
Allison B, Luth T, Valbuena D, Teymourian A, Volosyak I, Graser A. BCI demographics: How many (and what kinds of) people can use an SSVEP BCI? IEEE Trans. Neural Sys. Rehabil. Eng. 2010;18(2):107–16.
Allison BZ. Trends in BCI research: Progress today, backlash tomorrow? ACM Crossroads. 2011;18(1):18–22.
Allison BZ, Leeb R, Brunner C, Muller-Putz GR, Bauernfeind G, Kelly JW, et al. Toward smarter BCIs: Extending BCIs through hybridization and intelligent control. J Neural Eng. 2012;9(1):013001.

Allison BZ, Dunne S, Leeb R, Millan J, Nijholt A. Recent and upcoming BCI progress: Overview, analysis, and recommendations. In: Allison BZ, Dunne S, Leeb R, Millan J, Nijholt A, editors. Towards practical BCIs: bridging the gap from research to real-world applications. Berlin: Springer; 2013. pp. 1–13

ALSA. ALS association. What is ALS? http://www.alsa.org/about-als/what-is-als.html. (2013a). Accessed Feb 2013.

ALSA. ALS association. Facts you should know. http://www.alsa.org/about-als/facts-you-should-know.html. (2013b). Accessed Feb 2013.

Andersson P, Ramsey NF, Pluim JP, Viergever M.A. BCI control using 4 direction spatial visual attention and real-time fMRI at 7T. In Engineering in Medicine and Biology Society (EMBC), 2010 Annual International Conference of the IEEE. 2010; pp. 4221–4225.

Babiloni F, Cincotti F, Marciani M, Salinari S, Astolfi L, Aloise F, et al. On the use of brain-computer interfaces outside scientific laboratories: Toward an application in domotic environments. Int Rev Neurobiol. 2009;86:133–46.

Birbaumer N. Breaking the silence: Brain–computer interfaces (BCI) for communication and motor control. Psychophysiology. 2006;43(6):517–32.

Birbaumer N, Cohen LG. Brain-computer interfaces: communication and restoration of movement in paralysis. J Physiol. 2007;579:621–36.

Birbaumer N, Piccione F, Silvoni S, Wildgruber M. (2012). Ideomotor silence: the case of complete paralysis and brain–computer interfaces (BCI). Psychol. Res. 2012;1–9.

Blain-Moraes S, Schaff R, Gruis KL, Huggins JE, Wren PA. Barriers to and mediators of brain–computer interface user acceptance: Focus group findings. Ergonomics. 2012;55(5):516–25.

BRAIN. BCIs with rapid automated interfaces for nonexperts (BRAIN) http://www.brain-project.org. (2013). Last accessed Jan 2013.

Brainable Autonomy and social inclusion through mixed reality Brain-Computer Interfaces: Connecting the disabled to their physical and social world, http://www.brainable.org/en/Pages/Home.aspx. (2013). Last accessed Jan 2013.

Brain Communication. Brain Communication Foundation, http://www.braincommunication.org/. (2013). Last accessed Jan 2013.

BrainGate. BrainGate amazes again: Paralyzed woman moves thought-controlled robotic arm, http://neurogadget.com/2012/05/19/braingate-amazes-again-paralyzed-woman-moves-thought-controlled-robotic-arm-video/4370. (2012). Neurogadget, 19 May 2012.

BrainGate. BrainGate™ http://www.braingate.com/. (2013). Last accessed Jan 2013.

Brunner C, Allison BZ, Altstatter C, Neuper C. A comparison of three brain-computer interfaces based on event-related desynchronization, steady state visual evoked potentials, or a hybrid approach using both signals. J Neural Eng. 2011a;8(2):025010.

Brunner C, Andreoni G, Bianchi L, Blankertz B, Breitweiser C, Kanoh S, Kothe C, Lecuyer A, Makeig S, Mellinger J, Perego P, Renard Y, Schalk G, Susila IP, Venthur B, Müller-Putz G. BCI Software Platforms. In: Allison BZ, Dunne S, Leeb R, Millan J, Nijholt A. editors. Toward practical BCIs: Bridging the gap from research to real-world applications. Berlin: Springer; 2013. pp. 303–331.

Brunner P, Bianchi L, Guger C, Cincotti F, Schalk G. Current trends in hardware and software for brain–computer interfaces (BCIs). J Neural Eng. 2011b;8:025001.

Bauby JD. The diving bell and the butterfly: A memoir of life in death. Vintage (1998).

Calvo A, Chiò A, Castellina E, Corno F, Farinetti L, Ghiglione P, Vignola A. Eye tracking impact on quality-of-life of ALS patients. Computers Helping People with Special Needs, 2008. pp. 70–77.

De Laar B, Guerkoek H, Plass-Oude Bas D, Nijber F, Nijholt, A. Brain-computer interfaces and user experience evaluation. In: Allison BZ, Dunne S, Leeb R, Millan J, Nijholt A. Towards Practical BCIs: Bridging the Gap from Research to Real-World Applications, Berlin: Springer; 2013. pp. 223–237.

Ekandem JI, Davis TA, Alvarez I, James MT, Gilbert JE. Evaluating the ergonomics of BCI devices for research and experimentation. Ergonomics. 2012;55(5):592–8.

Emotiv. Emotiv company website. http://www.emotiv.com/. (2013a). Accessed Feb 2013.

Emotiv. Emotiv company website, EEG features. (2013b). http://www.emotiv.com/eeg/features.php. Accessed Feb 2013.

Ensrud E. Can computers read your mind? Neurology. 2005;64:E30. doi:10.1212/WNL.64.10.E30.

Future BNCI. Future Directions in BNCI Research, http://future-bnci.org/ EU funded: http://cordis.europa.eu/search/index.cfm?fuseaction=proj.document&PJ_RCN=11158040. (2013). Last accessed Jan 2013.

GARD. Genetic and rare diseases information center. (2013). http://rarediseases.info.nih.gov/GARD/Condition/5786/Amyotrophic_lateral_sclerosis.aspx. Accessed Feb 2013.

Gargiulo G, Calvo RA, Bifulco P, Cesarelli M, Jin C, Mohamed A, van Schaik A. A new EEG recording system for passive dry electrodes. Clin Neurophysiol. 2010;121(5):686–93.

Gehrig. Lou Gehrig Biography web page. (2013). http://www.biography.com/people/lou-gehrig-9308266. Accessed Feb 2013.

Gomez-Rodriguez M, Grosse-Wentrup M, Hill J, Gharabaghi A, Scholkopf B, Peters J. Towards brain-robot interfaces in stroke rehabilitation. IEEE Int Conf Rehabil Robot. 2011;2011:5975385.

Graimann B, Allison BZ, Pfurtscheller G. A gentle introduction to brain—computer interface (BCI) systems, In: Graimann B, Allison BZ, Pfurtscheller G editors. Brain-Computer interfaces: revolutionizing Human-Computer Interaction. Berlin: Springer; 2010. pp. 1–28.

Guger C, Daban S, Sellers E, Holzner C, Krausz G, Carabalona R, et al. How many people are able to control a P300-based brain? computer interface (BCI)? Neurosci Lett. 2009;462(1):94–8.

Guger C, Allison BZ, Großwindhager B, Prückl R, Hintermüller C, Kapeller C, Bruckner M, Krausz G, Edlinger G. How many people could use an SSVEP BCI? Front Neurosci. 2012;6:169.

Guger C, Krausz G, Allison BZ, Edlinger G. Comparison of dry and gel based electrodes for P300 brain–computer interfaces. Front Neurosci. (2012b):6.

Hayashi H, Oppenheimer EA. ALS patients on TPPV: totally locked-in state, neurologic findings and ethical implications. Neurology. 2003;61(1):135–7.

Hochberg LR, Bacher D, Jarosiewicz B, Masse NY, Simeral JD, Vogel J, Haddadin S, Liu J, Cash SS, van der Smagt P, Donoghue JP. Reach and grasp by people with tetraplegia using a neurally controlled robotic arm. Nature. 2012;485(7398):372–5. doi:10.1038/nature11076.

Huggins JE, Wren PA, Gruis KL. What would brain-computer interface users want? Opinions and priorities of potential users with amyotrophic lateral sclerosis. Amyotrophic Lateral Sclerosis. 2011;12(5):318–24.

Intendix. 2013. http://www.intendix.com/. Accessed Jan 2013.

Intendix. IntendiX®SPELLER Video. 2013a. http://www.youtube.com/watch?v=NlUPFpZswJk. Accessed Jan 2013.

Indendix. Domestic Control Video. 2013b. http://www.youtube.com/watch?v=bFwNi_M32cE&NR=1. Accessed Jan 2013.

Indendix. Painting by thoughts. 2013c. http://gtecmedical.wordpress.com/2012/10/22/paint-by-thoughts-only/. Accessed Jan 2013.

Intendix. Gaming using BCI. 2013d. http://gtecmedical.wordpress.com/2012/03/20/intendix-soci-breakthrough-in-gaming/. Accessed Jan 2013.

Jones M, Grogg K, Anschutz J, Fierman R. A sip-and-puff wireless remote control for the Apple iPod. Assistive Technol. 2008;20(2):107–10.

Kübler A, Nijboer F, Mellinger J, Vaughan TM, Pawelzik H, Schalk G, McFarland DJ, Birbaumer N, Wolpaw JR. Patients with ALS can use sensorimotor rhythms to operate a brain-computer interface. Neurology. 2005;64:1775-7.

Mak JN, McFarland DJ, Vaughan TM, McCane LM, Tsui PZ, Zeitlin DJ, et al. EEG correlates of P300-based brain–computer interface (BCI) performance in people with amyotrophic lateral sclerosis. J Neural Eng. 2012;9(2):026014.

Mason SG, Bashashati A, Fatourechi M, Navarro KF, Birch GE. A comprehensive survey of brain interface technology designs. Ann Biomed Eng. 2007;35(2):137–69.

Mattia D, Picchiori F, Molinari M, Rupp R. Brain computer interface for hand motor function restoration and rehabilitation. In: Allison BZ, Dunne S, Leeb R, Millan J, Nijholt A editors. Towards Practical BCIs: Bridging the Gap from Research to Real-World Applications, Berlin: Springer; 2013. pp. 131–153.

McCullagh P, Ware M, McRoberts A, Lightbody G, Mulvenna M, McAllister G et al. Towards standardized user and application interfaces for the brain computer interface. Universal Access in Human-Computer Interaction. Users Diversity. (2011):573–582.

McFarland DJ, Wolpaw JR. Brain-computer interface operation of robotic and prosthetic devices. Computer. 2008;41(10):52–6.

Mihajlovic V, Molina GG, Peuscher J. To what extend can dry and water-based EEG electrodes replace conductive gel ones? A steady state visual evoked potential brain-computer interface study," In: Proceeding of the 5th International Joint Conference on Biomedical Engineering System and Technologies (BIOSTEC 2012). (2012).

Millán JR, Rupp R, Muller-Putz GR, Murray-Smith R, Giugliemma C, Tangermann M, et al. Combining brain-computer interfaces and assistive technologies: state-of-the-art and challenges. Front Neurosci. 2010;4:161.

Millán JR, Carmena JM. Invasive or noninvasive: understanding brain-machine interface technology. IEEE Eng Med Biol Mag. 2010;29(1):16–22.

MindWave. MindWave Mobile from Neurosky. http://www.neurosky.com/Products/MindWaveMobile.aspx. 2013. Accessed Feb 2013.

Miranda ER, Magee WL, Wilson JJ, Eaton J, Palaniappan R. Brain-computer music interfacing (BCMI) from basic research to the real world of special needs. Music Med. 2011;3(3):134–40.

Molina GG, Mihajlovic V. Spatial filters to detect steady-state visual evoked potentials elicited by high frequency stimulation: BCI application. Biomedizinische Technik/Biomed Eng. 2010;55(3):173–82.

Müller-Putz GR, Leeb R, Millán JDR, Horki P, Kreilinger A, Bauernfeind G, Scherer R. Principles of hybrid Brain–Computer interfaces. Towards Pract Brain-Comput Interfaces. (2012):355–373.

Mulvenna M, Lightbody G, Thomson E, McCullagh P, Ware M, Martin S. Realistic expectations with brain computer interfaces. J Assistive Technol. (2012);6(4):233–244.

Münßinger JI, Halder S, Kleih SC, Furdea A, Raco V, Hoesle A, Kübler A. Brain painting: first evaluation of a new brain–computer interface application with ALS-patients and healthy volunteers. Front Neurosci. (2010):4.

Nam CS, Schalk G, Jackson MM. Current trends in Brain–Computer interface (BCI) research and development. Intl J Human–Comput Interact. 2010;27(1):1–4.

Neuper C, Müller GR, Kübler A, Birbaumer N, Pfurtscheller G. Clinical application of an EEG-based brain–computer interface: a case study in a patient with severe motor impairment. Clin Neurophysiol. 2003;114(3):399–409.

Nijboer F, Furdea A, Gunst I, Mellinger J, McFarland DJ, Birbaumer N, Kübler A. An auditory brain–computer interface (BCI). J Neurosci Methods. 2008;167(1):43–50.

Nijboer F, Broermann U. Brain–computer interfaces for communication and control in locked-in patients. Brain-Comput Interfaces. (2010):185–201.

Nijboer F, Clausen J, Allison BZ, Haselager P. The Asilomar survey: researchers' opinions on ethical issues related to brain-computer interfacing. Neuroethics. 2011:1–38.

Nijholt A. BCI for games: a 'state of the art' survey. Entertainment Comput ICEC. 2009;2008:225–8.

Orphanet. Orphanet, the portal for rare diseases and orphan drugs. http://www.orpha.net/consor/cgi-bin/index.php?lng=EN. 2013. Accessed Nov 2013.

Ortner R, Irimia DC, Scharinger J, Guger C. A motor imagery based brain-computer interface for stroke rehabilitation. Stud Health Technol Inform. 2012;181:319–23.

Ourand PR. FYI brain computer interface technology, ALS association, October 2004. Reviewed and Updated by Melody Moore, PhD, Feb 2005. http://www.alsa.org/als-care/resources/publications-videos/factsheets/brain-computer-interface.html. 2004. Accessed Jan 2013.

Patient. Motor neurone disease. http://www.patient.co.uk/health/Motor-Neurone-Disease.htm. 2013. Accessed Jan 2013.

Pfurtscheller G, Muller-Putz GR, Scherer R, Neuper C. Rehabilitation with brain-computer interface systems. Computer. 2008;41(10):58–65.

Pfurtscheller G, Solis-Escalante T, Ortner R, Linortner P, Muller-Putz GR. Self-paced operation of an SSVEP-based orthosis with and without an imagery-based "brain switch:" a feasibility study towards a hybrid BCI. IEEE Transactions on Neural Systems and Rehabilitation Engineering: A Publication of the IEEE Engineering in Medicine and Biology Society. 2010;18(4); 409–414.

Sellers EW, Vaughan TM, Wolpaw JR. A brain-computer interface for long-term independent home use. Amyotrophic Lateral Scler. 2010;11(5):449.

Sugiarto I, Allison B, Graser A. Optimization strategy for SSVEP-based BCI in spelling program application. Computer Engineering and Technology, 2009. ICCET'09. International Conference on, 1. 2009. pp. 223–226.

Tangermann M, Krauledat M, Grzeska K, Sagebaum M, Blankertz B, Vidaurre C, et al. Playing pinball with non-invasive BCI. Adv Neural Inf Process Sys. 2009;21:1641–8.

Thomas R. Diving Bell movie's fly-away success, BBC News Channel, http://news.bbc.co.uk/1/hi/entertainment/7230051.stm. (2008). Accessed Feb 2013.

TOBI. Tools for brain-computer interaction. 2013. http://www.tobi-project.org/. Last accessed Jan 2013.

Vaughan TM, McFarland DJ, Schalk G, Sarnacki WA, Krusienski DJ, Sellers EW, Wolpaw JR. The wadsworth BCI research and development program: at home with BCI. Neural Systems and Rehabilitation Engineering, IEEE Transactions on. 2006; 14(2):229–233.

Velliste M, Perel S, Spalding MC, Whitford AS, Schwartz AB. Cortical control of a prosthetic arm for self-feeding. Nature. 2008;453(7198):1098–101. Epub 2008 May 28.

Volosyak I, Valbuena D, Malechka T, Peuscher J, Gräser A. Brain–computer interface using water-based electrodes. J Neural Eng. 2010;7(6):066007.

Wadsworth. The wadsworth center brain-computer interface system. 2013. http://www.wadsworth.org/bci/faq.html. Accessed Jan 2013.

Ware M, McCullagh P, Mulvenna M, Nugent C, McAllister H, Lightbody G. A universal command structure for multiple domotic device interactions. TOBI Workshop. 2010; p. 41.

Wolpaw JR, Birbaumer N, McFarland DJ, Pfurtscheller G, Vaughan TM. Brain–computer interfaces for communication and control. Clin Neurophysiol. 2002;113(6):767–91.

Wolpaw J, Wolpaw EW, editors. Brain-computer interfaces: principles and practice. USA: Oxford University Press; 2012.

Zander TO, Kothe C. Towards passive brain-computer interfaces: applying brain-computer interface technology to human-machine systems in general. J Neural Eng. 2011;8(2):025005.

Zhu D, Bieger J, Molina GG, Aarts RM. A survey of stimulation methods used in SSVEP-based BCIs. Comput Intell Neurosci. 2010;2010:1.

Zhu H, Sun Y, Zeng J, Sun H. Mirror neural training induced by virtual reality in brain–computer interfaces may provide a promising approach for the autism therapy. Med Hypotheses. 2011;76(5):646–7.

Vignette: The Wilderness

Jeneva Stone

I was certain, when my son became ill in 1998, that giving him a vitamin each day, then gradually becoming inconsistent, was connected. No one else thought this. You know, Enfamil infant vitamins, the ones with bunnies on the label. After the first hospitalization, I gave them to Robert Henry again. He was also given PediaSure, which is chockfull of vitamins. He improved steadily. Still no one thought this. Then I became inconsistent with the vitamins, and he crashed again.

J. Stone (✉)
Maryland, USA
e-mail: jenevastone@gmail.com

What had happened was dramatic: with warning only in hindsight, Robert lost the majority of his voluntary motor function over the course of a long weekend. No medical findings presented themselves until 2003: no positive or abnormal test results, no data. No strange MRIs or CTs. Just lots of chaotic presentation, lots of static. And he was really, really ill. As though he might not survive. Eventually, Robert was confined to a wheelchair and tube-fed. His neuromuscular problems caused acid reflux and vomiting. Robert is non-verbal, but he listens to everything and his eyes belie a lively, unexpressed intelligence.

Robert wasn't diagnosed until 2012 when genome sequencing revealed one rare inherited and one *de novo* variant on each copy of his PRKRA gene, associated with dystonia. The diagnosis yields no definitive treatments, Dystonia 16 being so rare that Robert is only the 9th reported case in the world. His motor difficulties were always complex to describe, and drug therapies often seemed to help over the years we pursued treatment without a diagnosis. Let's just say that once you go the drug route, there are times you can't see the forest for the trees.

As I struggle to determine what image describes this 'route', the first thing that pops into my mind is a scene from the movie *Last of the Mohicans*, the one with Daniel Day-Lewis. His character, Hawkeye, helps Alice and Cora Monro escape—I don't know—the wilderness, other Indian tribes, the French, the smoking remains of Fort William Henry. They try to return to civilization, whatever that is. In the movie's version of 1757, civilization consists of British people drinking tea out-of-doors on parlor furniture in the middle of a field. This reminds me of our lives: pretending to some sort of order against a backdrop of benevolent or malevolent mayhem, faking decorum all the way.

Hawkeye leads the women into the Appalachian woods and a cave beneath the 'Falls of Glen,' or Glens Falls in modern-day upstate New York where businesses named after this period in American history still stand: a motel and a bar named the Montcalm and the Portage.

Hawkeye tells Cora, 'Stay alive, I will find you,' just before he jumps into the falls and her small party of British folks dressed in clothing inappropriate for wilderness junkets is captured by Magua and his band of evil French-influenced Native Americans. Hawkeye reasons that if he is captured with Cora and Alice, he will be unable to save them. Cora then does everything in her power to leave a trail he can follow: breaking branches, leaving footprints in soft soil, and so on.

The endless loop of medications we've tried is a bit like tracking something through an Appalachian forest. The terrain feels familiar, but the signs we're looking for are elusive and easily misread. It takes some kind of powerful love or crazy to keep pushing back the low-hanging branches, watching for the poison ivy and oak, and parting the undergrowth and ferns without crushing them. Whatever those old-growth east coast forests looked like, they were cluttered with deciduous vegetation thick enough to conceal a clearing or a sudden precipice.

At times, Robert has been on seven medications at once. This requires actually writing things down or having a spouse who can also remember doses and times so you can continually cross-check the day with him. When did he get that Artane dose? Was it the last one or only the second one he's supposed to have

Vignette: The Wilderness

had today? Should we give the Prevacid early because we've gotten behind on his feeds? Some medications are given once a day, some b.i.d., some t.i.d. We never had a drug regimen in which all were given on the same schedule. Try adding an innocuous antibiotic to that.

When away from Robert, I look at my watch for the first 2–3 days, involuntarily keeping track of when he should be getting medicine and each feeding bolus.

Medications and vitamins Robert has taken over time: Sinemet 10/100, Sinemet 25/100, Artane, Prilosec, Prevacid, Zantac, Reglan, Ritalin, Zoloft, Celexa, 5-HTP, Glycolax, Botox, Klonopin, Baclofen, intrathecal Baclofen, B-vitamin complex, biotin, mitochondrial vitamin cocktail. I'm not sure I remember all the minor trial medications or hospital-only drugs.

The extra-added complication with Robert, of course, is the g-tube. Many of these medications are not available in liquid form. Sinemet, for example, is only soluble in vast quantities of liquid fat—say, a cup of olive oil. For years, the proton-pump inhibitor drugs, Prilosec and Prevacid were only available in time-release capsules he could not swallow. If the gel cap were pulled apart to release the little beads to mix in something that could be pushed through a g-tube, no one guaranteed the medication would work effectively.

A lab at the University of Missouri, Columbia saved Robert's life with a solvent that stabilized either reflux drug for about a month. But it required refrigeration. Most stuff we crush into powder or release as powder from tiny capsules, one by one, and mix with applesauce, a great binder and suspension-maker. We thin it with a bit of water and draw the mess up into a syringe.

When I look back on this decade of wending our way through the medication wilderness, all I want to do is tear my hair out. Because it always felt as though we were getting somewhere but never arriving. Yes, that was the same tree we passed a few hours ago, and it is getting dark. But we keep walking by other stuff that looks different, don't we? I would yell at these previous incarnations of myself to "get an effing compass," but, unfortunately, they do not make compasses for administering and monitoring medication that are g-tube compatible.

But, I kept telling myself: we appear to be getting somewhere. Yes, we are in a clearing and the Falls of Glen are near. I can hear the water rushing. The air smells damp. I don't know who will find us though.

By 2008, tantalizing biochemical clues manifested themselves as MRI and other findings: odd neurotransmitter patterns in his cerebrospinal fluid, issues with metabolites of medium-chain triglycerides. Broken branches, torn leaves. Could he have a mitochondrial disease? Because it had produced results in another patient with a similar MRI, our neurologists suggested giving Robert high doses of the vitamin biotin.

So locked in that, most days, he could barely move, within 15 minutes of his first dose, Robert made sweeping, fluid motions with his arms. He tapped a plate repeatedly with a spoon and pushed two objects together with his fists. We changed his Sinemet formulation from 10/100 to 25/100. We left Zoloft lying by the side of the road—we were using it off-label for serotonin replacement. We waved good-bye to Artane and never looked back.

We upped the amount of biotin, a B vitamin. You can buy it at GNC. The biggest commercially available capsule for human consumption is 5,000 µg, or 5 mg. At one point, Robert took 150 mg/day. That's 30 capsules, going through a 120 count bottle in 4 days. Each bottle cost $29.99. This being America, insurance didn't cover it.

But he could move. And raise his hands and play with toys just a little. And pick up objects with two fists. And grab at stuff. Biotin cleared up his chronic vomiting syndrome within a month. And these seemed markers on a trail that appeared to be going somewhere.

A bottle of "Stress-Complex B Vitamins" I had been taking on a daily basis called to me from our kitchen shelf. I know what my hunch had been initially. And human metabolism is really, really complex. The leg bone's not the only thing connected to the hip bone, there are muscles, tendons, ligaments, you get the drill. If one B vitamin was doing something, perhaps he needed little helper B vitamins.

These did help. Robert's muscle tone relaxed, and he began making vocalizations with differentiated sounds. He could move his head independently and showed increased trunk control.

Unfortunately, not all of these improvements lasted, and none of our questions about human biochemistry have been answered by Robert's diagnosis. Yet, to this day, no one knows why biotin helps Robert because PRKRA has nothing to do with biotin response.

So off we go again to look for broken branches, branches broken with intent. For footprints in soft soil. Stay alive, I will find you.

(USA)

Opportunities and Challenges for Supporting People with Vascular Dementia Through the Use of Common Web 2.0 Services

Kyle Boyd, Chris Nugent, Mark Donnelly, Raymond Bond and Roy Sterritt

> *Alice: ... would you tell me please which way we should go from here?*
> *The Cat: That depends a good deal on where you want to get to*
> Alice in Wonderland, LEWIS CARROLL

Abstract The aim of this chapter is to present how Web 2.0 technologies can be adapted for Healthcare (Health 2.0). Specifically this will involve explaining the technology and real world scenarios of how this technology could be adapted to be usable for Vascular dementia patients and their carers. We consider Web technology that would aid the reduction of social isolation and to help dementia patients maintain social connectedness.

Introduction

The aim of this chapter is to highlight how Web 2.0 technologies can in general be adapted to help those with Vascular Dementia (VaD). Particularly, those subsets of VaD that are classed as orphan diseases. Many tools from the collection of

---

K. Boyd (✉) · C. Nugent · M. Donnelly · R. Bond · R. Sterritt
School of Computing and Mathematics, University of Ulster, Jordanstown Campus,
Shore Road, Co, Antrim, Newtownabbey, Northern Ireland BT37 0QB, UK
e-mail: boyd-k5@email.ulster.ac.uk

C. Nugent
e-mail: cd.nugent@ulster.ac.uk

M. Donnelly
e-mail: mp.donnelly@ulster.ac.uk

R. Bond
e-mail: rb.bond@ulster.ac.uk

R. Sterritt
e-mail: r.sterritt@ulster.ac.uk

Web 2.0 technologies may be suitable, however, many users may not know what is available. Put straightforwardly, people would benefit from being educated to allow them to discover what is possible. This can be through healthcare professionals or the patients themselves. Only then will the positives of using these tools will be fully realised. To quantify this vignettes will be presented to show how Internet technology could be used to help those with orphan diseases.

Dementia

A worldwide ageing population brings an increase in the number of people with dementia. Dementia does not have social, economic, ethical or geographical boundaries. With more than 35 million people already having dementia, the figures estimate that new cases are currently arising one every seven seconds (Alzheimers Disease International 2012). At this rate, figures are set to double by 2030 and triple by 2050, to approximately 115.4 million on a global scale (World Health Organisation 2012). Typically, people with dementia experience symptoms ranging from memory loss, mood changes, depression and anxiety to problems with communicating and reasoning (Alzheimers Association 2012).

Alzheimer's disease (AD) is the most common form of dementia (World Health Organisation 2012). The risk of developing AD doubles every five years in adults aged over 65, with those aged over 85 having a 50 % chance of developing the condition Babusikova et al. (2011). Studies (Xiao et al. 2012) have indicated if it were possible to delay the onset of AD for 5 year that it would be possible to half the number of people who die with the disease each year and also save the NHS necessary care costs. The second most common form of dementia is Vascular Dementia (VaD) (Xiao et al. 2012). A number of the causes of VaD are classed as being a type of Orphan Disease.

Orphan Disease

The name 'orphan diseases' is given to rare conditions (Azie and Vincent 2012). An orphan disease is a regulatory definition of a rare disease that only affects a small percentage of the population. According to Eurordis (2012) there are 6,000–8,000 orphan diseases in the world. An orphan disease is classified as such when at most five out of every 10,000 people have the condition (Merck 2012). The alternative term "orphan disease" singular is used to represent a combination of the insufficiency of treatment availability, lack of resources and severity of disease as well as the disease itself (Azie and Vincent 2012).

The diseases sometimes can be well known to the public like Cystic Fibrosis and Haemophilia, however, others are less well known. The diseases can be rare within a particular region of the world, however, be more prevalent in another, for example, Malaria. They can also move across countries and continents as with tuberculosis in Europe (Azie and Vincent 2012).

Asbjørn Følling discovered the first orphan disease, Phenylketonuria (PKU) in 1934 (Hanley 2012). He discovered that healthy babies had suffered from progressive brain and nerve damage, which resulted in seizures and a reduction in intelligence. Upon examination of the children's urine, he discovered large amounts of phenylpyruvate. This is one of the ketone bodies, only present in the body during sustained fasting (Merck 2012). If it is not detected after birth, PKU can impair the development of the brain. PKU is caused by an enzyme deficiency in the liver and can be treated with a low protein diet to control intake of phenylalanine.

Orphan diseases used to receive less attention from pharmaceutical companies because they were either not in a profitable market due to their rarity or the condition was only widespread in developing countries that could not afford to pay the drug prices set by a drug manufacturer. To address this issue, in 1983, the orphan drug act was created in the US(Van Weely and Leufkens 2004), which subsequently spread to other countries and eventually to the EU in 1999. This allowed companies to drive forward research and development into orphan disease solutions making it much easier to work in this area. As a result the number of orphan drugs has risen and currently stands at around sixty (Tejada 2013).

Vascular Dementia as an Orphan Disease

VaD makes up between 20 and 30 % of all cases of dementia (Dementia Guide 2012). It occurs when there is damage to the brain resulting from a low blood supply or a lack of oxygen. This occurs after a major stroke or after a series of 'mini strokes'. It can happen simultaneously with AD and when this happens, it is called mixed dementia. In contrast to the steady decline, cognitive function and abilities in AD, VaD can occur suddenly without warning. Changes are noticeable and they can get worse, then better and then stay the same or worse again. The risks for VaD are similar to stroke, older age, high cholesterol, obesity, hypertension and smoking, all of which can increase the risk of VaD (Holmes 2012).

Persons with VaD often have changes in personality, lack of activity and emotional spells are common. Confusion with simple tasks is also common, due to problems with functions and attention. The most common of VaD being multi-infarct dementia, caused after mini strokes, which go unnoticed but as more parts of the brain are damaged VaD appears. Cognitive impairment may occur suddenly with abilities decreasing over time, fluctuating with recurrent strokes. According to work by Korczyn et al. (2012) depression is observed in those with VaD frequently and is witnessed after strokes.

At this point it is worth noting that VaD in itself is not an orphan disease but it is the causes of the disease that allow it to be considered an orphan disease. Some of the causes are (Jellinger 2008) Binswanger's Disease, Vasculitis, Cerebral Haemorrhage and Cerebral Autosomal Dominant Arteriopathy with Subcortical Infarcts and Leukoencephalopathy (CADASIL) (Brown and Rossor 1998). CADASIL, in particular, is a recognised rare orphan disease (Orphanet 2012 Health on the Net Foundation 2012).

Causes of Vascular Dementia

Some of the most common causes of vascular dementia are explained in the following sections.

1. **Cerebral autosomal dominant arteriopathy with subcortical infarcts and leukoencephalopathy (CADASIL)**

CADASIL syndrome is a genetic form of VaD (Dotti 2012). It can occur in the young and during adulthood presenting itself as migraines or strokes. It is caused by dominant mutations in the NOTCH3 gene on chromosome 19. Salloway and Hong (1998) have found that it causes dementia and depression in 80 % of patients.

2. **Cerebral haemorrhage**

This is caused by bleeding in the brain from a ruptured blood vessel which causes inflammation and swelling. If one of the surface vessels of the brain ruptures, this can cause blood flow into the membrane that results in pressure on the brain. It can come on suddenly without warning and results in complications that can include difficulty with memory, thinking and talking (Federico et al. 2012).

3. **Sub-cortical Vascular Dementia (Binswanger's Disease)**

This is sometimes called progressive small vessel disease of which Binswanger's disease is a specific type. It is caused by damage to blood cells that lie deep within the brain resulting in difficulties with mobility and speech. Symptoms are not always present and at times come and go (Alzheimer's Society 2012).

Symptoms with Vascular Dementia

VaD can effect people in many ways. Depending on the severity of the strokes, these can appear suddenly or gradually over time, some are shown in Table 1. In general, VaD typically progresses systematically. With VaD dysexecutive syndrome is particularly common. Individuals with dysexecutive syndrome will have problems formulating goals, planning and organisation, which results in a lack of judgement (Stacpole 2011). They also are more likely to suffer depression and anxiety more so than in other types of dementia (Stacpole 2011) (Table 1).

Loneliness and Social Isolation

Social Isolation is a growing problem. According to Age Concern (2008) "1.2 million people over 50 years of age are severely excluded: 400,000 aged 50–64, 360,000 aged 65–79, and 400,000 aged over 80." With the increasing number of those with dementia, it is a certainty that this disease will become more

Table 1 Common issues of vascular dementia. Adapted from work by Alzheimer's Society 2012 and Block, Smith and Segal 2012 (Alzheimer's Society. 10/15/2012W; Block et al. 2012)

Common mental and emotional symptoms	Common physical signs and symptoms	Common behavioural signs and symptoms
Slowed Thinking	Dizziness	Slurred speech
Memory problems	Tremors	Language problems, difficulty finding right words
Mood changes	Balance problems	
Hallucinations and delusions	Moving with rapid, shuffling steps	Getting lost in familiar surroundings
Confusion, worsening at night		Difficulty following instructions
Personality changes and loss of social skills	Leg or arm weakness	Difficulty doing common tasks, like paying bills
		Reduced ability to function in normal life

widespread. It has been suggested that loneliness is a predictor of cognitive decline. Studies by Fratiglioni et al. (2000) and Wilson et al. (2007) have shown that loneliness has an effect on cognition and increases the risk of development of dementia. Murphy (2006) defines loneliness as a condition that describes the distressing, depressing, dehumanising, detached feelings that a person endures when there is an emptiness in their life due to an unfulfilled social or emotional life.

Two forms of loneliness can lead to depression subsequent to the onset of dementia: social isolation and emotional isolation (Wilson et al. 2007). Social isolation indicates that there is a minimal social integration and connectedness with others whereas emotional isolation refers to minimal or no attachment figures in the person's life. Many factors can contribute to social isolation. Findlay and Cartwright (2002) believe that this can be from a loss of health and function, hearing, communication abilities besides loss of relationships such as losing a partner through death, or if family moves away (particularly true of the case of emotional isolation.)

Not surprisingly, Wang et al. (2002) have found that frequent participation in social activities can decrease the risk of isolation, loneliness and therefore, dementia. Wang's findings suggested that the quality of social attachments and the quality of social interactions could positively affect the onset of dementia by slowing its development. This is similar to findings by Moyle et al. (2011) who carried out interviews with those with dementia. The study found that participants with dementia who took part in social activities with family were able to maintain interest, break up routine and minimise loneliness better than if they didn't.

Information and Communication Technologies (ICT) has been proposed as one method for alleviating some of the strain being placed on healthcare management and delivery by attempting to manage and monitor user's symptoms. Ambient Assisted Living (AAL) is a term used to describe systems that deliver help with tasks associated with daily activities, health monitoring, enhancing safety and access to medical systems (Alzheimer's Society 2012). AAL research has been used to monitor and support social communication, support with activities of daily living (ADL), enhance feelings of safety (Dementia 2012) and with memory support through regular reminders (Donnelly et al. 2010). A key component, which is common among most AAL solutions, is the reliance on the Internet. In general, the Internet has provided significant gains in terms of enhancing people's knowledge, entertainment, social communication and life-long-learning (Krishnamurthy 2008).

The most prevalent and readily available ICT solution is the World Wide Web. It is reasonable to hypothesise that the Internet could be used to help connect those socially isolated to enable them to stay connected with family and friends despite geographical location. World Wide Web technologies built on the platform of Web 2.0 technology allow the creation and exchange of user generated content. At this point we will define that the Internet is the underlying communication platform and the World Wide Web is the front-service that people use. Nonetheless these two terms are used interchangeably so the term Internet will be used hereafter.

The Internet and Social Media

The Internet population has doubled to 2.27 billion in 2012 from 1.15 billion in 2007. The meteoric rise of Internet usage began in the mid nineties when people could communicate via email and instant messaging services, beginning in the work place but gradually moving into the home. By 2004, the term 'Web 2.0' coined at a conference between O'Reilly and MediaLive International (O'Reilly 2009) was used to describe the Internet. It had gone from a static web to a dynamic and collaborative service into a growing phenomenon that was a collection of technologies, business strategies and social trends. (Murugesan et al. 2007) It allowed users to be more collaborative, supporting social interaction allowing developers to easily and quickly create applications that draw on data, information or services from the Internet.

The Internet is now a Social Web because of Social Media. Types and current examples of Social Media are shown in Table 2. Social Media allows people to create words, pictures, video and audio and share this with their friends through social interaction. This is facilitated through Web 2.0 Technology.

Over the past five years, there has been exponential growth of Online Social Networking products. Of these, Facebook is regarded as the most popular with more than 1 billion users, meaning that one in six people on earth now have a Facebook account. Of these 680 million are active users of Facebook mobile-based products (as of December 2012) (Facebook Statistics 2012). These direct and indirect interactions build up a social network were everyone is connected through friends and interests. This trend exists through all the latest Internet applications.

Online Social Networks

Online Social Networks allow users to setup a profile. This profile contains information about the user such as their name, age and where they live. It also contains a profile picture were a user can display a photograph of themselves. The profiles

Table 2 Types of Social Media (Facebook 2012; Twitter 2012; Yahoo 2012; Tumblr 2012; Pinterest 2012; YouTube 2012)

Facebook Social Networks www.facebook.com	Twitter Microblogs www.twitter.com	Flickr Photographs www.flickr.com

Tumblr Blogs www.tumblr.com	Pinterest Social tags www.pinterest.com	You Tube Video www.youtube.com

can connect or follow other profiles to build up a network (Fig. 2). Each member in the Online Social Networks can then view updates from virtual friends, which are typically displayed in a timeline or newsfeed (Fig. 1).

Online Social Networks allow updates to be made that contain text, images, videos or links. These can be shared with friends and followers. Friends and followers are essentially the same thing, just different terminology is used for different Social Networks. Followers receive your updates and can comment on these updates, which allows for communication and collaboration. Tagging is also used so content such as pictures can be shared with other friends and followers in the same location. Of the many Online Social Networks, Facebook (see Fig. 2) is the most popular. Nevertheless, there are many other online social networks like Twitter, LinkedIn, Flickr and YouTube (Facebook 2012; Twitter 02/09/2012; Yahoo 2012; YouTube 2012; LinkedIn 02/09/2012L).

In addition to Niche Social Networks that work specifically on Mobile Devices. Applications like Instagram (see Fig. 3), Foursquare and Path (Instagram 2012; Foursquare 2012; Path 2012) all have a Social Network platform were friends and followers share photographs, videos and locations, depending on which specialty the application is designed for.

Blogs

Blogs can be used as a powerful Web communication tool and they allow people to share thoughts, ideas and comments. Each entry into a blog is referred to as a post and these are displayed in chronological order. They can contain text, images, links and videos and allow comments and interactions from other Internet users if allowed by the administrator of the blog and whether it can be seen publicly or privately (Murugesan et al. 2007).

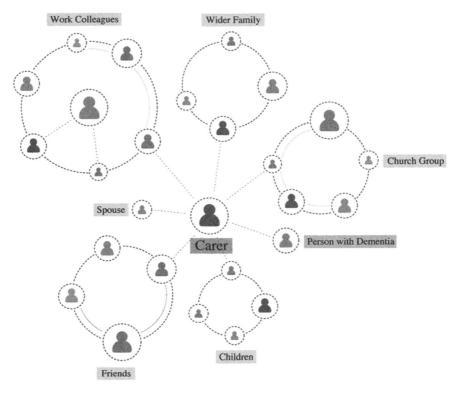

Fig. 1 How people on the internet can be connected through the online social networks via profiles

Blogs are now easier than ever to setup and can be performed with blogging software such as Wordpress. Wordpress has more than 55 million blogs (Wordpress 2012) worldwide and these blogs can either be hosted on the Wordpress server or you can setup your own blog on a dedicated server. It can be heavily customised either by using an existing theme it comes shipped with or you can easily create your own theme with some knowledge of HTML and CSS. Extra interaction and functionality can be added through ready made plug-ins created by developers for free. This makes Wordpress a powerful and stable platform.

Over the last five years there has been a rise of the Blogazine implemented by mainstream designers such as Dustin Curtis (2012) and Jason Santa Maria (2012). A Blogazine is used to create original layouts and designs customised to the content of each post. Making it more appealing and far reaching to the blogosphere (the blogging community). These posts, however, take time and specialised skills to implement.

Microblogging has seen a sharp rise since the introduction of services like Twitter (2012), and Tumblr (2012). Microblogging allows the subscriber to broadcast short messages to others using the same service. This can include text and links in addition to pictures and videos. Services like Twitter are limited to 140 characters, however, services like Tumblr allow you to share whatever you

Fig. 2 A screenshot of a Facebook profile

want in a more lightweight way than Wordpress. This can be shared with people who subscribe to your blog.

Really Simple Syndication

Never in human history have we had so much information available or in such an easily accessible format (through just a few mouse clicks). Frequently, we are so overloaded with information that we cannot make much sense of it. For example, YouTube, the online service that allows the creation and sharing of videos, receives 72 h of video uploads every minute (YouTube 2012) which, to a certain extent, makes it impossible to watch all of its content. RSS feeds provide a part solution to this by providing a high-quality way of gathering information online. They allow content from Webpages to be summarised with a title, its content, and a link to the content. Using an aggregator these feeds are checked displaying updated articles found. Of the many aggregators available, the most popular are Web Browsers such as Google reader (2012) or NetVibes (2012). Nevertheless, given the popularity of the Mobile Web it is now common to use applications specifically designed for devices such as Flipboard (2012) and Reeder (2012). These applications not only show the latest articles in a social magazine format, however, also incorporate Facebook and Twitter updates allowing content to be consumed from one application on the go.

Fig. 3 A screenshot of an Instagram profile

The new applications that allow you to 'save findings to read later' reinvigorated RSS technology. It allows users to consume content efficiently and helps keep them organised with the growing mass of information on the Internet. This shows in the popularity of applications such as Flipboard. At only two years old, it has 20 million users, 3 billion 'flips' per month and 1.5 million daily users (Flipboard Stastistics 2013).

Wikis

The most famous Wiki is Wikipedia (2012). It is a free collaboratively edited, multilingual Internet Encyclopaedia. The ideal Wikipedia article is well written, balanced, neutral and encyclopaedic, containing comprehensive, notable and verifiable knowledge. It can also include pictures, sounds and videos. A Wiki allows people to share knowledge whilst creating or contributing to new articles. These can be edited by anyone through a Web browser. The notion of a Wiki is based around the premise of collaboration. Wikis have now become popular as both educational tools between teachers and students and are also useful for technical documents as they are simple to use, flexible and expand as your content grows (Park et al. 2012).

Mashups

A Mashup is a website or service, which combines data from several sources to make it more useful. Many thousands of Mashups are available and can be found at WebMashup (2012).

Property Pal (2012) is a typical web Mashup. It allows the user to search for houses to buy or rent and displays where the houses are on Google Maps. YouPlayList (2012) is another service, which allows users to order videos found on YouTube. It is a platform to create, save and manage several playlists.

This is made possible due to the availability of APIs (Application Programming Interfaces). An API is an interface which allows users to respond or react to a product or service subsequently allowing for the exchange of data between the two. If for example, a user posts a photograph to Instagram on their mobile device they can decide whether to post this to Twitter or Facebook or both at the same time. This is only made possible through APIs. Spotify (2012), the online music library only lets you login to use the service if you have a Facebook account. This is achieved through the Facebook Social Graph and API.

Tagging, Folksonomy and Tag Clouds

Social Tagging has become popular with sites like Delicious and Flickr (Yahoo 2012; Delicious 2012). A user can assign a tag to the content, which is then grouped. On Delicious groups of URLs can be tagged and on Flickr groups of photographs can be tagged. Online Social Networks like Facebook allow users to tag photographs and 'Like' something such as a comment to tag.

People tag for a variety of reasons (Strohmaier et al. 2012):

- Future retrieval: A User finds something but cannot digest it at that point of time hence they tag it for future reference.
- Contribution: Allows the content to shared with others and categorised.
- Express Opinion: They allow users to convey opinions or judgments. The 'Like system in Facebook or the Like/Dislike system in YouTube.
- Organisation: Tags can be grouped into lists for ease of use and organisation for example a job search or to-do list.

Twitter uses hash tags within the content of the tweet which allows easier searching and categorising of the tweets. It is interesting that users created this ability and once it had gained popularity it was adopted as a recognised feature within Twitter. Tag clouds can be used to organise tags and have grown in popularity with Wordpress, which allows categories to be added blog posts with a tag cloud being automatically generated showing the most popular posts.

Pinterest (2012) is one of the newest Social Tagging platforms. It was the first independent website to cross the 10 million unique visitors per month mark over the shortest period from May 2011 to January 2012. It allows users to 'Pin' things they find on the Web to a board. It is a visual process and allows users to easily browse, find and share lists by curating what they find.

Health 2.0

The nature of the current Internet is such that many technologies can be used together in a mash up. Health 2.0 is part of Web 2.0 with the focus on special health applications, where people can take a more active role in their own healthcare. Indeed Eysenbach (2008) has stated that Health 2.0 applications and services that are Web 2.0 based can be used by caregivers, patients and health professionals to facilitate social networking, participation and collaboration. In the context of the current chapter it is postulated that these technologies can be used singularly or together to aid those with dementia. The Internet provides an alternative method for increasing social interactions. Previous research by Hong et al. (2001) has shown that increased support from networks has lead to increased psychological well-being. A study conducted by Small et al. (2009) between older users who were comfortable with Internet use and those who were Internet naïve found that those who could perform simple Web searches and read the results on screen could control vision, complex reasoning and decision making better than those who couldn't. They concluded that Internet searching could influence the responsiveness of older brains in a positive way and provide a benefit for those with cognitive impairment.

It has already been noted in research that there are benefits in using the Internet as a social support tool. It loses the implication of lack of participation due to transport and logistical or time constraints of people living in different areas. Everyone can participate with rapid responses arriving instantaneously. This is a benefit to caregivers (Perkins and LaMartin 2012). Traditionally persons with dementia (PwD) are cared for by family and formal carers from health and social services. As of 2012 there are 1.25 million carers in the UK who provide over 50 h of unpaid support, of those 670,000 act as primary carers to PwD (Alzheimer's Society 2012). Changes in memory and thinking means that PwD can struggle with everyday life and it can become difficult to undertake tasks such as paying bills and cooking. Carers end up providing emotional and social support in addition to dressing and bathing. This can become difficult logistically given that two

Table 3 Positive outcomes of using the internet for health 2.0 in those with VaD

Social Media Type	Solution	Proposed positive outcomes of usage
Online social networks	Facebook	Free of charge, not bound by time or location. Share photos and thoughts. Supports inclusion and communication and support (Hanson et al. 2011; Norval et al. 2011; OFCOM 2012; Zickuhr and Madden 2012; Xie et al. 2010).
Blogs	Tumblr	Increased social interaction as a result of enhanced and prolonged conversations. Allows for discussion (Cotelli et al. 2012; Peesapati et al. 2010). Increased quality of life, enhanced information sharing and social closeness
Tagging	Pinterest	Supports reminiscence by allowing them to catalogue their lives and curate what they find

thirds of PwD live at home meaning family members have to split time between them and their own families. Due to the dementia's progressive nature it means that support can become more intensive as time goes on.

The work of carers should not be underestimated. According to the Alzheimer's society they save the UK economy around £8 billion pounds a year (Alzheimer's Society 2012). Put simply, if the burden was left to our existing social care systems it would collapse with the inability to cope because of the strain.

It is clear that ICT is a potential solution to social isolation. A review by Lauriks et al. (2007) lists many ICT solutions, which includes aids for memory, robots to stimulate communication, decision makers, cooking reminders and GPS tracking devices for wanderers. Nonetheless, it is clear that using the Internet can not solve all problems associated with PwD. It would be futile to suggest that cooking or dressing could be completed in the Web browser, however, in some cases it could be viewed as an appropriate outlet to help prevent social isolation by providing social and emotional support for both the PwD and their carers that they stay in contact with family members and vice versa without actually physically being present.

Carers' burden in the provision for 'loved ones' would be relieved as they would have more time to focus on their own needs rather than the needs of the PwD. In the Case Studies previously discussed all three solutions were available free of charge via the Internet. No form of specialist hardware was required, only a device with Internet connectivity. In Table 3 the positive outcomes of why Health 2.0 solutions could be used over traditional methods is presented.

The positive outcomes can show that these solutions and the ability to provide an Online Social Network could provide emotional support and social closeness that a person with VaD would need assistance with. Considering they are more like to suffer from depression, anxiety and suffer from emotional liability. Supporting the work by Goswami et al. (2010) who reported report that the characteristics of Online Social Networks makes it easier and cheaper for the elderly to keep social ties active and therefore enhance feelings of social connectedness and Online Social Networks usage has psychological benefits, such as increased social capital, reduced loneliness and improved psychological well-being. Shapira et al. (2007) also reported that that older people have a greater sense of empowerment through online interpersonal interactions. They help them maintain cognitive functioning and have a greater independence in their lives.

This sociable Web has the potential to augment those with VaD and carers with a sense of belonging to a supportive kindred community reducing social isolation. Minimal costs of services mean that there is a better chance of adoption meaning more connectedness. This is something which is denied to many who are confined to their home through illness.

Case Studies

In the next Section, a series of hypothetical case studies that aim to illustrate how these technologies could be implemented and used by older people with VaD and their families are presented. Each case study aims to put into context how this would be achieved and managed exemplifying how the attributes of Health 2.0 are aligned with the needs of those suffering from VaD.

The Use of Online Social Networks Within VaD

Muriel a 68-year-old woman, lives in Harrogate near Leeds, and is originally from South Africa. She married Don in South Africa after he had come from Leeds in 1963 to work as an engine fitter for the RAF. They met while he had been stationed there. Muriel trained as a nurse and secured a job at the local hospital. After Don's tour ended in 1974, it was decided that they would move back to England with their three children, Geoffrey, Rachel and Alison. They bought a house at 49 Arncliffe Road and setup family life in their new home. Muriel held a post in the NHS as a nurse and was eventually promoted to a sister nurse on the ward. She said of her job:

> Coming to work in the UK was such a transition to South Africa but it was one I am embraced I loved my job, helping and caring for others

After Don died 6 years ago, Muriel has lived on her own. All three children are now married have moved away and live in other parts of the world. Geoffrey lives in Bermuda as a risk model developer for a Reinsurance Company. Rachel works as a vet in South Africa and Alison owns a fashion boutique in the leafy suburbs of west London. Muriel has five grandchildren, one in Bermuda, two in South Africa and two in London.

Muriel had a fall when she was 64. Her neighbour noticed that afterwards she began to repeat questions. After another fall a few months later she began to notice a decline in cognition. After going to see her GP with her daughter Alison, they conducted a Mini Mental Status Examination (MMSE) with a score of 21 and on the Geriatric Depression Scale (GDS) it indicated 3, which showed a mild impairment concluding that Muriel had a form of VaD referred to as Binswanger's disease due to the mini strokes caused by bleeding deep in the brain during her fall due to not being in regular contact with her children and grandchildren because of logistics it was believed Muriel's condition would not be helped with the continued social withdrawal.

Muriel's children are all ardent users of Social Media technology such as Online Social Networking, most notably Facebook. All of them used it beside regular media such as telephone and e-mail, however, this was a useful outlet for quick status and photograph updates through their Internet enabled mobile phones. Facebook's timeline allowed them to record these events automatically in a chronological order. They decided that this could be a novel way of keeping their mum, Muriel up to date as much as possible. They setup an account for her setting the privacy to private so her only friends were her children. Muriel had used a PC in the past, however, this had become more difficult for her. Therefore, they purchased an Apple iPad and installed the Facebook Application.

With the release of Apple's iOS 6 software guided access can be set (Apple Accessibility 2012). Particular features of an application can be disabled which prevent a user from exiting an application by disabling hardware buttons and touch screen sensors. Allowing the device to run an application, like Facebook as a stand-alone application without the risk of navigating to another part of the operating system. This therefore eliminates issues with starting the application as it is always on. It means that Muriel can either interact with her family or merely watch her family's daily life from across the globe from the comfort of her home. Helping her become socially involved and connected.

As the dementia increases, one of the effects is that those with the disease will exhibit a decline in communication skills. This is particularly relevant to seniors because isolation resulting from health issues, retirement or less involvement with your family, which increases cognitive decline. Cornwell et al. have found that online social networks have been associated with successful ageing and that social connectedness has a positive affect on the rate and extent of cognitive decline.

The Use of Blogs Within VaD

Charlie is a 56-year-old landscape gardener from Glossop in Derbyshire who is married and has two daughters. Originally from the West Country, Charlie grew up on a dairy farm and continued to farm the land after his father's death in 1980. Charlie was successful as a farmer for more than twenty years until the outbreak of Foot-and-Mouth disease in 2001. Unfortunately, Charlie's farm was affected and his total heard had to be culled. Given his age Charlie decided to retire from farming, as it would be impossible to return the farm to its former state. He sold the land and farm and moved with his wife to Glossop. Surprisingly Charlie had a keen interest in landscape gardening. He decided to start this as his full time job. It started slowly as garden maintenance like cutting grass, trimming hedges and planting flowerbeds. Nonetheless, as the property boom increased Charlie had contracts for many new housing developments in Glossop and the surrounding areas eventually building up a strong portfolio within three years of setting up the business.

In his thirties and forties Charlie started to have bad migraines, however, he put this down to the stress of working long hours, then losing the farm stock because of the foot-and-mouth outbreak. Nevertheless, these soon became more common with mood swings. Following this, Charlie suffered a stroke. Although he recovered, his memory was affected and he now has the early stages of dementia. He has been diagnosed with an orphan form of VaD called cerebral autosomal dominant arteriopathy with subcortical infarcts and leukoencephalopathy (CADASIL), the most common heritable cause of stroke and VaD in adults.

Due to his condition, he could not work full-time meaning he has to spend much time at home and at times, this makes him depressed, as he is not as outgoing as he used to be.

His two daughters are now married and have left home. One daughter is a journalist and the other is an interior designer. Both daughters keep blogs about their fields of work. They believed this could be a good outlet for their father to continue to talk about gardening sharing his experience with a wider audience. Charlie had never been a prolific user of ICT or the Internet in general but could send and receive e-mails in addition to purchasing items on the popular auction website, eBay.

With the help of his daughters Charlie setup a Tumblr blog called 'Charlie's Garden'. Charlie now blogs throughout the seasons providing advice for keen gardeners about planting, pruning and designing for spring, summer, autumn and winter gardens. He has amassed a strong following of more than 100 followers who regularly comment on his posts and advice. Charlie said of the Tumblr blog:

> Writing for my blog 'Charlie's Garden' has changed my outlook. It feels like I get out more even though I am still in the house—because I get to meet and share with people from all over the country and the world, not just Glossop!

This tool certainly hasn't cured Charlie's condition and more than likely he will lose all mobility before he is 60 and may not live beyond 65 depending how rapidly or slowly his condition progresses. Nonetheless, it allows him to continue to be engaged with the world and reduces his depressive symptoms.

The impact of using this technology means that patients are able to retain a sense of self. It helps them to remember things that they did in the past that can be supported by relatives and friends. For patients with dementia the past is often disconnected from the present meaning that relatives and carers often lose track of a patients mind-set therefore severing the ties of those around us. Blogs have helped others to share their experiences and share what they are interested in. Connecting with others to discuss interests can remove the loneliness associated with the disease, both with relatives and people with the same condition. It allows also acts as a personal marker of cognitive abilities that a patient will inevitablely lose at a later stage. The blogs allow patients to comprehend and preserve a sense of self.

The Use of Tagging with VaD

Joan is a 62-year-old grandmother and she lives with her husband George in Larkhall just north of Glasgow. Joan has been married to George for 38 years. They have one son called Jonathan. Joan worked as a telephonist for the University of Glasgow, however, after she had Jonathan, she gave up work and became a homemaker. George had a well-paid secure job as the post office postmaster in Larkhall.

The family was well ahead of the curve of having a computer in the home with George buying his son a Macintosh computer and later a Windows PC with Internet access. Joan was never interested in the technology and never used it much. When Jonathan went to university, she progressed to using e-mail for communication. She now enjoys shopping online, searching for interior design ideas and looking for potential places to visit. George and her always have one holiday a year.

A year ago, Joan suffered a Cerebral Haemorrhage, which caused bleeding in the brain. She now struggles to remember how to complete tasks and things that she had done. To aid this she now uses Pinterest. This allows Joan to pin things that she has found on the Internet and pin them to a board. It means that she can instinctively curate what she finds efficiently and easily making her in essence a curator. George has also included boards about him and Jonathan so she can look at these to aid the reminiscence process. Reminiscence therapy (RT) is a popular intervention for those with dementia. It is based on discussion of past events, activities and experiences with a person or group (Cotelli et al. 2012). It helps improve cognitive abilities by reminiscing about memories from the past. It can help prevent depressive feelings and it encourages social involvement resulting in a positive effect on the participant's quality of life.

Summary

This chapter presents an overview of how those with orphan diseases can be helped through the use of Web 2.0 technologies. An introduction to dementia and its particular types such as VaD was summarised. Then how this can be classed as an orphan disease through subtypes like CADASIL was presented. A summary of the particular social media technologies, how they work, and how they are currently being used has been detailed. Finally, three hypothetical case studies were detailed to help explain how those with VaD could use these technologies and the positive outcomes that might be seen.

Nonetheless, it is worth noting that some challenges still remain. Many researchers have studied the main barriers as to why older people are not using the Internet. These are as follows (Norval 2011; Mellor et al. 2008; Age Concern 2010; Lehtinen et al. 2009; Gibson et al. 2010; Sundar et al. 2011; Sayago and Blat 2011; Xie et al. 2012; Chou et al. 2010):

- Lack of understanding of and confidence with 'how it works'. A number of fears and anxieties were expressed about 'doing something wrong'.
- The only way to handle their lack of confidence is to dismiss the technology, however, they would be encouraged to use it more if they were given help and support.
- Many feel it is too hard and they are too old to learn.
- Affordability—it is too expensive.
- Don't understand why they would need to use it.
- Usability, including interface issues that arise from age-related disability and lack of experience.
- They couldn't grasp the purpose or benefits of technology like Online Social Networks.
- Daunted by the technology due to a complete lack of awareness of digital products.
- They were worried about breaking it.
- They were worried about privacy and security concerns.
- Language was a problem, and terminology used.

Some of the problems are that designers of these Web sites do not think about older people. In the UK ages between 25 and 34 is the biggest user group of Facebook at 25.6 % (Fanalyzer 2012) and Facebook's average employee (Wealthwire 2012) age is 26. These younger designers don't consider the needs of older people or those with VaD regarding the user experience and usability of the interfaces they use. For example small buttons, complex controls and complicated interface walkthroughs are a barrier to someone who hasn't grown up with this technology and coupled with a lack of manual dexterity this can cause problems (Independent Age 2011).

To conclude, we have proposed solutions using current Internet technology to help those with VaD, which was caused by orphan diseases. Barriers and challenges are clear, however, solutions are becoming clearer which would enable people to look after themselves through Health 2.0.

References

Age Concern. Older people and digital inclusion,http://Intelli.ageuk.org.uk/e/d.dll?m=1292&url=http://Www.ageuk.org.uk/documents/en-gb/for-professionals/computers-and-technology/older%20people%20and%20digital%20inclusion.pdf?dtrak=true. Accessed 07/04/2011. Age Concern, 2010.
Alzheimers Association. 2012 facts and figures alzheimer's disease. 2012.

Alzheimers Disease International. 09/09/2012 Dementia statistics. http://www.alz.co.uk/research/statistics.
Alzheimer's Society. 11/16/2012. Carer support. http://www.alzheimers.org.uk/site/scripts/documents_info.php?documentID=546.
Alzheimer's Society. 10/15/2012. What is vascular dementia? http://www.alzheimers.org.uk/site/scripts/documents_info.php?documentID=161.
Apple Accessibility. 13/11/2012iOS 6 Accessibility features: guided access. http://www.apple.com/ios/whats-new/#accessibility.
Azie N, Vincent J. Rare diseases: the bane of modern society and the quest for cures. Clin Pharmacol Ther. 2012;92:135–9.
Babusikova E, Evinova A, Jurecekova J, Jesenak M, Dobrota D. Alzheimer's disease: definition, molecular and genetic factors, 2011.
Block J, Smith M, Segal J. 10/15/2012. Vascular dementia: signs, symptoms, prevention, and treatment. http://www.helpguide.org/elder/vascular_dementia.htm.
Brown J, Rossor M. 5 vascular and other dementias. In seminars in old age psychiatry; 1998. p. 73.
Chou W-H, Lai Y-T, Liu K-H. Decent digital social media for senior life: a practical design approach. In Computer Science and Information Technology (ICCSIT), 2010 3rd IEEE International Conference, 2010. pp 249–253.
Concern Age. Out of sight, out of mind: social exclusion behind closed doors. London: Age Concern; 2008.
Cotelli M, Manenti R, Zanetti O. Reminiscence therapy in dementia: a review. Maturitas. 2012;72:203–205, 7.
Delicious. 02/09/2012. Delicious Home Page. http://www.delicious.com.
Dementia AT. 13/11/2012. Assistive technology dementia. http://www.atdementia.org.uk/editorial.asp?page_id=25.
Dementia Guide. 15/10/2012. Vascular dementia. http://www.dementiaguide.com/community/dementia-articles/vascular-dementia.
Donnelly M, Nugent C, McClean S, Scotney B, Mason S, Passmore P, Craig D. A mobile multimedia technology to aid those with Alzheimer's disease. IEEE Multimedia. 2010;17:42.
Dotti MT. CADSAIL: a hereditary cerebrovascular disease as a model for common vascular ischemic dementia. J Siena Acad Sci. 2012;1:34–5.
Dustin Curtis. 02/09/2012. Dustin Curtis Blogazine. http://www.dustincurtis.com.
Eurordis. 09/09/2012. About rare diseases. http://www.eurordis.org/about-rare-diseases.
Eysenbach G. Medicine 2.0: social networking, collaboration, participation, apomediation, and openness. J Med Int Res 2008
Facebook. 24/01/2012. Facebook Home Page. http://www.facebook.com.
Facebook Statistics. 24 January2012. http://tinyurl.com/mwf76j.
Fanalyzer. 11/16/2012. Demographics/Facebook Users in the UK. http://www.fanalyzer.co.uk/demographics.html.
Federico A, Di Donato I, Bianchi S, Di Palma C, Taglia I, Dotti MT. Hereditary cerebral small vessel diseases: a review. J Neurol Sci. 2012.
Flipboard. 02/09/2012. Flipboard Homepage. http://www.flipboard.com.
Flipboard Stastistics. 02/122013. Inside flipboard: the official flipboard blog. http://inside.flipboard.com/2012/08/28/flipboard-at-two-20-million-users-one-new-user-per-second/.
Foursquare. 20/08/2012. Foursqaure Home Page. http://www.foursquare.com.
Fratiglioni L, Wang H, Ericsson K, Maytan M, Winblad B. Influence of social network on occurrence of dementia: a community-based longitudinal study. The Lancet. 2000;355:1315–1319, 4/15.
Gibson L, Moncur W, Forbes P, Arnott J, Martin C, Bhachu AS. Designing social networking sites for older adults. BSC HCI, 2010.
Google. 02/09/2012. Google Reader. http://www.google.com/reader/view/.
Goswami S, Köbler F, Leimeister JM, Krcmar H. Using online social networking to enhance social connectedness and social support for the elderly,2010.

Hanley WB. Phenylketonuria (PKU): a success story. In: Tan U, editor. Latest findings in intellectual and developmental disabilities research. 2012.
Hanson V. Social inclusion through digital engagement. Berlin: Springer; 2011. vol. 6765, p. 473–477.
Health on the Net Foundation. 10/15/2012. List of rare diseases. http://www.hon.ch/HONselect/RareDiseases/.
Holmes C. Dementia, Medicine. 2012;40:628–631, 11/01.
Hong J, Seltzer MM, Krauss MW. Change in social support and psychological well-being: a longitudinal study of aging mothers of adults with mental retardation*. Fam Relat. 2001;50:154–63.
Independent Age. Older people, technology and community: the potential of technology to help older people renew or develop social contacts and to actively engage in their communities. Calouste Gulbenkian Foundation, 2011.
Instagram. 02/09/2012. Instagram Home Page. http://www.instagram.com.
Jason Santa Maria. 02/09/2012. Jason Santa Maria Blogazine. http://jasonsantamaria.com.
Jellinger KA. The pathology of "vascular dementia": a critical update. J Alzheimer's Dis. 2008;14:107–23.
Korczyn AD, Vakhapova V, Grinberg LT. Vascular dementia. J Neurol Sci. 2012.
Krishnamurthy B. Key differences between Web 1.0 and Web 2.0. First Monday. 2008;13:6.
Lauriks S, Reinersmann A, Van der Roest HG, Meiland FJ, Davies RJ, Moelaert F, Mulvenna MD, Nugent CD, Droes RM. Review of ICT-based services for identified unmet needs in people with dementia. Ageing Res Rev. 2007;6:223–246.
Lehtinen V, Naassanen J, Sarvas R. A little silly and empty-headed: older adults' understandings of social networking sites. In: Proceedings of the 23rd British HCI group annual conference on people and computers: celebrating people and technology, Cambridge: United Kingdom, 2009. p. 45–54.
LinkedIn. 02/09/2012. LinkedIn Home Page. http://www.linkedin.com.
Mellor D, Firth L, Moore K. Can the internet improve the well-being of the elderly? Ageing Int. 2008;32:25–42.
Merck. 09/09/2012. Less sometimes means more. http://magazine.merckgroup.com/en/Life_and_Assistance/orphan_drugs/orphan_diseases1.html?wt.srch=1.
Moyle W, Kellett U, Ballantyne A, Gracia N. Dementia and loneliness: an Australian perspective. J Clin Nurs. 2011;20:1445–53.
Murphy F. Loneliness: a challenge for nurses caring for older people. Nurs Older People. 2006;18:22–25.
Murugesan S. Understanding Web 2.0. IT Prof. 2007;9:34–41.
Netvibes. 02/09/2012. Netvibes Home Page. http://www.netvibes.com.
Norval C, Arnott JL, Hine NA, Hanson VL. Purposeful social media as support platform: communication frameworks for older adults requiring care. In 5th international conference on pervasive computing technologies for healthcare (PervasiveHealth), 2011. p. 492–494.
O'Reilly T. What is Web 2.0: design patterns and business models for the next generation of software; 2009.
OFCOM. Internet use and attitudes. OFCOM, London, UK, 2012.
Orphanet. Prevalence of rare diseases: bibliographic data, Orphanet, Tech. Rep. 1, May 2012.
Park S, Parwani A, MacPherson T, Pantanowitz L. Use of a wiki as an interactive teaching tool in pathology residency education: experience with a genomics, research, and informatics in pathology course. J Pathol Inf. 2012;3:32.
Path. 02/09/2012. Path Home Page. http://www.path.com.
Peesapati ST, Schwanda V, Schultz J, Lepage M, Jeong S, Cosley D. Pensieve: supporting everyday reminiscence. In: Proceedings of the 28th international conference on human factors in computing systems, Atlanta, Georgia, USA, 2010, p. 2027–2036.
Perkins EA, LaMartin KM. The internet as social support for older carers of adults with intellectual disabilities. J Policy Pract Intellect Disabil. 2012;9:53–62.
Pinterest. 20/08/2012. Pinterest Home Page. http://www.pinterest.com/.

Property Pal. 02/09/2012. Property Pal Home Page. http://www.propertypal.com.
Reeder. 02/09/2012. Reeder Home Page. http://reederapp.com.
Robyn Findlay Colleen Cartwright. Social isolation older people : a literature review. Brisbane QLD Australia: Ministerial Advisory Council on Older People, 2002.
Salloway S, Hong J. CADASIL syndrome: a genetic form of vascular dementia. J Geriatr Psychiatry Neurol. 1998;11:71–7.
Sayago S, Blat J. An ethnographical study of the accessibility barriers in the everyday interactions of older people with the web. Univ Access Inf Soc. 2011;10:359–71.
Shapira N, Barak A, Gal I. Promoting older adults' well-being through internet training and use. Aging Mental Health. 2007;11:477–84.
Small GW, Moody TD, Siddarth P, Bookheimer SY. Your brain on Google: patterns of cerebral activation during internet searching. Am J Geriatr Psych. 2009;17:116.
Spotify. 09/02/2012. Spotify Home Page. http://www.spotify.com/uk/.
Stacpole M. Caring for people with vascular dementia. Nurs Residential Care. 2011;13:228–30.
Strohmaier M, Körner C, Kern R. Understanding why users tag: a survey of tagging motivation literature and results from an empirical study. Web Semant Sci Serv Agents World Wide Web.2012;17:1–11, 12.
Sundar SS, Oeldorf-Hirsch A, Nussbaum J, Behr R. Retirees on facebook: can online social networking enhance their health and wellness? In: Proceedings of the 2011 annual conference extended abstracts on human factors in computing systems, Vancouver, BC, Canada, 2011, p. 2287–2292.
Tejada P. 02/122013Survey: patients' access to orphan drugs in Europe. http://www.eurordis.org/content/survey-patients'-access-orphan-drugs-europe.
Tumblr. 02/09/2012. Tumblr Home Page. http://www.tumblr.com.
Twitter. 02/09/2012. Twitter Home Page. http://www.twitter.com.
Van Weely S, Leufkens H. Orphan diseases, Priority Medicines for Europe and the World: A Public Health Approach to Innovation.Geneva (Switzerland): World Health Organization, 2004. p. 95–100.
Wang HX, Karp A, Winblad B, Fratiglioni L. Late-life engagement in social and leisure activities is associated with a decreased risk of dementia: a longitudinal study from the Kungsholmen project. Am J Epidemiol. 2002;155:1081–1087.
Wealthwire. 11/16/2012. Young Tech Rising: Average Facebook Employee is 26. http://www.wealthwire.com/news/equities/1258.
WebMashup. 02/09/2012. Web Mashup Homepage. http://www.webmashup.com.
Wikipedia. 02/09/2012. Wikipedia Home Page. http://www.wikipedia.com.
Wilson RS, Krueger KR, Arnold SE, Schneider JA, Kelly JF, Barnes LL, Tang Y, Bennett DA. Loneliness and risk of Alzheimer disease. Arch Gen Psychiatry. 2007;64:234–240.
Wordpress. 02/09/2012. Wordpress Stats. http://en.wordpress.com/stats/.
World Health Organisation. Dementia: A public health priority. 2012.
World Health Organisation. 09/09/2012. Dementia cases set to triple by 2050 but still largely ignored. http://www.who.int/mediacentre/news/releases/2012/dementia_20120411/en/index.html.
Xiao Y, Wang J, Jiang S, Luo H. Hyperbaric oxygen therapy for vascular dementia, The Cochrane Library, 2012.
Xie B, Huang M, Watkins I. Technology and retirement life: a systematic review of the literature on older adults and social media. In: Wang M, editor. The Oxford handbook of retirement (in press). Oxford: Oxford University Press; 2010.
Xie B, Watkins I, Golbeck J, Huang M. Understanding and changing older adults' perceptions and learning of social media. Educ Gerontol. 2012;38:282–96.
Yahoo. 02/09/2012. Flickr. http://www.flickr.com.
You Playlist. 02/09/2012. You Playlist Home Page. http://www.youplaylist.com.
YouTube. 11/09/2012. Site Statistics. http://www.youtube.com/t/press_statistics.
YouTube. 20/08/2012. YouTube Home Page. http://www.youtube.com.
Zickuhr K, Madden M. Older adults and internet use. Pew Internet & American Life Project. 2012.

Vignette: Recessive Dystrophic Epidermolysis Bullosa (RDEB): Sibling Experiences

Jason Barron

I was born in late 1987, but my rare disease story really began early in 1991 when my little sister was born with *Recessive Dystrophic Epidermolysis Bullosa (RDEB)*.

There are so many stories in the rare disease community. Each of us who has been touched by a rare disease—whether a patient or family member or any other person who would self-identify as an ally to those struggling with these kinds of diseases—each of us has a unique perspective. While certainly there can be no substitute for the first-person perspective of any illness, all too often the picture we have of a rare heritable disease remains incomplete until we also include the extended impact on family members, friends and loved ones.

In my case, everything changed the day my sister was born. It's hard to characterize just how our lives changed, but suffice to say it was a very different kind of childhood for both of us. When my little sister was born, the very friction—yes friction—of birth itself was enough to cause blisters and superficial wounds on her hands & feet. I remember the story vividly as my Mom has described it—

> All of a sudden there were all of these doctors in the delivery room, pediatricians, geneticists, other experts, and they kept cropping up throughout the day, taking shifts three to four at a time. It wasn't 24 h before the 'special child', still fighting for her life in the NICU was placed up front and center for grand rounds. Everyone was curious about her because most had never seen a live birth of someone with RDEB.

RDEB is a genetic disease that affects about 5 newborns out of every 1 million live births in the United States where individuals lack a fully functional copy of the gene that codes for a very specific type of collagen fiber. In my sister's case (and as a recessively inherited disease), she inherited two copies of a collagen gene that each had a 'misspelling' for lack of more expedient descriptor. My parents

---

J. Barron (✉)
Associate Director of Public Policy, National Organization for Rare Disorders (NORD),
1779 Massachusetts Avenue NW, Suite 500, Washington, DC 20036, USA
e-mail: jbarron@rarediseases.org

were both carriers and so have no physical expression of the disease. Each of any children they might have had, including myself, had a 25 % chance of having RDEB. I was born a healthy boy by any measure. My sister wasn't so lucky.

Physically, the disease is devastating, chronic, and extreme. Collagen type VII, the kind my sister lacks, is essential for the body's tissues to bind to one another. In most people there is a network of fibers at the exterior surface of the cells that make up skin and membrane tissues (think of the fuzz on the surface of a peach). These fibers bind to one another; that's why our skin behaves as if it's made up of one continuous layer—dermis contiguously binding to epidermis. However, in the absence of Collagen VII, the network lacks integrity and only loosely binds. The layers of tissue can slide against each other and shear apart at the slightest bump or rub, creating blisters, wounds, and scarring.

Now I know that might seem pretty dense so here's the bottom line: imagine that your skin is so fragile that wearing clothes causes blisters, especially at points of friction like the armpit, neck lining, and ankles. The pressure from walking on your feet causes massive blisters that get trapped under layers of scar tissue. The backs of your eyelids can become scarred and can sometimes scratch the surface of your eyes. Your throat and mouth get blisters and break down if you try to eat sharp foods like a potato chip, even if you are exceptionally vigilant about over-chewing. Most cruelly, the repeated scarring in the webs of your fingers and toes eventually fuse together, eventually denying you the use of your fingers and thumbs. Indeed, RDEB is one of the most challenging rare disorders.

'Treatment' mostly consists of cleaning and dressing the wounds and open areas as needed. This requires daily changes of sterile bandages and medicated gauze both to promote healing and to create a second barrier against infection.

Have you ever gotten a fresh cut wet? Imagine what it would be like to get into a bathtub with about a quarter of your skin open and exposed. The whole process can take up to three hours or more on a bad day. Obviously this kind of care is an example of doing the best you can; the wounds aren't going to fully heal and stop recurring, so it's a tremendously frustrating situation to be in on many levels.

As I experienced it, at first there were nurses that would come to our house to help with the daily bandage changes. I was a curious and concerned young boy at the time, and did my part with jokes and clowning around—anything that could make her smile and help her through the pain.

As we grew older and started school, additional support was needed, including an aide that could carry her backpack for her from class to class and help her with all the logistics. Most people have never tried, without fingers mind you, to twist a doorknob on a heavy door.

Growing up at school and with people our own ages, as hard as it was for us just to process how different our experiences were together, it was equally difficult to express that difference in a way that can really translate and resonate. Children, by and large, are not known for their empathy and they can be awfully cruel to others who are different from them. Different can mean having a rare disease that you couldn't even hope to explain in a sentence or a short moment. Different can also mean being thrust into a situation where you need to be uncommonly kind and giving just to survive.

Complicating matters, the stress of all of this took my parents' marriage as collateral damage. I was nine and she was six when they divorced. I remember distinctly how that felt, how betrayed I felt by my parents. If my sister could get on waking up every day and face such a tragic illness with a smile and sparkle in her eye, how dare they quit on each other and on us? For years we traveled back and forth for 5 h in a car each way every other weekend. Five hours to Dad's for the weekend. Five hours back to Mom's on Sunday afternoon so we could go to school the next day. My role as a caregiver throughout this period was initially modest. As we became older I consistently had to teach home health nurses how to properly debride a wound where the remaining skin was as fragile as tissue paper. I just wanted to make sure it was done correctly. Every detail counts. I had come to know her and her care so well and there was just so much need, so much endless detail from counting supplies to cutting bandages and preparing them for use, that every little bit of help I could give really mattered. I knew that instinctively. She would have done more herself if she could have, but with such little manual dexterity, how could I not try to help?

You know it's odd—as I write this I realize that this may not make a lot of sense to you. To be honest, it doesn't make a lot of sense to me either. It's very hard to describe an experience like this. Somehow the words seem so hollow compared to the truths that lay behind them—the reality of it all. To this day, it's a challenge for me to remind myself that most people don't recognize the necessity behind it all seeing as they haven't lived through this. People talk about 'unmet medical needs'. It's remarkable how such a small phrase could have such *diverse* **urgency** behind it. What's clear is that, we need new options to intervene right at the root cause of these diseases, to correct genetic errors and develop new strategies toward real and lasting cures.

Health Policies for Orphan Diseases: International Comparison of Regulatory, Reimbursement and Health Services Policies

Durhane Wong-Rieger and Francis Rieger

> *The Minority is always right*
> HENRIK IBSEN (An Enemy of the People)

Abstract Access to drugs for rare diseases varies considerably, not only in terms of which drugs are reimbursed but also which patients get access. Five-year-old Luc has been diagnosed with MPS II (Hunter's Syndrome), a genetic disorder with progressive debilitating symptoms affecting the joints and major organs and, in some children, cognitive functioning. The good news is that there is an approved drug treatment that is effective in "slowing the progression of symptoms"; however, the pivotal clinical trials were only conducted with children over the age of five who had no cognitive impairment. While disease severity is hard to diagnose at an early age, his physician believes Luc has some cognitive impairment. Luc's ability to get access to treatment depends on where he lives.

Prologue

Access to drugs for rare diseases varies considerably, not only in terms of which drugs are reimbursed but also which patients get access. Five-year-old Luc has been diagnosed with MPS II (Hunter's Syndrome), a genetic disorder with progressive debilitating symptoms affecting the joints and major organs and, in some

---

D. Wong-Rieger (✉)
Canadian Organization for Rare Disorders (CORD), 151 Bloor Street West, Suite 600, Toronto, ON M5S 1S4, Canada
e-mail: durhane@sympatico.ca

F. Rieger
University of Windsor, Windsor, Canada

children, cognitive functioning. The good news is that there is an approved drug treatment that is effective in "slowing the progression of symptoms"; however, the pivotal clinical trials were only conducted with children over the age of five who had no cognitive impairment. While disease severity is hard to diagnose at an early age, his physician believes Luc has some cognitive impairment. Luc's ability to get access to treatment depends on where he lives.

The drug is expensive, costing an estimated $500,000 per patient per year (Simoens et al. 2013). In many low- and middle-income countries, Luc has no access unless his family has private drug insurance coverage. In France or Belgium, Luc would not get access to treatment right away. If Luc were in Great Britain, he would be assessed against the access guidelines; he would be likely to be treated but his physician and parents would be consulted.[1] In Germany, Luc's cognitive impairment would disqualify him. In Canada, Luc's treatment access would depend on which province he lived in. In some provinces, the drug is not funded under the public drug plan, while in Ontario, Canada's largest province, Luc would not get access until he turned 6 years old and then only if he did not demonstrate cognitive impairment. He demonstrated significant cognitive decline, he would be taken off therapy.[2] In Taiwan, Luc would also need to demonstrate that his disease did not affect his cognitive functioning, but if he qualified, he would not only have his drug treatment covered but would also have access to education and social support.[3] In Korea, Luc would have access to a very well resourced MPS clinic where his treatment would be individualized and his response monitored. In Australia, Luc could apply for treatment under the Life-Saving Drugs Program; however, if he demonstrated significant neurological decline, his therapy would likely be discontinued.[4]

Introduction

The lack of consistency in national policies for rare diseases results in not only inequities and uncertainties for patients and their families but also disincentives for advancing research and development in new treatments, thereby undermining

---

[1] Vellodi A. Mucopolysaccharidosis type II: Guidelines for assessment, monitoring and enzyme replacement therapy (ERT). http://www.specialisedservices.nhs.uk/library/23/Guidelines_for_Mucopolysaccharidosis_Type_II.pdf.

[2] Ontario Public Drug Programs Exceptional Access Program Elaprase (idursulfase)—Reimbursement guidelines. Ontario Ministry of Health and Long-Term Care. http://www.health.gov.on.ca/en/pro/programs/drugs/pdf/elaprase_reimburse.pdf.

[3] New hope for MPS II, Enzyme replacement therapy lauches in Taiwan. http://www.tfrd.org.tw/english/news/Cont.php?kind_id=61&sid=27&top1=NEWS%20AND%20EVENTS.

[4] Guidelines for the treatment of Mucopolysaccharidosis Type II (MPS II) disease through the life saving drugs program. Australian Government Department of Health and Aging, July 2012. http://www.health.gov.au/internet/main/publishing.nsf/Content/lsdp-info/$File/FINAL%20MPS%20II%20Guidelines%20-%20July%202012.pdf.

the purpose of the orphan drug policies. It is not surprising that health policies for rare conditions vary considerably across jurisdictions, given the disparities among policies for common diseases. And just as there is no single best healthcare system, there is no "best practice" when it comes to rare disease policies. This review provides a snapshot of current policies and practices but given the rapidly evolving environment is most likely out of date even while being researched and written. The discussion is organized around three policy arenas: regulation of orphan drugs, patient access to drugs, and health services policies.

The regulatory policy arena governs safety and efficacy of drugs for market approval, and in many jurisdictions this also addresses support for clinical trial design (specific to small populations) and incentives for research and development. Access policies address how drugs will be made available to patients, including procurement, funding, reimbursement, and conditions for access. A formalized health technology assessment process in some jurisdictions defines appropriate use and value for money but assessment is complicated by the fact that there are publicly and privately funded insurance schemes (often with differential access), as well as limited long-term data and high per-patient costs. The health services arena encompasses newborn screening, centres of expertise offering centralized diagnosis and treatment guidance, patient registries, programs of care and support, and empowerment of patient groups. Research policies and programs cut across all of these arenas. Finally, it is important to note that in almost every country, the two major proponents of rare disease and orphan drug policies have been the patient organizations and the drug manufacturers, with researchers and policy makers playing instrumental roles in shaping and assuring their acceptance.

Orphan Drug Policy

The starting point for most rare disease initiatives is regulation for orphan medicinal products. While Japan has the distinction of being the first country (in 1972) to designate research funding and medical expense support for "rare and intractable diseases",[5] the real origins of the proliferation in rare disease treatments was the USA Orphan Drug Act. Passed in 1983 due to strong advocacy from the National Organization for Rare Disorders (NORD) and the support of key legislators in the House and Senate, the Act's explicit purpose was to remove barriers to "commercially viable" research and development into "rare and neglected" diseases. To that end, it provided incentives in the form of research and development grants, tax rebates for clinical testing, protocol assistance, fee waivers for registration,

---

[5] Orphan drugs in Japan. Orphanet: About orphan drugs. http://www.orpha.net/consor/cgi-bin/Education_AboutOrphanDrugs.php?lng=EN&stapage=ST_EDUCATION_EDUCATION_ABOUTORPHANDRUGS_JAP.

expedited regulatory review, and market exclusivity for a period of 7 years.[6] Japan followed with similar regulations in 1993, although using a much narrower definition of rare (1 in 30,000) persons as compared to the USA (fixed number of 200,000 persons or approximately 1 in 1,600) and market exclusivity of 10 years. In 2000, the European Union implemented the Orphan Drug Act, applicable to all 27 Member States, setting the now commonly accepted prevalence of a rare disease as 5 in 10,000 persons with the additional criteria that the disease be considered life-threatening, seriously debilitating or a serious and chronic condition and having no satisfactory diagnosis, prevention or treatment. Market exclusivity was set at 10 years, but the tax incentives for research and development costs are at the discretion of individual Member States and therefore quite varied.[7] Despite differences, the current US Food and Drugs Administration (FDA) and the European Medicines Agency (EMA) orphan drug frameworks so similar that a single application often suffices in filing for orphan designation from both regulatory agencies.

Several other countries have adopted or are in the process of developing rare disease/orphan drug policies.[8] Australia has announced the intention to update its Orphan Drug Program, initially put in place in 1997 for the purpose of importing rare disease drugs. It relies primarily upon the US FDA designation and approval and also offers waiver of filing fees, expedited review and a five-year market exclusivity (but no R&D incentives). In 2012, Rare Voices Australia was registered, creating an umbrella organization for rare disease groups. Canada has been very slow to recognize rare diseases. In 1996, Health Canada released an official position stating that orphan drug legislation was not necessary since market authorizations were being granted under current regulations. However, in 2012, after nearly a decade of advocacy led by the Canadian Organization for Rare Disorders and the biotechnology industry, Health Canada announced the establishment of an Orphan Drug framework, adopting the European criteria for orphan designation as diseases with a prevalence of less than 1 in 2,000 and also being seriously debilitating or life-threatening with no effective therapy. At the time of the announcement, the regulations included protocol assistance, fee waivers for registration, and expedited review but no provisions for tax incentives or market exclusivity.

In Asia, the countries with orphan drug initiatives are Singapore (1991), Taiwan (2000), and South Korea (2003) Wong-Rieger (2012). Singapore, similar to Australia, has developed a policy directed towards importing drugs for patient access and, to that end, requires only that the drug be approved in the country of origin. Taiwan's

---

[6] Orphan drugs in the United States of America. Orphanet: About orphan drugs. http://www.orpha.net/consor/cgi-bin/Education_AboutOrphanDrugs.php?lng=EN&stapage=ST_EDUCATION_EDUCATION_ABOUTORPHANDRUGS_USA.

[7] Orphan drugs in Europe. Orphanet: About orphan drugs. http://www.orpha.net/consor/cgi-bin/Education_AboutOrphanDrugs.php?lng=EN&stapage=ST_EDUCATION_EDUCATION_ABOUTORPHANDRUGS_EUR.

[8] 2012 Report on the state of the art of rare disease activities in Europe. Part 1: Overview of rare disease activities in Europe. http://ec.europa.eu/health/rare_diseases/docs/eucerd2012_report_state_of_art_rare_diseases_activities_1.pdf.

Rare Disease and Orphan Drug Act defines a rare disease as affecting fewer than 1 in 10,000, genetically based, and difficult to treat. The Act is very comprehensive, covering R&D, expedited review and 10-year market exclusivity, as well as treatment, care and supportive services. South Korea's regulation applies to rare diseases with a prevalence of less than 20,000 (fewer than 2.5 per 10,000) and with no available treatment, offering priority review and six-year market exclusivity. In China, no explicit orphan drug or rare disease policies have been enacted, but companies are already doing research and development on drugs for rare diseases. Similarly, despite the lack of Orphan Drug legislation, several rare disease drugs have been discovered by Canadian-based organizations, though most relocate to another country (with financial incentives) once clinical trials are begun. In 2012, the Canadian government announced its commitment to implement an Orphan Drug Policy (Gupta 2012). In addition, Brazil, Argentina and several Latin America countries, are in the process of developing policies for rare diseases but few provide R&D incentives at this time.

Drug Reimbursement Policies

In addition to policies that provide incentives for companies to develop orphan drugs and regulatory policies that support the appropriate clinical testing and approval of drugs with small patient populations, in order for patients to have access there must also be policies for funding and reimbursement. In the USA, the reimbursement of drugs is determined primarily by market forces and prices negotiated between the manufacturers and multiple public and private payers with complex arrangements that include discounts, rebates, subsidies, grants, service agreements, support initiatives, and compassionate access programs, just to name a few. Government-funded and government-mandated Medicare drug coverage usually applies to rare disease but the eligibility and actual coverage appear to vary considerably with no clear process for vetting medicines for inclusion in the drug listings. Patients with private insurance have generally been able to get coverage for approved drugs but face hurdles in terms of deductibles or co-pays that typically range from 10 to 50 %, putting many orphan drugs out of reach. The market price for orphan drugs (in Europe) is estimated to range from $1,500 to $520,000 (US) with a median price of $42,000. Recently introduced orphan drugs tend to be at the higher end of the scale, especially for ultra-rare diseases (incidence of fewer than 1 in 100,000) and genetic-based therapies that benefit only subtypes with specific genetic mutations (examples of cystic fibrosis, lung cancer, and myelofibrosis) (Simoens 2011).

While the EU has one centralized regulatory agency, the European Medicines Agency that provides market authorization for the sale of drugs, reimbursement is the responsibility of the 27 Member States and, therefore, patient access varies considerably. According to one report, France, Netherlands, and Denmark have the highest number of orphan drugs available (93–87 % of approved drugs) followed by Sweden, Italy, Belgium and Hungary (71–65 %). Substantially fewer drugs are available in Spain, Romania, and Germany, which list 33–25 % of the available drugs. However, it is important to note that even within a jurisdiction there is a lack of consistency in the approach.[9]

France, one of the leaders in orphan drug access, considers only the "clinical" added value of the drug and negotiates a price based on projected improved health outcomes. Other jurisdictions subject rare disease drugs to health technology assessment (HTA), although the lack of alternative therapies and the relatively high costs of orphan drugs make traditional HTA methods difficult to apply (Garau and Mestre-Ferrandiz 2009). Countries such as the Netherlands, Italy, Sweden, and Scotland calculate "value for money" either at the time of launch or later. For example, the Netherlands funds innovative drugs at 80–100 % reimbursement while setting up a post-market process to collect data for subsequent cost-effectiveness assessment. If the drugs are determined not to be cost-effective, the funding is stopped. However, in 2012, when the Dutch Healthcare Insurance Board tried to discontinue funding for two rare diseases, Fabry Disease and Pompé, based on a CEA of an incremental cost-effectiveness ratio (ICER) of €3.3 million per quality-adjusted life-year (QALY), there was a huge outcry from the patient and clinical community, pointing out that sub-group analysis would provide more accurate outcomes.

Belgium's conditional reimbursement program applies to innovative medicines, including drugs for rare diseases, which are eligible for a "price premium" based on factors not demonstrated at the time of launch. The company has 18–36 months to confirm added value with the possible outcomes of maintaining the price, changing the listing, or withdrawing the product. There is no record of withdrawal of any rare disease drug for lack of evidence.

Across Europe, while respecting national sovereignty and differential ability to pay, there is growing pressure to operationalize a "single" access model. This is being driven by many factors, including the EU directive for cross-border healthcare, international Centres of Expertise that are producing international guidelines for diagnosis and treatment, and the necessity to collect consistent post-market surveillance data on long-term safety and effectiveness. To that end, the European Committee of Experts on Rare Diseases has endorsed a process that provides a continuum between pre-market authorization practices (clinical development) at the EU level and post-marketing authorizations practices at member state level, entitled, Clinical Added Value of Orphan Medicinal Products-Information Flow (CAVOMP-IF).[10] This framework, in many respects, encapsulates the "coverage with evidence developing" approaches that have been developed by funding and reimbursement bodies to allow for patient access to medicines, including orphan drugs, approved without "robust" safety and clinical outcomes measures because they address unmet needs, intervene with severe or life-threatening conditions, or offer significant advantage over existing therapies. On the one hand, in situations where there are gaps in data, a premarket economic assessment may not provide valid or useful information; on the other hand, it is important to

---

[9] Le Cam Y. Inventory of access and prices of orphan drugs across Europe: A collaborative work between national alliances and EURORDIS. ERTC_13122010_YLeCam_Final.pdf.

[10] EUCERD Recommendation for a CAVOMP information flow. September 2012. www.eurordis.org/sites/default/files/cavomp.pdf.

make the therapies available to the broader patient population while monitoring patient response and collecting patient outcomes in real-world settings. Under CAVOMP-IF, the appropriate value (price) of the medicine is not determined until sufficient data have been collected over time.

There are many examples of this approach but the best documented has been the United Kingdom program for ultra-rare diseases managed by the Advisory Group for National Specialized Services (AGNSS). Patients are placed on therapy if they meet defined "start" criteria (biometric, functional, and other indicators) based on evidence from clinical trials); they stay on therapy if they meet pre-defined "benchmarks" indicative of response to therapy (biomarkers, functional ability, or clinical outcomes) and they "stop" treatment if they show evidence of adverse effects or no effects.[11]. Beginning in 2013, AGNSS is being subsumed under the National Institute for Clinical Excellence, which also conducts the HTA for therapy (drugs, devices and procedures) for common conditions.

This approach includes "performance-based risk-sharing" models whereby payers are seeking to offset "performance uncertainty" at the time of the market introduction with greater "budget certainty", either at the time of introduction or in the future. Examples include dose capping, that is, a maximum number of doses reimbursed regardless of the number required by the patient to achieve clinical outcome (Avastin in Germany), price adjustments and paybacks based on actual clinical results (drug for Multiple Sclerosis in the UK), access based on subgroups (Revlimid for multiple myeloma in the UK), and pricing based on outcomes (Velcade in UK). There are many challenge with these performance-based reimbursement schemes, most notably data collection.

Underlying all of these policies is the question of "fairness" or equitable access. The ultimate goal of rare disease and orphan drug policies is to provide principles (and guidelines) for "consistent" decision-making that leads to "equitable" access. An access policy needs to consider the public health benefits of diagnosing and treating a rare disease whenever possible, the human rights of all patients to access treatment, and the ethical imperative to save a life or reduce the impact of a debilitating condition whenever possible. Clearly, reimbursement policies for orphan drugs will continue to evolve and become even more complex with new therapies that are targeted to even smaller rare disease subtypes, reliant on genetic testing to determine likely responders, and used in combination or sequentially.

Comprehensive Rare Disease Policies

Despite the emphasis on orphan drugs, only 5% of rare diseases actually have an approved drug therapy, with many more diseases using drugs off-label, often without clear policies or guidelines for access or reimbursement. Clearly, comprehensive

---

[11] Appraising orphan drugs. National Institute for Health and Clinical Excellence. http://www.nice.org.uk/niceMedia/pdf/smt/120705item4.pdf.

rare disease strategies need to encompass more than innovative drugs. In 2009, the European Union called upon all Member States to develop national plans for rare diseases. According to a 2010 status report, the 27 EU members as well as Norway, Switzerland, Croatia, Turkey, and Israel have all consulted on and developed plans addressing some or all of the recommended categories, including:

- Definition of a rare disease
- National plan/strategy for rare diseases and related actions
- Centres of expertise
- Registries
- Neonatal screening policy
- Genetic testing
- National alliances of patient organisations and patient representation
- Sources of information on rare diseases and national help lines
- Best practice guidelines
- Training and education initiatives
- Rare disease events
- Research activities and E-Rare partnership
- Participation in European projects
- Orphan drugs (Orphan drug committee, Orphan drug incentives, Orphan drug availability, Orphan drug reimbursement policy, Other initiatives to improve access to orphan drugs, Orphan drug pricing policy)
- Orphan devices
- Specialized social services.

The French National Plan for Rare Diseases was implemented in 2004 and serves as the standard for the recommendations in the European.[12] While France continues to invest in its rare disease strategy, it is not clear how other countries will proceed with implementation of the proposed plans, given the economic constraints and competing interests.

Other International Rare Disease Policies

Asia

Much of the credit for Taiwan's rare disease policies goes to the Taiwan Rare Disease Foundation, which was started by parents as a mutual help association. The Foundation does fundraising to provide public awareness as well as a host of medical, educational, and social services, including medical equipment, patient scholarships, camps, and cultural events. According to the Foundation, the 37 articles of

---

[12] The French national plan for rare disorders. EURORDIS. http://www.eurordis.org/content/french-national-plan-rare-diseases.

the Rare Disease and Orphan Drug Act were negotiated among the patients, manufacturers and government to include not only the acquisition of orphan drugs but also diagnosis and treatment, prevention, and access to specialized medicines and nutritional supplements. Patients who are recognized by the Department of Health as having a rare disease are eligible for reimbursement for medical and nutritional expenses (70–100 % depending on income) Song et al. 2012.

Latin America

In Latin America, several countries have passed orphan drug or rare disease laws but there appears to be little enactment. In many cases, policies and facilitating regulations are still being negotiated 2–3 years later (Wong-Rieger 2012).

In Argentina, both levels of Parliament endorsed a national rare diseases law, which adopts the European definition of 1 in 2,000 persons and mandates coverage under the public and private insurance schemes. In addition, there is reference to patient registries, neonatal screening, and educational, social and support activities. Moreover, to coordinate these actions, the act proposed a central multidisciplinary committee, including patient organizations. Indeed, a driving force behind the legislation has been the Geiser Foundation (Grupo de Enlace, Investigación y Soporte—Enfermedades Rares), a Latin American regional network comprised of patients, families, and professionals, founded in Argentina to advocate collaboratively across diseases and borders.

In 2011, with support from Geiser and Peruvian patient groups now under the banner of the Peruvian Federation of Rare Diseases (Federación Peruana de Enfermedades Poco Comunes, or FEPEPCO), Peru passed a law that addressed both orphan drug access as well as a national strategy for diagnostics, surveillance, prevention, care, and rehabilitation. Without even a definition of rare disease, the legislation still requires considerable detail although it has provided a platform for patient activities.

Colombia's rare disease law affects about 3.8 million Colombians, but, according to advocates, patients feel "abandoned and neglected" because there are neither procedures nor funding to allow people to access the promised healthcare services.[13]. These would have to come from the Congress and specifically the Finance Minister; meanwhile advocates continue to host discussions and engage in awareness activities.

While Mexico does not have a distinct rare disease law, provisions of the Mexican General Health Law (GHL), which came into effect in January 2012, defined, for the first time, rare diseases and orphan drugs. Under the Article, an orphan drug is used to diagnose, treat, or prevent a rare disease, affecting no more

---

[13] Govt lacks funds to treat 3.8 M Colombians. Colombia Reports. 25 Feb 2011. http://colombiareports.com/colombia-news/news/14608-govt-lacks-funds-to-treat-38m-colombians.html

than five in 10,000 individuals.[14] Moreover, the Health Ministry has the power to implement measures necessary to encourage and promote access to orphan drugs, including the making of recommendations to National Health Institutes with regard to R&D. Finally, the law allows the Federal Commission for the Protection against Sanitary Risks (COFEPRIS) to establish the provision to govern market authorization for orphan drugs; however, these do not currently provide the same data protection and market exclusivity as the USA, Europe or other developed countries.

Brazil's regulatory framework states that orphan drugs should have "fast track" review but provides no specific guidelines or procedures for doing so; rather orphan drugs are subject to the same requirements as drugs for common conditions, which negatively impacts on market approvals.[15] There are no other special provisions for rare diseases or orphan drugs. However, some rare disease patients have been successful in gaining access to drugs by bringing lawsuits against the government, based on the universal "right to health" guaranteed in the Brazilian constitution. Key to success of this strategy appears to be good clinical protocols, supported by physicians and lawsuits advanced by lawyers, many of which have provided work *pro bono or through the support of patient associations*. In some cases, the government has been willing to grant patient access even before the lawsuits are filed in the courts.

Conclusions

The modern era in rare diseases has been driven primarily by two important factors: engagement of patient organizations and orphan drug policies. Patient groups, whether family-based, disease specific, or cross-disease international alliances, have served as the focal point for raising awareness and generating support across public, private, and political stakeholders to galvanize research and development in treatments for rare diseases. While the sequencing of the human genome, completed in 2000, has greatly facilitated the identification of the genetic causes of rare diseases, it was patient advocacy led by NORD in the 1980s that led to the creation of the US FDA regulatory framework that has incentivized the transformation of genetic knowledge into effective therapies. Almost every other country that has implemented orphan drug policy has elements analogous to the US Orphan Drug Act, and these consistencies have allowed international R&D, including clinical trials, essential to small patient populations.

The lack of consistency across countries in post-market policies not only affects patient access but also undermines a critical element of R&D for orphan

---

[14] Mexico defends orphan drugs but questions remain; US patent deal. The Pharma Letter, 14 March 2012. http://www.thepharmaletter.com/file/111811/mexico-defines-orphan-drugs-but-questions-remain-us-patent-deal.html.

[15] Focus Report: Brazil http://www.pharma.focusreports.net/index.php#state=InterviewDetail&id=1465.

drugs, and that is the need to collect long-term data on safety and effectiveness in real-world settings with a critical number of patients. International patient registries pooling outcome measures from patients with similar access is an essential strategy for pre-market approval and equally vital for post-market evaluation. The barriers to collaboration on patient access policies are primarily political and economic, not clinical or technical. The recommendations of the European Committee of Experts on Rare Diseases for an uniform access framework based on the clinical added value of an orphan medicinal product is an important step forward and should be embraced by all countries.

Not all rare disease do and could benefit from drug therapies but all rare diseases can and should benefit from health services, at least equitable to those for more common conditions. While health services policies will reflect the realities of the health, social, educational, and economic systems of each country, there is much that can and is being adapted across borders. However, a commitment to a national policy is a necessary first step but not sufficient. Implementation requires resources and engagement at all levels and, again, much can be achieved through international collaborations and especially the engagement of the patient organizations.

References

Garau M, Mestre-Ferrandiz J. Access mechanisms for orphan drugs: a comparative study of selected European countries. OHE Briefing, Office of Health Econ Res. 2009;52. http://www.raredisease.org.uk/documents/OHEBriefingOrphanDrugs.pdf.

Gupta S. Rare diseases: Canada's "research orphans". Open Med. 2012;6(1):e23.

Simoens S. Pricing and reimbursement of orphan drugs: the need for more transparency. Orphanet J Rare Dis. 2011;6:42. http://www.ojrd.com/content/6/1/42.

Simoens S, Picavet E, Dooms M, et al. Cost-effectiveness assessment of orphan drugs. Appl Health Econ Health Policy. 2013;11:1–3.

Song P, Gao J, Inagaki Y, et al. Rare diseases, orphan drugs, and their regulation in Asia: current status and future perspectives. Intractable Rare Dis Res. 2012;1(1):3–9.

Wong-Rieger D. State of the art of rare disease activities around the world: overview of the non-European landscape. Orphanet J Rare Dis. 2012;7(Suppl2):A2.

Wong-Rieger D. State of the art of rare disease activities around the world: overview of the non-European landscape. Orphanet J Rare Dis. 2012;7(Suppl 2):A2. http://www.ojrd.com/content/7/S2/A2.

Vignette: Route 125 (October 2009)

Jeneva Stone

We were all, in my household, climbing slowly out of an H1N1 flu-induced fog, through a long weekend of rain oscillating heavy and light against the peaks and valleys of the rooflines, the skylights, and against the quiet black asphalt of the neighborhood street beneath my bedroom window.

I hadn't been that sick for a long time, and when I couldn't quite sleep and couldn't quite rouse myself, the road through the valley of the last year, from early March until the present, was the path on which my thoughts moved.

J. Stone (✉)
Maryland, USA
e-mail: jenevastone@gmail.com

Early in 2009, we were certain there would be a diagnosis, finally, for Robert: an Israeli lab would run a genetic analysis for nup62, a gene implicated in biotin-responsive basal ganglia disease. That Robert's disease improved with biotin was our most significant finding. In early March, waking to the ash-gray light of a freak snow day for the kids, I opened the lid of my computer inside the silence of our house and read the email: no test could proceed as an interruption in funding had caused the lab to suspend testing for nup62. It was the only lab in the world that did the analysis.

We had arrived at the right test. We had found the right lab. All the lights were green in front of us. Now the stop sign of an email telling me there would be no test.

The brakes went on then. Hard. And, as it happens when life is at a certain speed, our habits of thinking, immutable as objects, continued traveling at speed, freed temporarily from reality, until, one by one, they lost velocity and stopped or crashed, as the case may be. And that's really a heap of broken images shored against our ruin.

I spent a lot of time over the next few months, picking things up. Over and over again. Because it took me a long time to realize that there really had been a paradigm shift and that stuff was going to keep falling off the shelves, no matter how many times I put it back.

The bumper of a car is there to absorb impact. It bends or creases, gives with the incoming shock, and what it doesn't absorb reverberates through the frame, through the passengers, and the energy dissipates or goes somewhere else, unspecified. While I was lying in bed, sick, I started wondering if that's how an emotional shock works. An event comes at you at a certain speed—sometimes you're facing it, sometimes you're looking the other way—and it impacts the soft and hard matter of your body, your self. The way you see things. Such as how would I come to understand how Robert was constituted of his condition, and how the world would come to accommodate him or not. And whether I could stand that. And how I would come to reframe this life, our lives, even if I could.

Other people see the way an event hits you, but they don't see the way the shock reverberates through your frame. Because the impact has to go somewhere. It has to dissipate. And sometimes it takes a long time for that shock to travel through your psyche, animating different parts, and where it exits may surprise even you. And it may exit in many places over time.

Mostly, I lay there and thought that, since the impact of March, I had never been sure if I were moving or standing still. Moving under my own power, or moving with a forward momentum over which I had no control. Standing still as in still standing, or merely stopped.

To check my velocity back in March, I'd called our friend Andrew, a geneticist. The human genome is a vast forest, I think he said, or so I imagined him saying. Unknown, uncharted territory. The brain the organ most dependent upon the greatest number of variants, variables, and unfortunate mutations. He found online, without even blinking, it seemed, all of the relevant journal articles by the medical researchers who'd studied infantile bilateral striatal necrosis and biotin-responsive basal ganglia disease. Their research focused on familial and co-sanguinous relationships—the deepening of the genome in populations that recycle their genetic material. These groups are the foundation of genetic research, but often little of this is applicable to a larger population—the rest of us with our sometimes inherited and sometimes 'sporadic' mutations and variants.

And therein lies the problem. Even if Robert had an implicated SLC19A3 folate transporter or a nup62 defect responsive to biotin, Robert is not a descendent of Bedouin tribesmen. His genome is radically different. So, diagnosis would be only information, in the phrase my cousin and I frequently use, to file in the 'good to know' category of our minds. Robert's genome reflects Ashkenazi Jews, French-Canadian Catholics, the British middle class of the 17th century, the Swedes, and the general roil of American intermarriage from 1620 on.

The implications of such a finding for his treatment would still be guesswork, said Andrew in conclusion. However, the job of an essayist is to drive the mind of the reader through a variety of material and toward an end of some sort, if not a conclusion.

Let me take you on one of my favorite few miles of road: from Ripton, Vermont, on Route 125 into East Middlebury. I've driven this road often, under a lot of different conditions (both environmental and emotional) and in different seasons. The road descends sharply from an open plateau, broad meadows, through a forest and along a creek. As you descend, the creek is on the right and, on the left are steep banks—a mountain wall.

My memories of this road cast it mostly in dark and shadow, partly because the forest is quite thick, and partly because I've driven the road at night many times—once at 3 or 4 in the morning after one of the most significant realizations of my life. The road is steeply curved and sharply banked. One moment momentum draws the car hard to the right, toward the creek, and then the car is drawn rapidly back toward the mountainside. And again and again. In the dark, headlights catch the sharpest point of each curve, a series of reasonably harrowing hairpin shifts, illuminating whatever natural object could have been your demise, before shifting back toward the road and relative safety.

In the dark of my college years, I drove by watching the white line that marks the edge of the road, not the double yellow in the center. And braking judiciously, but not continuously. It was, at that point in my life, fun in its own way.

Let's just say I drove this road at intervals during a period when danger of all kinds seemed inviting rather than frightening. But I always felt a twinge of relief at the moment the road straightens and levels and pours itself into the village at the base of the mountain.

The road becomes an oscillation between the poles of what could still happen and what didn't. And it dissipates and the energy fades in a village of frame houses that never quite seem to change—a place I've never actually lived, but a place in which I always imagined I would be happy.

Rare Diseases Challenges and Opportunities

Rashmi Gopal-Srivastava and Stephen C. Groft

> *The best way to predict the future is to create it*
> PETER DRUCKER

Abstract Rare diseases present special challenges to the rare diseases community. The community includes research investigators, clinicians, funding organizations in the public and private sectors, patient advocacy groups, the biopharmaceutical and medical device industries and the regulatory agencies.

Introduction

Rare diseases present special challenges to the rare diseases community. The community includes research investigators, clinicians, funding organizations in the public and private sectors, patient advocacy groups, the biopharmaceutical and medical device industries and the regulatory agencies. The challenges that need to be met include expanding the knowledgebase of the approximately 7,000 rare diseases and conditions, patient recruitment for clinical trials, and training of clinicians to increase the pool of investigators, developing a research emphasis on rare diseases, gaining the interest of research investigators. The lack of patients in any one location requires novel approaches to conducting research and the development of orphan products for rare diseases. These approaches have resulted from the global distribution of patients requiring families, health care providers, the research community and patient advocacy groups across borders and continents.

R. Gopal-Srivastava (✉) · S. C. Groft
Office of Rare Diseases Research (ORDR), National Center for Advancing Translational Science (NCATS), National Institutes of Health, Democracy 1 Building, Room 1008, 6701 Democracy Blvd, Bethesda, MD 20892, USA
e-mail: Rashmi.Gopal-Srivastava@nih.gov

Challenges of Expanding the Knowledgebase of Rare Diseases

The major task is to increase understanding of rare diseases. Optimally this would be during the training programs of clinicians. However, with so many diseases under the rare diseases umbrella, it is unlikely that most physicians will be exposed to all of the rare diseases encountered in their practices. The rarity of the diseases precludes exposure of most physicians to many rare diseases. In recent years, tremendous emphasis has been placed on participation in patient registries, natural history studies and epidemiological studies. Only after a critical mass of data from larger cohorts of patients with a rare disease we will be capable of providing fundamental characterization and delineation of the heterogeneity in patient populations. This data will help identify the severity of diseases in all age groups throughout the lifespan. Results from these studies will establish reliable data on the prevalence of rare diseases. How we observe the efforts of interventions and better clinical care, standards of care should emerge for the treatment of rare diseases, including appropriate use in emergency and critical care settings. This is the basis of establishing the Global Rare Diseases Patient Registry Data Repository (GRDR) to capture data from patients around the world (Forrest et al. 2011). GRDR was initiated by the ORDR to collect de-identified patient clinical data for research and test data mapping, the data import/export processes of new and existing registries, and to promote data sharing (Rubinstein et al. 2010a, b).

With a majority of rare diseases having a genetic component, it is important to identify and understand the roles of genotype and phenotype expressions occurring in individual rare diseases. The promise of more accurate molecular diagnosis resulting from improving whole genome sequencing and analysis capabilities provides hope to many patients with rare diseases. We are observing a change in product development as we expand the knowledgebase of genetic predisposition of patients responding as we have seen with KALYDECO, a prescription medicine used for the treatment of cystic fibrosis (CF) in patients age 6 years and older who have a certain mutation in their CF gene called the G551D mutation (KALYDECOTM (ivacaftor) Tablets: highlights of prescribing information; Insights from the Kalydeco journey: a real life story of progress and possibility). We also realize that the rare diseases can inform about more common diseases.

One major need is the rapid transfer of new and useful information about individual rare diseases. The scientific and biomedical communities need to expand and make more accessible long-distance learning opportunities for the diagnosis and treatment of rare diseases for practitioners and trainees alike. Most major scientific conferences and rare disease-specific research workshops will have an emphasis on rare diseases. To increase participation on a global basis to the rare diseases communities throughout the world expanded access to the conferences needs to occur. The conclusions and outcomes, particularly improved diagnostic criteria and better treatment methods, need to be communicated utilizing available information technology resources. Organizers and sponsors of research-based conferences should utilize available resources to expand access and virtual attendance at these conferences. Many scientific conferences on rare diseases can be accessed from ORDR website (Scientific conferences).

Small clinical trials are appropriate and even a necessity in various contexts that include rare diseases. Being able to conduct them with scientific rigor is of increasing importance in the current regulatory and scientific environment. The Office of Orphan Products Development at the US Food and Drug Administration (FDA), in collaboration with the National Institutes of Health (NIH) Office of Rare Diseases Research (ORDR), National Center for Advancing Translational Sciences (NCATS), conducted several 2-day public courses titled The Science of Small Clinical Trials. This course presented an overall framework and provided training in the scientific aspects of designing and analyzing clinical trials based on small study populations. The goal of this course was to engage and educate FDA reviewers, NIH scientists, clinicians, academics and industry representatives with experience in human subject research, seeking to build upon their existing knowledge and to obtain a broader context of what is known about small clinical trials across medical products (e.g. drugs, biologics, and devices). The course brought together subject experts and stakeholders to discuss examples of small clinical trials and to identify strategies and trial designs that are conducive to overcoming the challenges of executing clinical trials using small study populations. Speaker presentations and webcast recordings can be accessed (The FDA_NIH science of small clinical trials course 1 website).

The value of and need for Natural History Studies have been recently recognized and emphasized. Natural History Studies provide useful information on the progression and etiology of the disease. In addition, these studies are useful in developing research hypotheses that can be tested in clinical trials. Natural History Studies also help in identifying genetic variability in rare diseases patients and in establishing clinical endpoints for trials. These studies provide patient cohorts willing to participate in clinical trials thus avoiding long delays between various phases of clinical trials usually encountered in recruitment of subjects. A workshop on Natural History Studies of Rare Diseases was recently organized by FDA and NIH and is available as a resource (Workshop on natural history studies of rare diseases).

The globalization of the pharmaceutical and devices industries has expanded to emphasize rare diseases in recent years. Recognition of access to niche markets for rare diseases, even with a small population in individual countries, now enables a research, development, and marketing access to patients around the globe. Patients and clinicians are learning of newer interventions more readily.

Challenges to Patient Recruitment

One of the major barriers to initiating and completing clinical studies in a timely fashion is recruiting a sufficient number of patients. For rare diseases this can be particularly problematic with patients distributed over a large geographical area or even in a number of different countries. When this occurs, investigators and sponsors must understand and consider the cultural, ethical, legal and social issues related to data gathering and sharing of information between multiple diverse populations. In one respect this could be considered a positive outcome as we observe the genetic similarities or diversity that can occur in more distributed global populations. With the increased development of genetic and molecular diagnostic tests, particularly from whole genome sequencing tests and analyses, identification of more homogeneous population is possible. These analyses will also provide access to a more heterogeneous patient population reflecting greater genetic diversity in the patient population enrolled in the study.

To address the fundamental challenge of patient recruitment for clinical studies in rare diseases the Rare Disease Clinical Research Network (RDCRN) was established by the ORDR. It is designed to address this problem by fostering collaboration among scientists and shared access to geographically distributed research resources.

The RDCRN supported by the ORDR in collaboration with eight other NIH research Institutes and Centers, requires active collaboration with patient advocacy groups (PAGs) (Griggs et al. 2009). The 17 research consortia comprising the RDCRN and one Data Management and Coordinating Center (DMCC) include active participation with more than 95 patient advocacy groups (PAGs). The direct involvement of PAGs in RDCRN operations, activities, and strategy is a major feature of this network. Each consortium in the RDCRN includes relevant PAGs in the consortium membership and activities. These PAGs representatives serve as research partners within their own consortia. These collaborations have enabled rapid accrual of patients into study protocols. These activities in addition to establishing partnerships with medical specialty organizations for referral purposes have been found helpful in speeding up the recruitment of patients into studies. Along with a globalization of society, and the pharmaceutical industry, a similar pattern is developing with PAGs as they extend their connections to patients around the world.

Likewise we have observed the addition of research investigators at numerous sites to eliminate barriers to patient recruitment. Many PAGs have grasped the expansion of their traditional roles of providing information to the patients providing counseling services to patients and families serving as a liaison to other clinicians and a referral source to local and national media. These PAGs have established their new role as a research collaborator. Their services include the review of clinical research protocols and informed consent documents. This movement is occurring in many countries around the world and is expected to increase.

Challenges to the Research Investigators

The research investigators of rare diseases confront many challenges. There are limited but growing resources for studies of rare diseases. For the most part the investigators must use existing grant and contract resources for their research. Their outcome requires identification and utilization of the existing research infrastructure of networks and consortia such as the RDCRN. The existing infrastructure enables the participation of researchers from different countries with a focus on the same diseases. For example, the 17 disease-specific consortia of the RDCRN have ongoing research studies at 225 clinical sites including more than 25 other countries.

Research of rare diseases requires the recruitment of a critical mass of investigators who are willing to work under common protocols, share data from clinical studies, share biospecimen samples, and co-author publication. Research of rare diseases requires a collaborative multidisciplinary approach. Access and accepting central Institutional Review Board has been shown to be useful to investigators from multiple sites working under a common protocol. The collaborative approach of the scientific community is critical to the direction of research with the greater emphasis on patient registries, longitudinal natural history studies and limited biospecimen samples. The multi-site approach is required to avoid recruitment fatigue for a limited number of patients for various studies. This concern has been expressed when multiple studies are planned or ongoing with access to a limited number of patients. It is also important to realize that that the data collected from multi-site studies become available for use by other research investigators.

The RDCRN is unique in its approach to addressing rare diseases as a group. Previously, the NIH's institutes and centers funded research on individual rare diseases in their respective disease-type or organ domains. The RDCRN established in 2003 is the first program that aims to create a specialized infrastructure to support rare diseases research. To increase the number of trained investigators, the 17 research consortia in the RDCRN have established training programs for clinical investigators who are interested in rare diseases research.

DMCC supports RDCRN by providing technologies, tools to collect clinical research data, and support of study design and data analysis. It also provides on-line protocol management system for patient enrollment/randomization, data entry and collection with data standards, works with the each consortium's Data and Safety Monitoring Boards in collaboration with NIH ICs to establish protocols for adverse events notification and reporting and provides protocol training for research staff for all multi-site studies.

The RDCRN's DMCC has also provided additional systems to address needs of individual studies, such as a laboratory data collection system, a specimen tracking system, and a pharmacy management system (to support blinded distribution of study agents and placebos). The DMCC created and maintains RDCRN's central public website, developed as a portal for the rare diseases community, including patients and their families and health care professionals, to provide information on rare disease research, consortium activities, RDCRN-approved protocols, disease information, and practice guidelines. In addition, it has developed a unique voluntary patient contact registry that provides on-going contact with approximately 11,000 individuals from over 90 countries representing 180 diseases, alerting them when new studies are opened in the RDCRN or when on-going studies expand to new sites.

Sharing of data is another challenge faced by rare diseases investigators. The data from multi-site well-characterized population samples constitute an important scientific resource and their full value can only be realized if they are made available, under appropriate terms and conditions consistent with the informed consent provided by individual participants, in a timely manner to the rare diseases community and the largest possible number of qualified investigators.

The ORDR has supported data collection from participants in numerous clinical trials and epidemiologic studies as part of approved protocols under the RDCRN. The data from RDCRN studies will be made available for sharing with the scientific community for the purposes of scientific research. To accomplish this, the DMCC coordinates with ORDR including registration with and data uploading of appropriate RDCRN studies to ORDR-governed data repository through dbGaP, a database for genotype and phenotype information available from the National Library of Medicine.

Challenges of Providing an Emphasis on Rare Diseases Research

As a result of numerous scientific discoveries and completion of the mapping of the human genome, the rare diseases present scientific opportunities to challenge the community to translate the laboratory discoveries into products to be evaluated in the clinic. With the failure rate of new product development very high, attempts are now underway to de-risk potential therapies from failure or, to put it in a positive light, to optimize the chance for success. Expanding USA and international interest along with a revived interest by the biopharmaceutical and device industries, has led to the development of more directed research agendas with a focus on the delivering to patients and clinicians more interventions and diagnostics. The expansion of research activities at NIH maintains the emphasis on basic research programs while expanding the emphasis on rare and neglected diseases. This emphasis has resulted in the establishment of novel programs and includes the NIH's National Center for Advancing Translational Sciences and programs such as Therapeutics for Rare and Neglected Diseases Program (TRND), Bridging Interventional Development Gaps Program (BrIDGs), and the Cures Acceleration Network (McKew and Pilon 2013). The TRND program offers collaborations, the opportunity to partner with TRND researchers and gain access to rare and neglected disease drug development capabilities, expertise, and clinical/regulatory resources in a collaborative environment with the goal of moving promising therapeutics into human clinical trials. BrIDGs makes available, on a competitive basis, certain critical resources needed for the development of new therapeutic agents. In general, synthesis, formulation, pharmacokinetic and toxicology services in support of investigator-held investigational new drug (IND) applications to the U.S. Food and Drug Administration (FDA) are available.

NIH institutes' translational research programs such as NINDS' Network for Excellence in Neuroscience Clinical Trials (NeuroNEXT), the Center for Accelerating Innovation (CAI) at NHLBI, NCI's Experimental Therapeutics Network (NExT), NHLBI's Science Moving Towards Research Translation and

Therapy Program (SMARTT), the NIH PhRMA Collaboration on Repurposing of Products, and the NIH-DARPA-FDA Collaboration on the development of tissue and organ systems on a microchip initiative to develop better predictors of safety and efficacy for testing compounds. Many of the novel translational research programs offer opportunities to assess the gaps in the product development stages of compounds with limited industry interest or support.

Considerable emphasis has been devoted to the identification of appropriate clinical endpoints to measure the safety and effectiveness of investigational compounds and medical devices. There is also growing interest in developing *In Vivo and In Vitro* models representative of human diseases. It is now possible to establish better definitions of possible patient responders with the development of appropriate biomarkers and clinical and surrogate endpoints for clinical testing. The research community also encourages pharmaceutical and biotechnology compounds to identify and make available with appropriate confidentiality and material transfer agreements the chemical libraries and compounds available for research and development.

Conclusion

Challenges to rare diseases research and orphan products development can be viewed as barriers or opportunities to advancements. The rare disease community is gaining momentum as more information is becoming readily available from an expanding cadre of research investigators with experience in conducting clinical trials with small patient populations. Increased emphasis by the patient and the research communities is leading to the generation of data from patient registries and natural history studies. These activities increased access to patients with improvements in recruiting patients for clinical studies. The biopharmaceutical industry provides an emphasis on the special populations offered by the niche markets associated with the rare diseases. In recent years, NIH developed translational research program emphasis to provide resources for research discoveries to be developed into interventions for rare diseases. Despite expanded program emphasis on rare diseases, numerous scientific research opportunities exist to study one of the 7,000 rare diseases. Collaborative research partnerships are required for advancements in rare diseases. Knowledge of existing resources facilitates translational research activities. With these resources and partnerships, we anticipate the increased and speedier development of orphan products.

References

Bridging interventional development gaps (BrIDGs) program. http://www.ncats.nih.gov/about/faq/bridgs/bridgs-faq.html. Accessed 18 Feb 2013 from NCATS.NIH.gov.
Forrest CB, Bartek RJ, Rubinstein Y, Groft SC. The case for a global rare-diseases registry. Lancet. 2011;377(9771):1057–9.

Griggs RC, Batshaw M, Dunkle M, Gopal-Srivastava R, Kaye E, Krischer J, Nguyen T, Paulus K, Merkel PA. Clinical research for rare disease: opportunities, challenges, and solutions. Mol Genet Metab. 2009;96(1):20–6.

Insights from the Kalydeco journey: a real life story of progress and possibility. http://www.cfri.org/pdf/2012CFRInewsFall.pdf.

KALYDECOTM (ivacaftor) Tablets: highlights of prescribing information. http://www.accessdata.fda.gov/drugsatfda_docs/label/2012/203188lbl.pdf.

McKew JC, Pilon AM. NIH TRND program: successes in preclinical therapeutic development. Trends Pharmacol Sci. 2013;34(2):87–9.

Rubinstein YR, Groft SC. Driving interest in consolidating resources for the creation of a global rare disease patient registry. Contemp Clin Trials. 2010a;31(5):393.

Rubinstein YR, Groft SC, Bartek R, et al. Creating a global rare disease patient registry linked to a rare diseases biorepository database: rare disease-HUB (RD-HUB). J Contemp Clin Trials. 2010b;31(5):394–404.

Scientific conferences. http://rarediseases.info.nih.gov/Scientific_Conferences.aspx. Accessed 18 Feb 2013 from rarediseases.info.nih.gov.

The FDA_NIH science of small clinical trials course 1 website. (http://www.fda.gov/ForIndustry/DevelopingProductsforRareDiseasesConditions/OOPDNewsArchive/ucm312575.htm).

Workshop on natural history studies of rare diseases 16–17 May 2012 Bethesda, Maryland. http://www.fda.gov/ForIndustry/DevelopingProductsforRareDiseasesConditions/OOPDNewsArchive/ucm292294.htm. Accessed 18 Feb 2013 from FDA.gov.

Epilogue

WE HOPE THAT this book has met its key objectives which were to be clear and simple. The plethora of perspectives from such a wide range of stakeholders (including patients, their families, health professionals, researchers, NGOs, Pharma, policy makers and charities) underline the difficulties faced by the rare disease community. Those working in the area already are all too familiar with its inherent frustrations—we envisage that this book will act as a guiding beacon to new arrivals to the field whilst offering support to others. Any newcomers should be under no illusions: this is an extremely challenging area littered with multiple obstacles (be they financial, legal, regulatory, clinical, technical and perhaps even ethical and moral).

With challenges, come opportunities. The proliferation of contemporary IT tools (the web, email, social media—the core of Health 2.0) has allowed patients and their families to be connected in a way that could simply not be imagined a few years ago. The diagnosis of a rare disease is harrowing enough—to hear that "sorry, we have no answer" must come as a hammer blow to patients and their families. Patient-led advocacy is perhaps strongest when answers are not forthcoming from the (professional) healthcare community and the power and passion that radiates from our vignettes attests to this perspicacity of spirit. "No" certainly does not mean "no".

In discussion with many of the chapter and vignette authors, there was an overwhelming consensus regarding rare diseases: patients and families could not care less about regulation, policy and procedure. They want answers, they want a cure and they want it now. Stigmatization, lack of awareness, ignorance and other difficulties are not limited to any one rare disease—it could be argued that many of the challenges of one rare condition are replicated among the others. It is almost a case of "insert name of rare disease here".

The very nature of rare diseases means that there is a limited number of patients and a paucity of expertise. The use of Health 2.0 has been very important in connecting patients and their families with one another. As exemplified by several of our chapters, patient-led advocacy and information groups can help drive new research avenues as they act as an invaluable knowledge base. The rarity of a disease often means that patients themselves know more about their condition than the healthcare professionals, especially the challenges of living with it on a daily basis. Patients and their families are all too willing to be involved in research activities once they learn of new initiatives.

As communication and dissemination of promising rare disease research is of paramount importance, we trust that this book has offered new insights and a reassurance that multiple stakeholders are working on fundamental issues in the field. We look forward to increased patient-centric approaches and coordinated efforts in order to make better sense of the situation, which unify currently dispersed knowledge centers and which push for reform and action.

Rajeev K Bali, PhD
r.bali@ieee.org

Lodewijk Bos, MA
lobos@icmcc.org

M Chris Gibbons, MD, MPH
mgibbons@jhsph.edu

Simon R Ibell
simon@ibellieve.com

August 2013